Located Research

Angela Campbell • Michelle Duffy
Beth Edmondson
Editors

Located Research

Regional places, transitions and challenges

palgrave
macmillan

Editors
Angela Campbell
Arts Academy
Federation University Australia
Ballarat, VIC, Australia

Michelle Duffy
Centre for Urban and Regional Studies
University of Newcastle
Callaghan, NSW, Australia

Beth Edmondson
Humanities and Social Sciences
Federation University Australia
Churchill, VIC, Australia

ISBN 978-981-32-9693-0 ISBN 978-981-32-9694-7 (eBook)
https://doi.org/10.1007/978-981-32-9694-7

Cover illustration: Penny Dunstan

This Palgrave Macmillan imprint is published by the registered company Springer Nature Singapore Pte Ltd.
The registered company address is: 152 Beach Road, #21-01/04 Gateway East, Singapore 189721, Singapore

Angela, Michelle and Beth dedicate this volume to the regions that nurture us, support us and continue to enrich our lives and the lives, both human and non-human, around us. We hope that this investigation into what is particular, and at times peculiar about regions, inspires further located research into place and change, transition and resilience, and our shared challenges for the future.

Contents

Notes on Contributors

Chloe Benson is Lecturer in Screen and Media Studies at Federation University Australia. Her current research interests include bisexual representation, regional film festivals and paratextuality.

Simon Bradley joined the University of Highlands and Islands (Lews Castle College) in 2012 and teaches on the HNC Music, BA Applied Music, and is the Program Leader for MA Music and the Environment. Simon was awarded Fellow of the Higher Education Academy in 2017 and has extensive knowledge of designing and teaching on Networked Blended practical degrees and Master's delivered from remote locations. He is a distinguished Celtic Fiddle player and composer. As artistic director of Albastur Cultural Exchange, he integrates staff, students and musicians from Scotland and Asturias (Northern Spain) in collaborations, field trips, performances and teaching opportunities.

Angela Campbell is a Senior Lecturer at the Arts Academy, Federation University, where she teaches into the Performing Arts program. Her research and published work in theatre and performance has been both practical and theoretical and has investigated performance from the archives, site-specific theatre, the politics and poetics of place, intercultural theatre, Indigenous theatre, contemporary paradigms and practices in theatre and performance, and practice-led research.

Uncle Lex Dadd is a Darug Traditional Custodian who leads a program called 'Yanama Budyari Gumada'—'walking in good spirit'. He teaches the fundamental principles of how to be a humble, respectful and responsible role model, through taking care 'as' country and looking after one another.

Michelle Duffy is Associate Professor in Human Geography in the Centre for Urban and Regional Studies, University of Newcastle, Australia. Her research interests include community resilience, wellbeing and sustainability; the significance of emotion and affect in creating notions of belonging and exclusion; the role of art and sound practices in creating and/or challenging notions of identity and belonging in public spaces and public events; and an exploration of the body as a means of embodied, emotional and affective communication.

Penny Dunstan is a Newcastle-based artist, agronomist and environmental scientist working with land rehabilitation after open cut coal mining in the upper Hunter Valley. Her art practice employs analogue and digital photography, topographic and stratigraphic drawing, writing and soil exploration to derive lived encounters with terraformed lands.

Beth Edmondson has lived and worked in regional locations for most of her life. As a Senior Lecturer in the School of Arts, Federation University Australia, she engages in co-designed research in Gippsland Victoria. She is series editor of the Palgrave Studies in Environmental Transformation, Transition and Accountability Series.

Pia Eiringhaus is Research Associate at the Martin-Luther-Universität Halle-Wittenberg, Germany. She is writing her PhD thesis on the history of West German adoption after 1945. She received her MA and MEd degrees in March 2017 at Bochum Ruhr University, Germany. Her research interests include contemporary history, family history, deindustrialisation and industrial heritage research.

Susan Fayad is the coordinator, Heritage Strategy at the City of Ballarat, Victoria, Australia. She is managing the roll out of UNESCO's

Recommendation on the Historic Urban Landscape (HUL) pilot program in Ballarat through the signing of a Strategic Cooperation Agreement with the World Heritage Institute of Training and Research in Asia and the Pacific (WHITRAP) in Shanghai. Susan has been an active member of UNESCO's global historic urban landscape program, contributing to international discussions in France, China and Ecuador as well as in Australia.

Paul Glass is a passionate Kamilaroi man working within the National Parks and Wildlife Service in the Blue Mountains and Hawkesbury Area and alongside the Darug Community, Merana Aboriginal Association Hawkesbury Inc., Muru Mittigar, Blue Mountains Aboriginal Cultural Resource Centre (ACRC), Green Army, Hawkesbury Young Land Carers, Friends of the Colo Hawkesbury Environmental Network, Yarramundi Rural Fire Service, Bass Fishers, Firesticks Network, Gundungurra Heritage Archaeological Institute and many Aboriginal and community groups.

Marnie Graham is a Formas post-doctoral Research Fellow at Stockholm Resilience Centre, Stockholm University, Sweden, and honorary associate at the Department of Geography & Planning, Macquarie University, Australia. Her post-doc examines Indigenous tourism initiatives in Australia, Sweden and southern Africa, focussing on the role such initiatives play in reconciliation.

Michelle Graymore is an Honorary Senior Research Fellow in Environmental Sociology at Federation University Australia and Board Member for the Wimmera Catchment Management Authority. Her research is focused on the social dimensions of natural resource management, particularly water, in a changing climate. She works with environmental scientists, natural resource managers and community helping to enable local knowledge, community needs and values, and best available science are considered in planning, policy and decision making.

Monica Green is a Senior Lecturer in the School of Education at Federation University Australia (Gippsland campus). Monica's teaching and research focuses on place- and community-based pedagogies and curriculum. She is an active committee member of the Regional Centre of Expertise in Education for Sustainable Development (RCE Gippsland) and lead editor/author in Green, M., Plowright, S. & Johnson (2019) *Educational researchers and the regional university: Agents of regional-global transformation*, Dordrecht: Springer.

Fern Hames is a leader in communication, collaboration and engagement at the Arthur Rylah Institute for Environmental Research; Department of Environment, Land, Water and Planning, Melbourne, Australia. She works in the contexts of biodiversity, natural resource management and encouraging people to act for nature. Fern is passionate about wild places and their future, and the people who care for them. She works with scientists, natural resource managers, decision makers and communities to connect knowledge, partnerships and action, with the aims of improved biodiversity and people strongly connected to, valuing and actively caring for nature. She is committed to diversity, acknowledging multiple perspectives, connecting multiple ways of knowing and encouraging multiple ways of experiencing place.

Paul Hodge teaches at University of Newcastle in Geography and Environmental Studies. His research focuses on Indigenous-led environmental management, particularly caring-as-Country practices on/with Darug Ngurra, Yarramundi, western Sydney and Gumbaynggirr Country on the mid-north coast of NSW.

Carole L. James is Associate Professor in the Discipline of Occupational Health and Safety at the University of Newcastle. Her research interests have included supervising honours projects (in Occupational Therapy, Physiotherapy and Occupational Health and Safety at Bachelor's and Master's level), RHD students in Occupational Therapy and Occupational Health and Safety and research related to Functional Capacity Evaluations, Work and Occupation. She has been involved with a research team investigating occupational rehabilitation of NSW nurses, a team studying biomechanical analysis of musculoskeletal injury risk in an industrial setting, a team investigating health professional interventions for patient weight management and musculoskeletal injuries in academics. She is also engaged in a team investigating mental health in the coal mining industry and another investigating task rotation in coal mining.

Geraldine Jones leads the Business Partnerships Unit within the Research and Innovation portfolio at Federation University Australia. She has significant senior management and portfolio development experience from working and leading teams within the utilities, construction and education sectors. She specialises in researching and writing on leadership, transition, strategic thinking and organisational psychology. Geraldine is a Certified Project Director and holds a Master of Organisational Leadership as well as qualifications in business, humanities and teaching.

Sara Judge is a Geography PhD candidate at Macquarie University. Her research focuses on collaboration and deep relationships with more than just human Earth Kin. Sara works with the bats and birds of the Cumberland Plain woodlands, Greater Blue Mountains, and lower Hunter regions of NSW, and their human allies.

Brian Kelly is Professor of Psychiatry, Head of School of Medicine and Public Health, University of Newcastle and dean of the Joint Medical Program, University of Newcastle and University of New England. He has completed a number of mental-health related research studies including the NHMRC funded Australian Rural Mental Health Study. Prof Kelly is co-director of the Priority Research Centre for Brain and Mental Health Research, University of Newcastle and Chair (Aspen Medical) of the Centre for Resources, Health and Safety, within the University of Newcastle's Institute for Energy and Resources.

David Kelly is a cultural geographer that works at the social intersections of political controversies around housing and home in remote, regional and urban contexts. He has a theoretical focus on affect, difference and encounter.

Lisa Kendal is a senior planning and sustainability professional, with specialist skills and experience in land use planning, environmental planning, climate adaptation planning, sustainable development, project management and community engagement. She holds a master's degree in Planning and a master's degree in Environmental Science, with a focus on rural and regional planning and application of tools for developing sustainable communities.

Jessica Lemire is an honours student in Human Geography at the University of Newcastle. Her research explores the relationship between Indigenous cultural dance practices and the healing and rejuvenation of the environment at Yarramundi Regional Park.

Michele Lobo is a social and cultural geographer at Deakin University who explores race, affect, encounter and co-belonging in cities. More recently her work focuses on Anthropocene urbanism, multispecies worlds and climate change. She is the author of three books and her articles have appeared in *Geoforum* (2019), *Area* (2019) and *Urban Studies* (2018).

Judith Mair is an Associate Professor and leader of the Tourism Discipline Group of the UQ Business School, The University of Queensland, Australia. Her research interests include the impacts of tourism and events on community and society, consumer behaviour in tourism and events, the relationship between events and climate change, and business and major events.

Sherie McClam is Associate Professor of Science Education and Education for Sustainability in the School of Education at Manhattanville College in Purchase, New York, USA. With an extensive background in social and ecological justice teaching and research, she created and directs an advanced certificate program in education for sustainability designed to generate educational leaders for a sustainable future.

Kathleen A. Moore is an Associate Professor in the School of Health and Life Sciences, Federation University Australia. Her major areas of research are related to people's mental and physical health, but she is also interested in organisational psychology and the factors that contribute to staff and organisational wellbeing. She is particularly interested in social anxiety, the relationship between anxiety and depression, obesity and exercise behaviours, chronic illness and people's psychosocial adjustment (coping) to these and other life stressors, as well as people's help-seeking behaviours.

Damian Morgan is a Senior Lecturer at the Federation Business School, Federation University Australia. His research interests in outdoor recreation focus on risk management for visitors and businesses servicing visitors. He has a specific interest in safety management for aquatic and coastal activities. Research projects include development of a risk management training program for small tourism businesses, epidemiological investigation of surf bather drowning, appraisal of sustainable workplace practices for employees across power generating, water utility and education industries, development of a regional resilience monitor, evaluation for a rock fisher safety campaign, estimation of beachgoer's exposure to poor water quality based on water quality forecasting, and measuring visitor satisfaction across Gippsland.

Darug Ngurra Darug Country, is beauty and complexity, enabling everything to be, to become together. She gifts her presence in loud and gentle ways: double rainbows, light rain, flooding waters, fallen tree branches, sandstone rock shelves, sea eagles circling, owls shrieking, axe heads within basalt boulders waiting to emerge.

Aunty Corina Norman-Dadd is a Dharug/Dharawal woman. She is a singer/songwriter, dancer, teacher and language consultant. Corina is the Secretary of Dharug Ngurra Aboriginal Corporation and is a cultural advisor for Yanama Budyari Gumada. She facilitates cultural knowledge and practices, weaving in Traditional culture and protocol into a contemporary framework.

Anne Poelina is a Yimardoowarra marnin, an Indigenous woman from the Mardoowarra, Fitzroy River, who has been appointed as an Adjunct

Senior Research Fellow by the Nulungu Research Institute, University of Notre Dame, in Broome, WA, Australia. She has been a research consultant and research fellow for various universities for more than 28 years. She is an Adjunct Fellow of Charles Darwin University and an affiliate of Gaia University and has been a staff member of several other universities including the University of Queensland and the Southern Cross University.

Her PhD addresses the topic of Nyikina Indigenous Australian tradition and the importance of heritage and language to both Nyikina people's social, cultural and economic wellbeing. Her current work focuses on building culture and conservation economies on country, which is on the National Heritage Listed Fitzroy River, in the West Kimberley region of Western Australia.

Jessica Reeves is Senior Lecturer in Environmental Sciences at Federation University Australia, Gippsland Campus, and co-chair of the Impact Ecology Research Group. Her research revolves around long-term climate and environmental change. In particular, she is passionate about the interplay between humans and aquatic environments through time—including recent industrial pollution and long-term modification of land and waterscapes. She takes great pleasure in engaging students in debate around our environment and how to engage with it more sustainably.

Jane Louise Rich is a qualitative health research academic with the Centre for Resources Health and Safety. She is a human geographer with special interests in how psychosocial and environmental factors impact health and wellbeing. Ms Rich's PhD explored women's experiences with drought in Australia and the mental health and ageing implications of this. Jane has authored many articles on mental health of rural people, young people, workplace risk factors and gender and these have appeared in many publications.

Rebecca Scott is a PhD candidate in Human Geography at the University of Newcastle. Her research focuses on Indigenous-led environmental management practices within urban places. She works with National Park rangers and Darug custodians in western Sydney.

Naomi Smith is a digital sociologist in the School of Arts, Federation University Australia. She examines how technologies like social shape our understandings of and relationship to space and place. Naomi also researches the intersection of bodies, health practices and digital technologies

Anna Wendy Stevenson is a Senior Lecturer and programme leader for the innovative BA (Hons) Applied Music and is responsible for delivery across the University of the Highlands and Islands and beyond. For the past decade, she has led the traditional music programmes for Lews Castle College in Benbecula, designing and delivering dynamic courses which have produced many of today's award-winning traditional musicians. She has performed and delivered workshops in traditional Scottish music world-wide and curated and directed events for a wide range of organisations, communities, media groups and festivals in the UK, Canada, USA, Australia and Europe.

Sandie Suchet-Pearson is Associate Professor in Human Geography at Macquarie University. Her research is in the area of Indigenous rights and environmental management. Her work focuses on Indigenous self-determination in the context of cultural tourism in North East Arnhem Land, northern Australia, and caring-as-Country in western Sydney.

Leigh Sullivan is Deputy Vice Chancellor (Research & Innovation) at the University of Canberra. His research has addressed significant regional environmental management issues including managing acid sulfate soil and water issues in coastal areas regionally, nationally and globally, and novel approaches to enhance carbon biosequestration using crops and pastures.

Gordon Waitt MA (Edinburgh) 1981, PhD (Edinburgh 1988), is a Professor and head of the School of Geography and Sustainable Communities, University of Wollongong, Australia. His current research interests focus on vulnerable bodies including cyclists, senior drivers and low-income households. Many of Gordon's articles have appeared in many journals including *Gender, Place and Culture, Social & Cultural Geography and Australian Geographer.*

Sally Weller is an economic geographer in the University of South Australia Business School. She is especially interested in processes of regional restructuring and labour market change. Since 2009, she has been working on various aspects of the challenges facing the Latrobe Valley as it transitions from coal-based electricity production.

Susan Yell is a communications scholar, and a Senior Lecturer at Federation University Australia, Gippsland Campus. She researches social practices, discourse and affect, and the role of communication technologies forming social relations. Her research focuses on the role of media and social media in community resilience and disaster recovery.

List of Figures

List of Tables

1

Introduction

Leigh Sullivan

Regional areas often possess a characteristic sense of place and purpose dependent on the specifics of their environmental, historical and societal nature. This sense of place and purpose is reflected in the activities and enterprises that occur in these locations, as well as by the collective and shared values and aspirations of those who inhabit these places. We could argue that, in comparison with regional areas, non-regional areas such as metropolitan cities exhibit far more strongly a "patchwork" of senses of place and purpose that reflect variations in societal factors, especially socioeconomic status. In any case, the health and wellbeing of our regions are undeniably critical to the health and wellbeing of our society as a whole.

Whilst in times of uncertainty, be they due to climatic, social or economic changes, regional areas are often the first areas that suffer from and exhibit adverse impacts from changes that will subsequently also sweep through other areas. Wider society has much to both learn and benefit

---

L. Sullivan (✉)
University of Canberra, Canberra, ACT, Australia
e-mail: Leigh.Sullivan@canberra.edu.au

A. Campbell et al. (eds.), *Located Research*,
https://doi.org/10.1007/978-981-32-9694-7_1

from studies aimed at understanding the characteristic vulnerabilities and strengths of their regional areas. By developing a deeper appreciation and understanding of the interdependencies between regional areas and metropolitan areas, for example, it is more likely that policies and practices will be developed and established for the benefit of all.

Research aimed at gaining a better understanding of how our regions function, how the functioning of regions may be enhanced further, and considering what lessons regional research can provide to help address broader societal issues, all need to be strongly encouraged. The use of both transdisciplinary and cross-disciplinary approaches are critical to the success of this research and indeed these approaches are advanced considerably by this important publication.

This book embraces at its core, a framework of co-design in the context of the appreciation of different ways of knowing, and of incorporating different values and perspectives from all stakeholders, and of a continuous search for balance between these dynamic aspects.

There are three themes advanced here: *Place and change*, *Transition and resilience*, and *Challenges for the future*.

Chapters in the *Place and change* theme provide case studies that begin in NSW with a cross-cultural interdisciplinary exploration of how involving ways of knowing and experiencing by the traditional owners of country are and will ever more increasingly be vital in the repair of our degraded landscapes, caused largely by recent and widespread adoption of European land-management practices ill-suited to our Australian landscapes. This case study illustrates how implementation of traditional knowledge will be essential to the healing of both our lands and, in turn, the healing of the land's people.

The chapters in this theme then move and focus largely on Victoria, especially the Central Highlands and Gippsland. The importance of letting the landscape represent itself in the making of art as distinct from an interpretational landscape art approach is explored. Other chapters explore the significant role of festival networks in creating a sense of community and place, the value of regional arts practice supported by wider communities, and of how the consequences of economic transition in regional areas are often dependent on a very narrow range of economic activities—something that is of stark difference to that of metropolitan areas and cities supported by highly diversified and inter-connected economies.

Chapters in the *Transition and resilience* theme provide examples of regions undergoing shifts away from European relations to place and explore the importance of specific place-based considerations in determining the pathways and trajectories of the social and economic wellbeing and health of those in our regional and remote areas. Issues addressed here include how to best approach problems of mental health and lifestyle behaviours in both the workers and communities that support our mining industries, especially during the frequent economic downturns that beset most of our resource-based industries.

The growing emergence and empowerment of women in demanding the adoption of sustainable land-use practices in rural landscapes are also examined in this theme. The roles of education, social media and collectives (such as citizen groups) to increase the sense of belonging and place in remote and regional areas via enabling enhanced social connections and collaborations is scrutinised. The utility of using social media (especially Facebook) to assess and explore regional identities is developed in an especially novel way in this theme.

This theme also contains a particularly insightful investigation of the role of employment and leadership in the workplace in creating and maintaining the sense of both belonging and wellbeing in regional areas (in this case the Latrobe Valley) that have been subject to sustained external forces that challenge and threaten to undermine the economic viability of the regional organisations that have long provided much of the employment in that community.

Chapters in the *Challenges for the future* theme set out exciting directions and opportunities to address the challenges that regional areas are facing currently and will face in the future. Fundamentally, the chapters argue that the development of solutions will require the jettisoning of long-held beliefs that have been found wanting. For example, the first chapter in this theme details how the parlous state of many of our waterways can rent apart the social, economic, as well as environmental fabrics of many of our regional communities. Rehabilitation of these waterways presents a significant opportunity to repair and indeed revitalise these communities: it is an opportunity that cannot be ignored without continued substantial societal cost.

Another critically important contribution to this theme emphasises the fundamental, yet too often over-looked, importance of collaboration with, empowerment of, and truly listening to all of the community—not only the loud and/or powerful in the community—for the development of initiatives and strategies that speak to, and are capable of, leading to meaningful improvements in the health and wellbeing of all in the community.

The need to successfully address empowerment and social issues from diversity and equity considerations is examined in the light of the unique challenges that regional communities face with a view to the opportunities that the employment of successful approaches will provide to the enhancement of the resilience and wellbeing of these communities.

This third theme also addresses successful agricultural approaches that a rural community in the climatically challenged Wimmera Mallee region of Victoria has driven to ensure their continued agriculturally based viability. These approaches have centred around taking a leading role in the development and translation of appropriate and innovative agricultural practices through targeted impactful research.

Finally, this thought-provoking and insightful book provides an important and timely contribution to our understanding of the meaning of "regional" in relation to issues of wider significance, with the promise of catalysing new ways of thinking about how we may approach the development of solutions to some of our most pressing regionally based issues.

2

Regional Contexts and Regional Research

Angela Campbell, Michelle Duffy, and Beth Edmondson

Contemplating the creative process, polymath musician Brian Eno reflects:

> So I came up with this word "scenius"—and scenius is the intelligence of a whole … operation or group of people. And I think that's a more useful way to think about culture, actually. I think that—let's forget the idea of "genius" for a little while, let's think about the whole ecology of ideas that give rise to good new thoughts and good new work. (Brian Eno, cited in Pattie & Albiez, 2016: 2)

A. Campbell (✉)
Federation University Australia, Ballarat, VIC, Australia
e-mail: a.campbell@federation.edu.au

B. Edmondson
Humanities and Social Sciences, Federation University Australia, Churchill, VIC, Australia
beth.edmondson@federation.edu.au

M. Duffy
University of Newcastle, Callaghan, NSW, Australia
e-mail: Michelle.Duffy@newcastle.edu.au

A. Campbell et al. (eds.), *Located Research*,
https://doi.org/10.1007/978-981-32-9694-7_2

The collection of research from regional places offered in this volume explores cultures of interconnected and responsive research shared between our authors and their communities. What is important is not so much the "genius" of the work outlined in the individual chapters (though no doubt, this exists), but as Brian Eno suggests above, the "scenius"—the ways in which each research culture and the assemblage of research cultures collected here—touch, shape and hopefully improve regional lives and environments. The deep investment, complexity and fine-tuning of the scholarship shown throughout this book is a testament to the significance of the projects being done in regions across the world and how such work informs research approaches and methodologies more broadly.

For many of us, when we imagine regional places and contexts, what comes to mind are those locations set apart from that of the metropolitan. From this is conjured a series of binaries that offer stark difference between places and communities, whereby the urban is positioned as global, connected, cosmopolitan, and central while the regional is often positioned as isolated, homogeneous, and peripheral. A contrasting myth offers the rural and regional as bucolic and wholesome while the urban is congested and dangerous. This deep imaginary divide is traced through English literature and poetry by Raymond Williams in his seminal work *The Country and the City* (1973). Williams suggests that 'the contrast of the country and city is one of the major forms in which we become conscious of a central part of our experience and of the crises of our society' (1973: 289). More recently, Taylor (2014: 3) argues that such a focus has important implications as it reinforces ideas about 'where things are "happening" or not', and leads to

> treating the meanings of places as given rather than as social constructions and therefore historically contingent [and thus] opportunities to reflect on the ideas accompanying these movement are few and often limited to thinking about consequences in empirical and mostly casual terms.

Nevertheless, in a sense, the whole world is comprised of regions. Spaces and places and intersections. Moods of light, shades of colour, and sound. Languages of country and people, topography and terrain. Rivers, trees, towns, villages, forests, and farms. Yet, regions are also intersections between pasts and presents and possible futures. Regions are ideas and experiences of what is real and what is imagined. Regions cannot be dismissed as "edges" or "hinterlands". They cannot be contained in descriptions that accord them status as "other" in discussions and decisions that prioritise cities. Instead, they create (and perpetuate) dynamic gravitational fields. They're relational spaces. They are neither centres nor peripheries. Their fluid (often unmarked, sometimes unmapped) perimeters do not create fixed boundaries. Nor do they establish borders or delineate territorial claims.

We inhabit space in the world in ways that trouble such spatially determined definitions. Regions are comprised of places and peoples and systems—interconnected in conversations with time, culture, reality, belief, nature, industry, and relationships. Regions are more than human settlements and societies. They are "horizon places" with global and local roles and importance for how we live and how we find (and produce) meaning as individuals and communities. Thus, researchers who explore regions deal with scale, fit and interplay in ways that differ from research that examines local and global challenges through an urban lens. Rather than existing as separate and confined entities, regional places and lives are stretched, interrupted and woven across places and times (Duffy, 2010).

While this is so, Judith Brett discusses the historic and political dimensions of the changing relationship between regions in Australia, tracing growing disconnect between the city and country as a part of a wider neoliberal trend, captured by policy changes of the 1980s under Prime Minister, Paul Keating,

> City and Country in Australia share a history, a long history both of interdependence and of watchful suspicion. The understanding of that interdependence was strong in the first two centuries of Australia's European settlement, and the attempt to build a vibrant and self-sustaining countryside was major political preoccupation ... This understanding has waned rapidly since the neoliberal 1980s... since then the country has seemed to be in a perpetual state of crisis. (2011: 42)

Regional areas certainly play key roles in supporting our increasingly globalised urban societies and economies. In regions, problems, solutions and policy contexts intersect in distinctive ways, and so, while "the urban" and "the region" are interdependent, regional areas in some ways play the "canary in the coalmine". In times of transition, this regional canary indicates what might be benign or toxic in the air that we all share. If the canary suffers, soon enough, so will we all. Thus, in times of transition, regional research matters.

Expertise in regional research enhances our understanding of how knowledge is constructed and emerges across a range of environments and communities, wherever they are found and however they might be defined. The knowledge holders who identify problems are often those who are likely to work to solve them while also striving to influence others from outside the region to help support these solutions. This "situated knowledge" (Haraway, 1991) does not mean a parochial vision of the region; rather, as Haraway argues, situated knowledge offers a means to critically examine categories of ideas that 'are dependent upon unequal power for their maintenance' (1991: 1). Consideration as to how our lives are embedded within place is integral to our approaches to regional research, and thus, our "situated knowledges" can help us question certain assumptions about the region and relationships beyond. As Haraway (1991: 196) argues, 'location is about vulnerability … the only way to find a larger vision is to be somewhere in particular'. Acknowledging the situatedness of knowledge invites an understanding of 'our location and the location of others within complex webs of relations, places, affects, experiences, thoughts and processes' (Fisher et al., 2015: 25). Such conversations flourish between and across disciplinary boundaries. From a transdisciplinary and cross-disciplinary base, we can identify an emerging "landscape of practice" in regional research that can help us translate "knowledgeability" from local to global contexts (Wenger-Trayner & Wenger-Trayner, 2014). Regions provide the *terroir* of non-metropolitan communities. Regional *terroir* flavours the relationships among people-nature-place-space-culture-livelihood-industry-past-present-future-beliefs-languages-knowledge.

This shared space also draws attention to the ethical role of the researcher within regional communities and environments, not as a separate observer, but as an embedded and active agent across the range of human and non-human elements. As researchers, we are affected by and affect the range of others—people, places, things, networks, and so on—that constitute regionality. This interdependence, accelerates regional trajectories, offering

> energised possibilities … of tensions that constantly challenge us to think differently about the relations between presupposed and fixed definitions of "region" and what lies beyond them and yet relates back to it in multiple ways. (Campbell, 2016: 5)

Recognising opportunity and acknowledging the challenges of research conducted outside of highly resourced major metropolitan centres is strategically important at a time when global connectivity and mobility are increasing and the complexity of "wicked" problems demands more than one approach or solution. Addressing these wicked problems increasingly requires transdisciplinary approaches to better understand and support adaptation and transformation (Galvin et al., 2016; Marin, Ely, & van Zwanenberg, 2016). Such work suggests more complex ways of being, engaging with, and even participating in the co-constitution of knowledge. To this end, the research presented in this volume exemplifies the value of diverse perspectives offered by researchers and their lived experience in regional communities and environments across disciplines spanning social sciences, environmental science, critical indigenous studies, art practice, pedagogy, tourism, health, and cultural geography.

Our contributors point to the potential of shared training and expertise in regional research methodologies to create and further enable active, research "citizens" (Omidvar & Kislov, 2014: 272) to build capacity and vitality in our regional communities and diversity and abundance in our regional environments, for the benefit of all. As such, an important structural framework underpinning the work presented in this volume is that of co-design through which we acknowledge the importance of developing forms of knowledge to enable and foster social change through the inclusion of different ways of knowing different kinds of stakeholders and diverse perspectives (Galvin et al., 2016; Marin et al., 2016). In addition, we recognise the dynamic ways in which different stakeholders come together, often via informal network clusters or socio-economic systems (UNIDO, 2013). Bringing together co-design principles and network clusters helps to address the concerns and issues within regions and facilitates ways to seriously consider knowledge building as an open-ended process in which 'new balance is continually being sought among multiple social, economic and environmental challenges and goals' (Grove, Chowdhury, & Childers, 2015: 6).

This collection brings together researchers and practitioners who seek to identify what is particular to regional research. Do certain methodological approaches work better in regional environments than others? Is research that is situated within a regional environment different from research into regional issues that originate from major metropolitan centres? In asking such questions and considering them in regional contexts of *Place and change*, *Transition and resilience*, and *Challenges for the future*, this book examines the diversity of practice through specific examples, it highlights the contribution of local research to national and global issues and, hopefully, contributes to the vibrancy of research practice, overall.

References

Brett, J. (2011). Fair share: Country and city in Australia. *Quarterly Essay, 42*, 1–99. Melbourne: Black Ink.

Campbell, N. (2016). *Affective Critical Regionality*. London and New York: Rowman & Littlefield.

Duffy, M. (2010). Sound Ecologies. *Cultural Studies Review, 16*(1). Retrieved April 23, 2019, from http://epress.lib.uts.edu.au/journals/index.php/csrj/index

Fisher, K., Williams, M., FitzHerbert, S., Instone, L., Duffy, M., Wright, S., et al. (2015). Writing difference differently. *New Zealand Geographer, 71*, 18–33.

Galvin, K., Reid, R., Fernández-Giménez, M., Kaelo, D., Baival, B., & Krebs, M. (2016). Co-design of transformative research for rangeland sustainability. *Current Opinion in Environmental Sustainability, 2*, 8–14.

Grove, J. M., Chowdhury, R., Childers, D. (2015, April). Co-design, co-production, and dissemination of social-ecological knowledge to promote sustainability and resilience. *GLP News*, pp. 6–11.

Haraway, D. (1991). *Simians, cyborgs, and women: The reinvention of nature.* London: Free Association Press.

Marin, A., Ely, A., & van Zwanenberg, P. (2016). Co-design with aligned and non-aligned knowledge partners: Implications for research and coproduction of sustainable food systems. *Current Opinion in Environmental Sustainability, 20*, 93–98.

Omidvar, O., & Kislov, R. (2014). The evolution of the communities of practice approach: Toward knowledgeability in a landscape of practice – An interview with Etienne Wenger-Trayner. *Journal of Management Inquiry, 23*(3), 266–275.

Pattie, D., & Albiez, S. (2016). Introduction, Brian Eno: A problem of organization. In D. Pattie & S. Albiez (Eds.), *Brian Eno Oblique music*. London: Bloomsbury.

Taylor, W. (2014). Introduction to Australian places, place-making and the politics of displacement in a transient society. In P. Goldswain, N. Sully, & W. Taylor (Eds.), *Out of place (Gwalia): Occasional essays on Australian regional communities and built environments in transition* (pp. 1–13). Crawley, WA: UWA Publishing.

United Nations Industrial Development Organization (UNIDO). (2013). *The UNIDO approach to cluster development key principles and project experiences for inclusive growth*. Retrieved April 13, 2019, from www.unido.org/fileadmin/user_media_upgrade/What_we_do/Topics/Business__investment_and_technology_services/CUP/UNIDOs_CLUSTER_APPROACH.PDF

Wenger-Trayner, E., & Wenger-Trayner, B. (2014). Learning in landscapes of practice: A framework. In E. Wenger-Trayner, M. Fenton-O'Creevy, S. Hutchinson, C. Kubiak, & B. Wenger-Trayner (Eds.), *Learning in landscapes of practice: Boundaries, identity, and knowledgeability* (pp. 13–30). London and Routledge.

Williams, R. (1973). *The city and the country*. London: Chatto & Windus.

Part I

Place and Change

3

Yanama Budyari Gumada, Walk with Good Spirit *as Method*: Co-creating Local Environmental Stewards on/with/ as Darug Ngurra

Darug Ngurra, Uncle Lex Dadd, Paul Glass, Aunty Corina Norman-Dadd, Paul Hodge, Sandie Suchet-Pearson, Marnie Graham, Sara Judge, Rebecca Scott, and Jessica Lemire

Introduction

'Thank you to Aunty Corina, Darug custodian, singer, songwriter, dancer, weaver and teacher for sharing her knowledge of Darug language as part of our research collaboration' Yanama budyari gumada, to "walk with good spirit", is a Darug phrase that embodies what we try to do together in our intercultural collaboration at the Yellomundee Regional Park, located in the Greater Blue Mountains area in New South Wales (NSW), Australia.

D. Ngurra • U. L. Dadd
Yilby, Blue Mountains, Sydney, NSW, Australia

P. Glass
National Parks and Wildlife Service, Glenbrook, NSW, Australia
e-mail: paul.glass@environment.nsw.gov.au

A. C. Norman-Dadd
Dharug Ngurra Aboriginal Corporation, Sydney, NSW, Australia

A. Campbell et al. (eds.), *Located Research*,
https://doi.org/10.1007/978-981-32-9694-7_3

Yellomundee, also known to many custodians and community members as Yarramundi is, and always has been, Darug Ngurra, Darug Country, a place of immense connection and meaning to its Aboriginal custodians, the Darug people. Yarramundi has always been a highly significant place for gathering and trade, with rock engravings, grinding grooves, knapping sites, and layers of inhabitation woven throughout Country. However, Yarramundi is also a heavily colonised place. It is bounded by intense residential development from the edge of western Sydney and is overrun by weeds and pollutants from colonial agricultural and mining practices. As a regional park within the NSW national park estate, Yarramundi is highly recreated, used for sanctioned and unsanctioned activities, including mountain biking, abseiling, fishing, four-wheel driving, and horse riding.

Our research collaboration, which we have called yanama budyari gumada, supports current efforts to nurture Darug Ngurra as a place in transition. As senior Darug custodian Uncle Lex Dadd explains: "We're walking our dreaming together now", trying to listen carefully to Darug Ngurra, so that we might heal Country. This is an important learning process of reconnecting—with the ancestors, with Mother Earth, Father Sky, with each other, with ourselves, and with all the sentient beings which make up, through their co-emergence and co-becomings, what is known in Aboriginal English as Country (Bawaka Country et al., 2016; Rose, 1996).

Our collaboration brings together Darug custodians (Uncle Lex and Aunty Corina), NSW National Parks and Wildlife Service (NPWS) (Paul

P. Hodge (✉) • R. Scott • J. Lemire
University of Newcastle, Callaghan, NSW, Australia
e-mail: paul.hodge@newcastle.edu.au; rscott2@uon.edu.au; jessica.lemire@uon.edu.au

S. Suchet-Pearson • S. Judge
Department of Geography and Planning, Macquarie University, Sydney, NSW, Australia
e-mail: sandie.suchet@mq.edu.au; sara.judge@hdr.mq.edu.au

M. Graham
Stockholm Resilience Centre, Stockholm University, Stockholm, Sweden

Glass, an Aboriginal man from Kamilaroi Country and ranger for Yarramundi), and Macquarie University and University of Newcastle researchers (Paul, Marnie, Sandie, Sara, Bec, and Jess) to develop, model, and advocate greater environmental stewardship. We are doing this by facilitating important connections between Darug and other Aboriginal people, including youth, environmental experts, management authorities, and users of Yarramundi. While our work is currently funded by a NSW Environmental Trust grant (2018–2020), we have all been slowly developing our relationships together with, through, and as Yarramundi since 2016, and Uncle Lex and Paul Glass have been developing their relationships since 2009.

In this chapter, we share some of the ways we yanama budyari gumada at Yarramundi *as method* to care-*as*-Country. Caring-as-Country acknowledges that humans and their actions are integral to environmental wellbeing, and that Country cares in return (Bawaka Country, Suchet-Pearson, Wright, Lloyd, & Burarrwanga, 2013; Darug Country et al., 2019). Our caring-as-Country practices at Yarramundi bring together different knowledge systems and ways of knowing and being (Ens Finlayson, Preuss, Jackson, & Holcombe, 2012). We reflect here on two particularly important and iterative caring practices embodied at Yarramundi—yarning and dance circles—which are aimed at healing Darug Ngurra and enhancing environmental stewardship of the area. We share herein the ways we are facilitating, reflecting on, and documenting this intercultural and more-than-human interdisciplinarity. Human and non-human entanglements at Yarramundi "co-fabricate worlds, spaces and encounters", asking that we acknowledge the ways "that non-humans are active in shaping research processes" (Bell, Instone, & Mee, 2018: 136; see also Rose, van Dooren, Chrulew, Cooke, Kearnes, & O'Gorman, 2012). Country teaches us to yanama budyari gumada through our relationships with Yarramundi—the birds, plants, river, weather, and more. If we listen to Country and know what to do with this learning, we might all have a chance to heal.

To yanama budyari gumada today at Yarramundi, on Darug Ngurra, is an incredible privilege, yet it is also complex and sometimes fraught. Co-existence in a place of immense cultural value on unceded Darug land, located at the regional fringe of Australia's most populous city, presents myriad challenges. Policy, bureaucratic, managerial, and organisational challenges abound given that many current institutions are built on assumptions of *terra nullius*. Transforming these systems and the relationships which sustain them presents an immense challenge. Yet, there *are* collaborations in the Greater Blue Mountains area that embody the kinds of intercultural and Indigenous-led relationships that exemplify what it is to yanama budyari gumada (see Yellomundee Aboriginal Bushcare Group; Yellomundee Firesticks http://www.nrm.gov.au/indigenous-nrm/greater-sydney/yellowmundee-bush-care-firesticks). Our collaboration is another example that hopes to build relationships that heal Country and communities. In the following section, we share how "we're walking our dreaming together now" (Uncle Lex in Scott, 2017) as part of this healing. We reflect on how we come together, our relationships with our human and non-human kin, how we co-emerge together, and what we co-create. We also share some of the challenges and how we attempt to negotiate these, the mistakes made, and dilemmas faced. We do this by sharing two yarns we have had over the last year, *in, with, and as* Country, and draw attention to the learnings as we embody and advocate local environmental stewardship on/with/as Darug Ngurra.

We begin our first yarn on the banks of the Deerubbin, the Hawkesbury-Nepean River, January 2019 at Yarramundi (see Fig. 3.1). While centred on this time and place, our yarn draws across and through time from the contributions of the present ancestors to ongoing yarnings as they have been drafted and re-drafted through the writing and revising of this chapter. For this yarn, we've taken a moment during one of our regular culture camps, where we bring people together on Country, to reflect on what it is to yanama budyari gumada. This is what we do: yarn in/with place on/as Darug Ngurra. It adds meaning to what we are doing together, now, and in ways that invite Country Herself to make Her presence/s known.

It's been a particularly wet camp characterised by rain, thunder, and mozzies. Following December flooding, stagnant pools have formed, gifting mosquitos the ideal conditions to be, and co-create, with and off us. "The flood was a cleansing and rebirthing", Sara reflects, "but now the mozzies are seeing what we're made of!" Their bites remind Sara of Val Plumwood's work that we're part of Country and that being fed on by

Fig. 3.1 The banks of the Deerubbin, Hawkesbury-Nepean River (Photo taken by: Suchet-Pearson, 2018)

other beings is part of that relationship (Plumwood, 1995). The mosquitoes, through their presence, brought ideas and reflections to our conversation.

There is a comfort and ease among the group today, as there always is. It's work, yes, but it's so much more. We are mudjin, family, sharing what it is to be together in *this* place. We offer some of ourselves with a familiar knowing. When it's time, we share in a way that reminds us that people have done this very thing, in this very place, for thousands of years, forever. Uncle Lex describes our collaboration, yanama budyari gumada, in this way:

> When we have our yarns we really come together with that respect. We never have any competition of egos… we come to each other as equals. We each have different expertise in different fields, in the framework of our old ways.

Sandie adds a different understanding of this mutual, respectful, coming together:

> It's love that's pulling us in. I feel like there's a force here that pulls people together and draws people and the energy in. History but also the place and very much Uncle Lex. I think there's something about Yanama Budyari Gumada that is about inclusivity—so a really generous opening up for anyone to be drawn in.

For Sandie, there is something about this area; the boulders, rocks, river system, layers and layers of pull. Deep time over centuries, right where we're sitting, here, right now, "it's vibrant and pulls people in". Marnie describes connecting to "a sense of deep time, and the present and the future" at Yarramundi, and also a connection to her own past. It's the rivers, swimming, and bush that remind her of growing up on the NSW Central Coast. Now living in inner Sydney, under a flight path, and "with no green bush to access—it's hard to connect with Country". Emile, Marnie's two-year-old son, doesn't have the same opportunities to be in nature, but Uncle Lex and this connection with Darug culture and Country has changed all that, and each visit to Yarramundi with mudjin leaves them feeling connected to each other and to Country in new ways.

Just as yanama budyari gumada opens up new relationships, it also exists because of enduring ones; relationships of trust with others beyond, but vital to, the collaboration. Paul H describes the way his strong relationship with Sandie only came about because of an existing relationship he has with a mutual colleague, friend, and once PhD supervisor, Sarah Wright. So it is the trust that Sandie and Sarah have cultivated over time that is transferred to Paul—because of his relationship with Sarah—thus making possible the invitation to meet Uncle Lex. Paul elaborates on this relationship of trust and intergenerational aspects of yanama budyari gumada:

> We have Sara here who has Sandie as a PhD supervisor. While Sandie was introduced to Uncle Lex by Richie Howitt, who was her PhD supervisor. Not only the pull in the here and now but also the histories, the intergenerational sharing and learning, humility and patience… Sara and I were

> talking how one day Sara and Bec will be supervisors of the next generation. Malakai, Archie and Daisy [some of our children] might go to university and on it goes... It's just profound. It's something that struck me.

The pull described by Sandie and Marnie is unmistakable in the younger mudjin too. Paul G talks about the enticing nature of Yarramundi camps for his two children:

> [Daisy and Archie] changed their holiday dates (with their mum) so they could come here (to this camp)... Daisy got up at 6 am because she was so excited! They had packed their bags two hours early!

Mudjin makes yanama budyari gumada possible. Sara refers to "the two Chris'" (Sandie's and Marnie's husbands), hinting at their centrality in enabling what we're doing by being present at the camps to share parenting responsibilities. The "blurring of personal, private, work", Sandie adds, is how it all happens, like at Bawaka (Lloyd et al., 2012; see Fig. 3.2). Aunty Corina (Uncle Lex's partner) at home with Eli (their youngest

Fig. 3.2 The blurring of work and family (Photo taken by: Suchet-Pearson, 2018)

child) allows Uncle Lex to be here now. For Marnie, having Chris here makes her "feel calmer", taking a break from running after Emile and allowing time for other things during the camps. Paul G talks about the difficulty of "switching off" from his role as park ranger as part of this blurring, yet having his children here deepens their relationship. With a smile he reminds Uncle Lex of what he told him at an earlier camp: "remember to play with the kids, it's what they remember".

So the "pulling in" of yanama budyari gumada enmeshes old ways, serendipity, deep time, new and enduring relationships, and intergenerationality. These things coalesce as we co-emerge and co-become together (Bawaka Country et al., 2016). Yet, there are always challenges. One challenge is that we're not always here. Mudjin misses mudjin. Sara describes the way she is feeling not having Bec or Jess here at this camp:

> I feel they are my sisters in a lot of ways so their absence hurts and reminds me how awesome it is to have those relationships—we are similar age, same point in our relationships, we've had amazing experiences together.

Sara reflects on the regular contact via Facebook with Jess who is in Fiji, and Jess missing being here with us. Yet, for each of us, they *are* here, and *always will be*. Sandie and Paul H recall a moment a few days earlier while tending to the dance circle, where Jess was so intimately connected through her thesis over the past year. Two butterflies made their presence known to us. With their gentle and familiar rhythms, they darted in and out around us. "Jess and Bec are checking in", we thought, making their own imprints in the air and soil, reconnecting across time and space.

While there are challenges, the "pulling in" continues, reverberates, as it has effects beyond the collaboration. Paul G talks about the way "the bosses" are "starting to come in". He is describing the way people within the NSW NPWS are getting a sense of the positive connections and events emerging through these camps; Darug community involvement, the co-creation of the dance circle, the making of clap sticks, the clearing of invasive grasses, and the sharing of gatherings on social media (see Fig. 3.3). So the "pulling in" has its own life and energy as others not directly connected to Yarramundi become deepened—and sometimes

Fig. 3.3 A camp participant sanding their clapsticks (Photo taken by: Suchet-Pearson, 2017)

distanced when relationships don't work—by what yanama budyari gumada is and does.

Uncle Lex speaks next in a sincere voice. "'This weekend we've had four traditional custodians of this Country, from this clanal Country, join us' and they are 'all loving it!' 'It's not my clanal Country', Uncle Lex adds, but 'none of them have pulled me up', in terms of creating friction over who may 'speak' for Country. They recognise what we're doing and how it's building".

But Uncle Lex reminds us that we "need to be patient" as we walk with good spirit. A reminder just as much to himself, it seems, as he recounts burning his toe and stubbing his feet over the weekend. Sara had told him it's Country "slowing you down". Wanting you to "get more grounded in Country". Uncle Lex elaborates on the conversation they had and what it means to him:

> I love that she can say that to me... Country is telling me something. That makes such sense. That we can have that understanding and patience with each other, we have the attitude where we can stop and listen.

And Country seems responsive, too, as the source of this pull. Sara describes the way the Deerubbin is "a huge pull" of Yarramundi, and "very much alive", changing and moving with each encounter, both "nurturing and fierce". Through such changing moods, Darug Ngurra tells us things. Gives us signs. Teaches us. When birds fly over the Deerubbin, we listen. "The pelican and sea eagle are our elders too", Sara reflects.

Yarning in/with/as Soil

Our second yarn unfolds in the soil of the dance circle at Yarramundi in September 2018. The dance circle was co-created during two culture camps in early 2018. Together we cleared, weeded, dug out roots, and prepared a soft-dirt dance circle on a plateau up from the Deerubbin in preparation for a major culture camp and corroboree in April 2018—the first gathering of its kind for five years and very much anticipated by local Aboriginal communities (https://youtu.be/0tXIAeQ-4sg). So several months on after this and other highly successful camps and corroborees, it is special to be here once again, yarning together (see Figs. 3.4 and 3.5). Jess is posing questions as part of her Honours research on dance and emotions. She is interested in teasing out what it means to us to connect with Darug Ngurra in this place, on the dance circle, over time, and what emotions or memories come in for us.

Jess: Being on the dance circle right now, what kind of emotions or memories are coming up for you?

Uncle Lex: For me knowing that we'd done it [the dance circle] so quickly, and it's still here… seeing the old fires, knowing that we had a corroboree, and knowing that we're going to have a corroboree here again very soon, and we're going to be all dancing here. You know? I'm excited about keeping it alive.

Paul H: I've been feeling like something's been missing actually, and so coming back here and even just walking up then to the circle, I couldn't wait to get my shoes off, and then Paul G saw the remnants of Country speaking in terms of the animals who'd been here and their footprints in the soil.

Fig. 3.4 The dance circle in its early emergence; working together to dig out lantana and African lovegrass (Photo taken by: Judge, 2018)

Sara: I feel such affection for the dance circle. A couple of weeks ago, I had the same experience that I did this morning: I got to the dance circle and felt reverent, I went very quiet. I remember the moment when I first stepped inside the perimeter of the dance circle, and it felt like being held. Like hugging the dance circle when you step into it. When I step into it, I have very clear memories of what it was when we started, how it looked. I have memories of the corroboree and the times that we met here and did things here. There are some parts of it you can't explain or fully share. The dance circle is a person, and you have a relationship with that person. I feel a lot of kinship with the dance circle.

Jess: How's the transformation of the dance circle affected you?

Uncle Lex: So, for instance, I've just travelled up to Queensland and travelled a lot of different Country and back into New South Wales and I see a lot of cleared land for sugar cane and things like that. They serve a

Fig. 3.5 Preparing the dance circle for a corroboree (Photo taken by: Suchet-Pearson, 2018)

purpose but a lot of it's just for pure greed… This is being cleared for a very special, spiritual essence of connecting a lot of people to Country. To get their shoes off, to feel mother, and it's through parallels or through polarisations to different ends. People clearing land for greed, and people clearing land for a spiritual purpose to get back to living to a closer essence of our old people.

Sandie: For me… it was almost a sense of it was emerging, like it was there already, and we were just helping it along. It's not that we created a dance circle, it's more like we were nurturing and looking after and helping it re-emerge. I don't know if there had been a dance circle before, but it was meant to be here now… It definitely belongs here now, I don't know how much longer or before, but definitely it's meant to be here now.

Uncle Lex: Well, Paul G had already actually picked this area out many, many years before, or maybe the area picked Paul G?

The conversation comes to the signs of Country speaking. What She is telling us.

Uncle Lex: And then the other thing that we have to remember too while we were digging, clearing the dance circle and preparing it, were that we had lots of signs. We had eagles flying over and different types of birds visiting us. And that, for me, was just a sign from our old people saying good on you, well done, you know, it was affirmation....

Sara: There was also the big double rainbow in the sky, there was the wind moving through just as we learned a particular dance, what else happened?

Bec: There was a bit of rain.

Paul G: Yeah, I remember when we were all dancing and learning those dances I just hadn't felt that good naturally for ages, like especially in the lead up to the camp I was going through all this personal stuff and it's like now it's come full circle. And you know? I feel strong again... I feel like that whole journey helped me get through it. And this group definitely. I've come out the other side, you know, from a personal thing.

Jess: So are there any feelings of—especially tangible, sensory-related feelings that you feel in your body on the dance circle, remembering your connection to this place?

Sara: Sweat! Both in making the dance circle and dancing.

Sandie: I remember goosebumps. I think when Uncle Pete was pointing out the rainbow and the wind going through and that sense of more than the normal, I think very physical goosebump-y, shivery, ancestors here, "part of what's happening" feeling. And that rake-hoe going under the lovegrass. And my sense that the lovegrass was this really deep-rooted thing that was not going to budge and the rake-hoe just going "nup, we can work together". And it's sad because you are killing the lovegrass, but it's also like I'm okay leaving here. It sort of felt like it's okay.

Paul H: The lovegrass then became the kindle, the sort of flare up for each of the dances during the major corroboree. It participated in the dance. It was part of it.

Paul G: It's alive.

Uncle Lex: So on that note brother, it wasn't wasted. We weren't just killing a weed for the sake of killing a weed, or what's perceived as a weed, it had a purpose and it gave us light for our dance. So it had a purpose.

Paul G: Yeah it's interesting, it was brought in actually after the mining to stabilise the soil. So it's done, its job [pause], but now it's time to move back to….

Marnie: Have another job?

Sara: When we were digging all the weeds to make the dance circle, I was aware of my heartrate increasing. And when I dance I'm also aware of my heartrate increasing. It's reminding you that you're alive, and that you're bringing something life-giving—you're dancing the dance circle alive, and you're weeding it alive. That sense of there being a heartbeat involved—collective heartbeats.

Sandie: And the texture is just so inviting, it feels beautiful. But it needs you to do it to become soft. You get here and it's not automatically soft, so the bit we haven't sat on, initially is hard, but it doesn't take much at all to work with it. So just your feet or your fingers really quickly turn it—the silkiness emerges.

Paul H: So it's the co-becoming, isn't it? It isn't until the human hits the non-human, and then it becomes more-than-human, or something….

Sandie: And I think that was one of the things you told us, Uncle Lex, really early on, we were softening it, going around and around. And the kids, that's why they love running on it as well. There's something really satisfying… physically, feeling that transformation from the hard earth into the silk.

Rhian (Sandie's daughter): The textures you might have to [rubs her hands together with some dirt between them]… they add another texture, they're not sticks which can hurt your feet. It just brings something. It's not just the dirt [hands rubbing].

Uncle Lex: I think for me too this is the closest I would come to for want of a better word—an ownership of Country—I never wanna really own land—and I don't believe you own it because it's your Mother but I suppose it's that connection to the dance circle cos we all achieved it together. So you know probably not ownership, and all the little weeds which are popping up, which is not a lot of them, but they're struggling we don't even have to pull them out. We should clear the sticks and dance, and we will weed through dancing.

Rhian's loving it that much that she's pouring the dance circle all over her jeans... How cool is that? A young fella just rubbing the dirt cos it feels so good all over yourself. From the dance circle. And there's been so many energies here, dancing and loving the place. And there's a collective love and energy in the dance circle together. A lot of people's spirit are here. A lot of people's beautiful energy. And I remember all the people sitting around and watching the corroboree (Fig. 3.6).

Marnie: For me it's not just about doing the dancing, it's about that you can jump in and out, and it feels good both participating or sitting on the side. And when we were watching at the major corroboree, I felt my heart swelling. I remember saying to Lex at the time that I hadn't quite understood the gravity of what we were doing when we made this [dance circle] until that moment. I totally cried during the corroboree. I was like "oh my god, this is so amazing".

Fig. 3.6 At the dance circle, watching the corroboree (Photo taken by: Judge, 2018)

Sara: During the corroboree, I had a really strong sense of smell from the dance circle. You could smell the soil and you could also smell people's bodies. There was a lot of energy going into the dance circle that night, really big, vibrant, energetic dancers. Especially when the men were dancing, you could really smell the sweat. But it wasn't in an unpleasant way. It was very powerful, it smelt very earthy. I remember watching the animal dances, and the combination of the visual dance and the smells coming from it—I had this moment where I thought, "we really are animals, they're really summoning up the animals in these dances, aren't they?" They really embodied the animals in such huge ways through the dances (Fig. 3.7).

Sandie: There was something about the sound and the singing and the stomping and the vibrations. There really is that person, land, Country, air, sky, trees, people, quite a physical connectivity between everything through the sound, and the vibrations, and the rhythms. You don't get that very much in general life….

Jess: Is it a different feeling of connecting to the dance circle when there's a lot of people on it compared to (now)… how does it feel in your body when you dance with other people?

Marnie: I feel connected to them. Whoever's in here, through the dancing… but it's also a different kind of dancing from what I've grown up with. You know, jazz ballet kind of dancing, which is really performative, but it's like you're always doing the performance for an audience, whereas it never feels like that here. It feels like you're doing it for yourself. So it lacks this self-consciousness, that I haven't had with the other

Fig. 3.7 Learning, dancing, co-becoming (Photo taken by: Suchet-Pearson, 2018)

dancing, you know, this experience. I find it connects me to the other people that I'm dancing with.

Sara: I agree, I don't feel comfortable in my own skin a lot of the time, so when there's a small group, I love dancing here. With the bigger group, it takes me a little while to feel comfortable doing that. But I feel like the dance circle does facilitate that comfort in a lot of ways. It's almost like—look at this little worm—sorry, I got distracted!

Suddenly we all notice a caterpillar making its way through the dance circle, right where we are sitting, and the conversation deviates a little as Country speaks.

Sandie: Oh wow!

Paul G: Yeah

Uncle Lex: Where?

Sara: He's right here.

Sandie: Covered by all our dirt. Where did he come from?

Sara: I don't know. Yeah, but the shape of the circle, and that when we're learning something we gather around the edges of the circle, really facilitates that comfort. Even though when there are more people, a lot of people you don't know, you get this immediate sense that it doesn't matter because we're all equal, we're all together. No one's in front of someone, no one is more important, and that really breaks down that discomfort.

Uncle Lex: So just on that note Sara, on exactly what you just said when you pointed out the little caterpillar that's cruising along through the dance circle now. I've been out west and I've seen a long line of probably a couple of hundred caterpillars and they're all following each other on this line, and old uncle showed me—he put a stick in their pathway and they go around the stick, okay, but they're still going on there, and they go back—so there's the line, perfect line, round the stick, then the line continues perfectly. So that just resonated with what you said Sara, that we're all trying to stay on that same pathway, or that line of yanama budyari gumada, to walk with good spirit. To stay on that path of goodness and what we're achieving here.

Paul H: And I guess those challenges represent the stick.

Uncle Lex: Yes.

Paul H: And we sort of do our best to go around them and try to stay on that path....

Uncle Lex: Instead of confronting the stick and beating the stick up or burning the stick or....

Paul H: Going over the stick....

Uncle Lex: We just move around. You're there, we respect you, but we're going to go around and still continue on our path.

Paul H: For me it's like they're still here, it's almost like the presence of the people at the big corroboree is still kind of here, that it's cumulative. That all their essences are still here and we're just another layer, and there's going to be more and more layers. So, for me there's a timelessness about being here and coming here. I can get a feel for them again, I can still see them, I can see where I stood, so for me whether there's human bodies in it or not, it doesn't matter, because it's just an accumulation of those memories and presences and experiences and sounds and smells and textures. There's no difference for me whether there's all these bodies here or not. There's been all these people using this space, and then a layer of colonialism and now there's this kind of decolonising. Or re-layering of important past practices coming back in new ways that are inclusive. That it is about connecting culturally, and like Uncle Lex often says, you're just coming back home. Everyone's just coming back home.

Jess: During the big corroboree did you feel connected at all to the animals that you danced and when you dance how does that connection relate to the environment around you?

Paul G: I really love doing the kangaroo, and feeling this space, the rocks and stuff. We see them moving across here all the time and so I just felt... like a kangaroo at the time.

Marnie: Someone who did feel connected who was on my hip when we were dancing the emus is young Emile who now has two emus of his own since that day, they're called Hink and Nyneen. We hear about them every single day since the camp. They speak together and they speak to him and...

Sandie: They're furry toy emus or are they...

Marnie: No, they're his hands.

Uncle Lex: Wow! That's amazing.

Marnie: Uncle Tim says that it's the old people speaking through him. It's just like they just came, and they haven't gone anywhere, and they have distinct personalities. It's amazing.

Sara: I really loved learning the robin dance because we learned that dance right after I had an encounter with the rose robin here. That encounter with the robin, then doing the dance, is pretty spiritual for me. Since that time the robins have become significant in making life decisions and seeing things unfold in my life. I feel like in doing that robin dance, I invited something into my life. So it's not just dancing. You really are putting out invitations.

Sandie: For me it's an invitation to stop and slow down and pay attention. I'm not good at doing that at all. We had to do the willie wagtail. I love the willie wagtail, so I often notice them—well, I noticed then that having to try and dance one takes it to a new level of connection of body and movement and brain and care and attention. Yeah, I just think that's so important.

Jess: So was there ever a moment where you really felt out of your body or connected to the larger group rather than a single individual?

Bec: At the end of the corroboree when we were stomping and had our hands up at the sky, that was such a moment. Everyone was dancing at the same time. It was beautiful.

Uncle Lex: Stomping for mother.

Sara: And the stomping was like that heartbeat too, a unified sense of "we're together, everyone's heart beating together".

Paul G: I feel when we do the stomping part, Bec mentioned it. I love just doing it all together, being—I guess I get lost in that. Like feeling it. Feeling together, and feeling that energy. And feeling as though you're sharing that.

Uncle Lex turns the conversation to Jess, reminding her that she'll be leading some of the dances with the university students at an upcoming student culture camp.

Uncle Lex: How do you feel about that responsibility?

Jess: Excited and nervous, nervous in the sense that I really want to do it justice and really want to show Country respect and appreciation—but excited at the opportunity of being able to do that and being able to show that appreciation for Country and for the dance circle. Yeah, and being

Fig. 3.8 The trees embracing the dance circle (Photo taken by: Suchet-Pearson, 2018)

able to share that experience, I think is really special, especially because I was able to help do the weeding and do the dancing and just all of those experiences together have just made it really personal and really connected.

Jess: Can you reflect on the trees outside of the dance circle?

Sandie: Yeah, for me they do shape the dance circle, like they're a part of it. The trees tower over it (Fig. 3.8). They're very relational, they move depending where you are in the circle and how you look at them. So right now those ones are really leaning over it, and then the fire and in the painting brings that alive for me. I think they're a key part of it. I can't imagine it without the trees actually.

Uncle Lex: I'll be interested to see how these trees grow over the next few years, because my feeling and my gut feeling is that, they can breathe easier, they've got room to grow. But not only that, we're here and we're loving them and giving them energy. I think it would just be interesting to see if they grow taller and bigger and stronger. Just from our love and our energy here. Old people talk about that—where we're powering up Country, we're giving it our energy, and that's what's happened throughout Australia for thousands of years. There's been a big shift in that since colonisation. We're not powering up Country like we should be, we're on Country, but we're just using Country, just for what it can give us, not from what we can give it, or share with it. And so, yeah, just want to see if there's any change. And it might just be that there is a big change because we want a big change. And we want it to have our love. You know, because it's loving us back.

Sara: It's all about your intentions. I remember when we first came out here. The lantana really demanded your attention. It was everywhere. And now these large trees around the dance circle, I really feel their presence a lot more. It's like the dance circle has allowed them space. Opened them up, revealed them again. The way the wind is moving through the trees at the moment, and Uncle Lex has always told us that the wind is the Old People moving through Country, they start to do this gentle sway. It's like a dance between the trees and the Old People around the dance circle. Everything's dancing together in subtle ways that are maybe different to how we dance, but it's still a dance and it's still part of this.

Creating Openings

This chapter focused on and is deeply embedded in our experiences working with/as Darug Country at Yarramundi in western Sydney. We recognise that in a time of global environmental crises and attendant cultural and ecological loss, we have important lessons to share with other peoples and places through yanama budyari gumada, walking with good spirit. As a research collective we are led by Darug custodians, and together we walk this good spirited walk with patience, humility, and respect, always trying our best to carefully listen, lead with love, and leave our egos behind. We know these research protocols are essential for the health of our collective, enabling us to learn from and with each other always.

We also know these protocols are fundamental for working towards our research aim of together healing Darug Ngurra, Darug Country. Our yarning circles show the power and reverberation of Country speaking. Country is always "pulling" us, inviting, gifting, loving, challenging, and trying to communicate with us—in, through, across, and as time and place. However, if we don't listen appropriately, recognise, and respond in embodied and multi-sensory ways, then we can't or won't hear what Country is trying to say.

Indeed, so much of conventional "modern" land management practices fail to listen to what Country is saying, which inevitably leads back in a cruel loop to our current time of ecological and cultural crisis and loss. We, therefore, have an ethical imperative to *really* listen and respond to Country—wherever your Country may be, to ensure that we are hearing all the messages—using all our senses, embracing diverse ways of knowing, and connecting to the ancestors, past, present, and future. We are faced with incredible uncertainty in this time of multiple crises. What is apparent is that we need to do something different. By opening ourselves up to listening to Country and walking in good spirit, we open ourselves to the possibilities of re-imagining and re-shaping responses, ideas, and perspectives on how to approach environmental degradation, loss, and uncertainty.

Acknowledgement This chapter is shaped by, and an intimate part of, Darug Ngurra and we acknowledge and thank Ngurra for nurturing and enabling this work. This includes our broader mudjin, our wonderful and supportive family—our partners and children, parents and friends—always an integral part of our yanama budyari gumada. We'd also like to acknowledge and thank other Countries which we live on, travel through, work on, and which are also a critical part of our walking. This project has been assisted by the New South Wales Government through its Environmental Trust.

References

Bawaka Country including Suchet-Pearson, S., Wright, S., Lloyd, K., Burarrwanga, L., & Bawaka Country. (2013). Caring as country: Towards an ontology of cobecoming in natural resource management. *Asia Pacific Viewpoint, 54*(2), 185–197.

Bawaka Country including Wright, S., Suchet-Pearson, S., Lloyd, K., Burarrwanga, L., Ganambarr, R., Ganambarr-Stubbs, M., Ganambarr, B., Maymuru, D., & Sweeney, J. (2016). Co-becoming Bawaka: Towards a relational understanding of place/space. *Progress in Human Geography, 40*(4), 455–475.

Bell, S. J., Instone, L., & Mee, K. J. (2018). Engaged witnessing: Researching with the more-than human. *Area, 50*(1), 136–144.

Darug Country including Dadd, L., Glass, P., Scott, R., Graham, M., Judge, S., Hodge, P., & Suchet-Pearson, S. (2019). *Yanama budyari gumada* – Reframing the urban to care as Darug Country in western Sydney. *Australian Geographer*. https://doi.org/10.1080/00049182.2019.1601150

Ens, E., Finlayson, M., Preuss, K., Jackson, S., & Holcombe, S. (2012). Australian approaches for managing "country" using Indigenous and non-Indigenous knowledge. *Ecological Management and Restoration, 13*(1), 100–107.

Lloyd, K., Wright, S., Suchet-Pearson, S., Burarrwanga, L., & Bawaka Country. (2012). Reframing development through collaboration: Towards a relational ontology of connection in Bawaka, North East Arnhem Land. *Third World Quarterly, 33*(6), 1075–1094.

Plumwood, V. (1995). Human vulnerability and the experiences of being prey. *Quadrant, 39*(3), 29–34.

Rose, D. B. (1996). *Nourishing terrains: Australian Aboriginal views of landscape and wilderness*. Canberra: Australian Heritage Commission.

Rose, D., van Dooren, T., Chrulew, S., Cooke, S., Kearnes, M., & O'Gorman, E. (2012). Thinking through the environment, unsettling the humanities. *Environmental Humanities, 1*, 1–5.

Scott, R. (2017). *Yanana Budjari Gumada: 'Walk with good spirit', Caring as Country at Yellomundee, Western Sydney*. Unpublished Honours manuscript. University of Newcastle, Australia.

4

Animating Frankenstein Lands: The Place of Landscape Art in Post-mining Lands

Penny Dunstan

Part 1: What Is Lost

> *The NSW mining industry is committed to rehabilitating mined land to a condition that supports an agreed post-mining land use. Mine owners do this, not only because this is a requirement under the lease, licence or other approvals granted, but because it is the right thing to do. This also helps to ensure that mine owners remain a responsible and accountable member of the communities in which they operate.* (NSW Mining, NSW Minerals Mine Rehabilitation Fact Sheet, n.d.: Accessed 2019)

Out there. Once. She sighs. The woman stands at the kitchen window looking out. *Out there, once, was Bates Hill.* Her sigh is deep and sad. She shows me photos going pink with age. *That's where it was.* She gestures to an empty space hanging with yellow mine dirt. Memories swarm. *My grandparents lived there. We spent holidays there.* Sheep, cattle, bus-stops,

P. Dunstan (✉)
University of Newcastle, Callaghan, NSW, Australia
e-mail: penny@telsa.com.au

A. Campbell et al. (eds.), *Located Research*,
https://doi.org/10.1007/978-981-32-9694-7_4

grass, trees, soils, bricks, siblings and earth. A memory kaleidoscope of entangled histories springs into speech. Only words and photos are left now. Even the roots of the hill are pulled out in the search for coal (Fig. 4.1).

Only the woman keeps a vigil. Others try not to remember. It makes them so angry that they give up. For them, the personal cost of remem-

Fig. 4.1 *Mined Hill remnant* (Digital Photograph: Penny Dunstan, February 2019)

bering is too high. And the old people have the pleasure of not being able to remember. Sometimes old age can be kind.

The woman bears the vigil as a mark of love for the hill, the last of a long line of carers. Bates Hill had shared stories with a millennia of carers. Respect and love had passed through generation to generation, from indigenous to coloniser, not that such classifications mattered to the hill. But now it is subject to a new invader, a conqueror of all. All that remains of Bates Hill is an edge and it too will soon be consumed. The woman has made it her work to be a witness, collecting the shards of memory. Photographs and words are a poor substitute but it's all she has to acknowledge the missing. And a pall of mine dust.

Tea? She asks me. Sharing tea is the welcome custom. A small meal of biscuits and fresh figs becomes a meal of remembrance for Bates Hill. Books of her family history show a long attachment to the place. Black and white photos of her distant ancestors look directly out at me, challenging me to write a memoir too. How should I remember Bates Hill?

On the computer are grainy colour shots of Bates Hill as it is slowly taken. Her photography is part of her witness. But how she feels is complicated. Entangled in the sorrow of losing land and losing history is the fact that her husband worked for the mine for many years. The house, the farm, their lifestyle are all derived from the vanishing hills of the Mt Thorley open cut mine. She understands that we are all complicit but knowing this does not relieve the pain. We trade our twenty-first century lifestyles for history, hills and land. Her acknowledged complicity only serves to underline my own. Bates Hill and many other hills are giving up their form, their space and their memories for me, at home, writing these words under an air-conditioner, while the daytime temperatures reach 40°C.

Given the depth of loss and mourning, how will it be possible for post-mining manufactured land to be taken back into relationship with people? Looking through the lens of traditional western land ownership, where land is seen merely as a resource for humans, there seems no satisfactory answer to the conundrum. The human gaze of ownership objectifies the land as resource: land has no intrinsic value of its own. If the land is not a commodity, but a place for living with earth-others, it becomes a repository for memory of those interactions, and the loss of land means the loss of belonging and a dispossession from personal and ancestral history.

How should we begin the long journey to accepting a fully engineered hill as country? Even more importantly, when should caring for country start, given it is stitched together from carcasses of the old? How can forgiveness begin for the destruction of belonging and place history?

It's a hard place to stand, stuck between complicity and mourning. For me, this is not a place to dwell, despite my ethical queasiness. It's a sadness that cannot make or create a future. My witness is to examine the future for the trucked-out remnants of Bates Hill, those sewn-together landscapes that appear on the other side of the mining void. I want to interrogate how we breathe life into landscapes without a history and the place of art in those first gasps of life. How can we re-establish love, care and respect for engineered lands? What is it that reanimates terraformed rock into earth? (Fig. 4.2).

On the drive away from Bulga along the Putty Road, on the left I see a vestigial agriculture of grape vines, olives, pastures and tree lots; on my

Fig. 4.2 Penny Dunstan Mine rehabilitation area showing frog dam and stag trees (apex bird habitat) (Digital Photograph: Penny Dunstan, 2015)

right are manufactured hills of grey overburden. Large, lumbering yellow mine trucks dump rock creating clouds of dust. A closer examination of the terraformed shape is prevented by banks of struggling, scraggy trees. Don't see. Don't look. Don't cry. This is big business. The country of the Indigenous creator spirit, Baiame, is being remodelled into vast chasms and Frankenstein hills.

A remnant of Bates Hill spikes upwards, its white and yellow horizontal stone stripes plunge downwards into a void of unseen depth. A silent, vacant space has replaced almost all of Bates Hill. As I drive across a bridge that straddles a mine access road, I encounter a newly forming hill. Its fully naked body shows beautifully engineered slopes approximating an engineer's understanding of an 'original' landscape. Water ways, dry and unconnected, have been freshly constructed out of angular stones broken from newly unearthed Permian rock. Dead trees and small piles of carefully placed rock form designer habitat on trucked-in soils currently bereft of grasses or trees (Fig. 4.3).

I park the car on the roadside and look. It is a replacement hill reforming 500 m across a mining void. Perhaps this new creation emulates a lost hill, like an anatomical model of a body showing muscle, veins and arteries emulates a person, but it is plainly not alive. All the things land does in a day are absent. There is no breathing, photosynthesising, nesting, walking, squawking, trickling, washing, decaying, germination or recycling.

Fig. 4.3 *A new creation—hill half made* (Digital Photograph: Penny Dunstan, 2019)

At what point does rehabilitated open-cut coal mine land pass from an engineered marvel to animate and agentic? When I have asked this question of mine staff they hold differing views. One mining engineer thinks that the transformative point is when cattle return to graze the land as they become a way of generating a return on investment: animation is not complete until it can be entered on an accounting balance sheet. A contrasting opinion comes from an environmental officer who suggests animation occurs when all the engineering work of remaking is done, and the rain arrives to transmute fertilisers and seeds into living green grasses and legumes. He sees this as the agentic point where, providing the conditions are right, nature asserts ownership over terraformed land. For myself, as an artist, I argue that there is one final (and often missing) step to mine rehabilitation, and it involves ordinary people who need to cross the threshold between mourning for the loss of personal and ancestral history, to begin the long journey of accepting engineered land as country to be cared for.

Part 2: Legacy

> *Think of Earth creatures and systems as poems in languages that are foreign but not entirely incomprehensible. Our task as humans is to translate: to find the meaning and the music, the ways of life and life's poetry. For we are part of the music of Earth and our capacity to join in harmoniously depends on both the accuracy of our knowledge and the skill of our translations.* (Rose, 2016: online)

To travel through the Hunter Valley is to bear witness to what Val Plumwood (2008: 146) calls 'shadow places… places that bear the ecological traces of one's passage, or that carry the ecological impacts of supporting your life'. In 2017, it was a sobering 21,000 hectares (NSW Mining, 2018) that bore the collective ecological scars of open-cut coal mining in the Hunter Valley. Through the years, the scars continue to expand. The Valley bounty is not yet exhausted. Such 'shadow places' of big cities, like Sydney, are the backyards of regional cities, like Newcastle, and smaller towns, like Singleton and Muswellbrook. Tiny towns, such as Bulga and Wollar, find themselves overcome by dust and noise and big money. The human tendency to bind to place is deemed to have a price.

As part of the mining restoration process, new land forms are created from the leftovers of ancient Permian rock strata. Some are designed to be the inversion of the mining void from which coal is taken. One long hill is created, topsoil is spread, drop structures are added to remove water from the new landscape and keep it dry. And finally trees and grasses are planted. There is very limited poetry in the new landscape. It continues to be a mine waste hill on which trees and grasses have been planted.

More nuanced ways of thinking about post-mining land design are contained in software programs to emulate the natural land form of the surrounding area. Hills and dales are created, with twisty waterways designed to take water through the landscape rather than off the landscape. Rock outcrops are planted, and trees are transported from the working edge of the mine to become log habitat. Such programs produce a more literal translation of previous landforms, but again, the storied sense of belonging is not considered. A new land with no history rests uncomfortably alongside much older places that have belonged-with people for 60,000 years (National Museum of Australia, 2016).

How is it that landscape art might help twenty-first century Australians to discover the sense of belonging-with land, to find the poetry to animate our lands without history, our Frankenstein lands? How can an artist's view transform the physical world? Barbara Bolt (2008: online) understands 'the performative act doesn't describe something but rather it does something in the world. This "something" has the power to transform the world'. She understands the work of artists, who make and create landscape art, as actions *with* their environment rather than examination of their environment from an external point of view. Landscape art is not the act of looking-at but the act of being-with. It is in the act of being-with that the poetry of belonging and caring animates the terraformed and re-created. As Deborah Bird Rose (2015: 5) wrote: 'Respect (for land) is a matter of knowledge—of knowing the connections so that one knows the many contexts in which respect is due'. Finding how we belong to manufactured lands is a difficult idea and acknowledging that respect is due is doubly difficult. Landscape art is a way of showing respect and acknowledging land as a co-creator in its own future.

Art-making, in particular landscape painting, has long reflected land and belonging, where artists find themselves not only reproducing a view but making what it feels like, how it affects them. There is a kind of magic that happens where the artist enters into a relationship of embodiment with landscape. John Wylie (2007: 17), in his examination of landscape from a European angle, quotes a letter by French post-impressionist painter, Paul Cezanne, '*the landscape thinks itself in me... and I am its consciousness*'.

Donna Haraway (2011), has arrived at a similar view from a different starting point, and conceptualised in the term, 'natureculture' to describe an embodied view of the world. She argues that the land is a living entity, that performs, and that we are entangled within the webs of relation whether we understand it or not.

Not far from the mines of Mt Thorley and Bulga, nestled in a sacred valley is the image of Baiame, Creator Spirit, from the Wonnaura Dreaming. Visiting the cave where Baiame is painted in ochres on the wall teaches me that place, people and dreaming are so tightly interconnected to be indistinguishable from one another. To view Baiame is to look at the creation of the world and all the ways the world functions, with people as co-creators with rights and responsibilities.

If we understand land to be a living entity in conversation with people, then it becomes necessary to understand landscape art practice as a collaboration, rather than generating from a single artist's mind. Conventional landscape art-making required a framed view. In the case of photography, mediated by the use of a lens, a processing file in a camera, post-processing in a computer and a printer. Landscape painting is also mediated by the choice of a view, compositional techniques and commercially decided paint colours. However, it is possible to step outside the traditional conventions of landscape art-making and make art that acknowledges the deeply entangled relations between humans and non-humans, between artists and their environments. In such a framework, the gaze is returned by the land and its inhabitants. It is risky business to hear the other and to make art with the non-human, however, it can make rewarding viewing/participating for artists and viewers.

Part 3: Animating Frankenstein Lands

Remember, I am not recording the vision of a madman. The sun does not more certainly shine in the heavens than that which I now affirm is true. Some miracle might have produced it, yet the stages of the discovery were distinct and probable. After days and nights of incredible labour and fatigue, I succeeded in discovering the cause of generation and life; nay, more, I became myself capable of bestowing animation upon lifeless matter. (Mary Wollstonecraft (Godwin) Shelley, Frankenstein; or, the Modern Prometheus, 1818)

The work of landscape art, in the context of reanimating Frankenstein lands, is to show the ways in which humans might connect and learn to care for lands without histories. Art cannot reconstitute a past any more than rehabilitation of post-mining sites can reconstitute the land before mining—land is so much more than the sum of its physical parts. The work of art is to bring land into relationship to people, to co-create stories of beginning and to enact conversations about futures. It is the function of landscape art to take the risk, to stand in the presence of the other and to trust co-creation. Christopher Orchard (2017: 143) would call it an 'eco-aesthetic' when artists are 'in no way separate from the living systems they aim to record, and indeed they are accountable to this complex symbiotic organisation of relationships'. It is the representations of these relationships that bestow animation upon what, under traditional western thought, is perceived to be lifeless matter.

While there are many artists who are making eco-aesthetic art, I have chosen three Australian artists who concern themselves with co-creating landscape art; Susan Purdy, John Wolseley and myself, Penny Dunstan. Susan Purdy, from the tall forests of the Dandenong ranges, has a practice in photograms, prints and drawing, and seeks to *contact something primal, ancient, persistent in the psyche and in nature* (Purdy, 2014). John Wolseley practices being-with land by camping around ancient billabongs and waterways, and using earths and charcoals, and occasionally bird carcasses, to find marks directly from nature, over which he makes his own contributions. He is particularly interested in the crimes of humanity against the environment in Australia. Finally, my own art seeks a relational way to interact with the world using a practice of walking as wayfinding to explore post-coal mining rehabilitated land in the Hunter Valley. My practice of collaborative art-making seeks to allow land to represent itself; works are made directly with earths, and the environment influences extreme exposures of photographic paper to make in-situ lumen prints.

Susan Purdy

Susan Purdy has a photographic art practice that is centred on photograms. (Photograms are made without the mediation of a lens or camera where images are produced using light-sensitive paper.) She makes lyrical works that relate to her bond to the land where she lives and studies. Purdy finds that *the photogram process still seems like big magic to me, like radiation. It enables me to explore the energy and structure of various forms* (Purdy, 2014: online). Representational photographs give way to a manifestation of life force. 'In an age when what was once certain has now become more fluid, the photogram seems a timely representation to present an enlivened understanding of the material world'.

Purdy's art acknowledges that there is conversation that we as modern humans should tap into to find our true place in our environment. This is particularly relevant in Australian ecology where old stories give us access to ancient occupations and interactions between land and human. It is the place where we find ourselves in conversation with earth-others. There is a centeredness and meditative quality to Purdy's work that indicates her great respect for the conversation she has with land.

John Wolseley

John Wolseley is an eco-artist who understands his work as collaborations with the natural world, rather than a traditional documentation view of disembodied observation.

Wolseley's interaction with the environment is recorded primarily using paint, ink and water-based media on paper. Collaborative mark-making with the environment means printing from leaves, sticks, moss and burnt trees. In the waterways, paper can be dissolved, eaten, stained, torn, folded by land, buried, dug up again, collaged and layered. Wolseley's ambition to provide a space for land to represent itself is enacted through the use of earths, direct printing from vegetation. His immersive methods include swimming in a billabong to retrieve dead birds that he gives a second life as a direct print, painting on the carcasses and slap-printing on paper on the bank. At exhibition, these works still had feathers embedding in the paint on the paper. All the while the artist keeps a light hand using minimal artistic interference.

Wolseley assembles and installs his work to replicate various natural shapes, such as trees or waterways, to give the audience an immersive environmental view to replicate the feeling of being in the wild to see the land, and the non-human dwellers of that land, through the artist's eyes. This choice drags his conventional gallery audiences into the uncomfortable zone of participators in the world, rather than external observers gazing from the safety of a white cube gallery space. Wolseley makes his art works collaboratively with the environment, giving at least equal space to the voice of the land, and ultimately shows a world where humans are fully integrated into nature.

Penny Dunstan

Central to my landscape art practice is the experience of being-with rehabilitated post-mining land using walking as wayfinding for a method of opening conversations with the non-human. Lesley Instone (2015: 138) defines wayfinding as *an experimental, non-linear mode of movement (that) helps us meet the challenge of unknowable futures in a changing world.* Walking as wayfinding allows for performative, action research where new knowledge is accreted, laid down layer by layer, sedimenting into an art practice. This new knowledge enables me to move from an understanding of nature as commodity, as an engineered item, to nature as co-existing in the same space as myself, where my relationship with the new land is *intrinsic, material and relational* (Instone, 2015).

In my art practice, I have walked my way through rehabilitated open-cut coal mine land in the Hunter Valley. The newly surfaced land **pushes back** at my feet. History is being told **to** me by lumpy gravel millennia old. Walking becomes a way to hear what the land is saying, a place to find a co-creator for landscape art and a way of transitioning land from an engineered solution for mine waste into country to be respected.

Co-creating with land is fundamental to my art practice. Two co-creating methods are described here: lumen prints (lens free photography made in-situ) (Fig. 4.4) and unfired earth bowls (Figs. 4.5 and 4.6).

Fig. 4.4 *Lucerne*, Lumen print with insects, plant matter, mine dirt and fungi (Lumen Photograph: Penny Dunstan, 2017)

Fig. 4.5 *Earth Bowl, From Dirt/Soil/Earth installation* in *Altarld*, Adamstown Uniting Church (Digital Photograph: Penny Dunstan, 2018)

Fig. 4.6 Penny Dunstan, Earth Bowls, Dirt/Soil/Earth installation in *Altar/d*, Adamstown Uniting Church, 2018 (Photograph: John Cliff, 2018)

Overexposed lumen prints form unexpected beauty. Like Purdy's photograms, they are made by direct printing, without a lens, on light-sensitive paper, but the exposure times are much longer. The specimen plants sweat during a three to four-hour overexposure, producing coloured prints from black and white paper. Working on site with a picnic blanket for a darkroom means there is very limited preparation time for each print, but the land has already put in months, possibly years, to grow the plant and create the soils. The images happen as they want. Pollen, microscopic insects, mine dust, plant sap, fungi and chlorophyll all become part of the print and the print paper itself begins to be colonised by moulds. The resulting prints, scanned from the original and barely processed, are romantic, tattered, unearthly and of the earth, an embodied presence of land in print (Fig. 4.5).

Unfired earth bowls form another part of my practice. The raw materials are collected samples of soils used for mine rehabilitation or local Hunter Valley agriculture close to mine sites. The collected earth is fashioned into a shape often associated with the provision of human sustenance to open a conversation about an ethical response to land. Some bowls have inclusions from the soils, stones, plastic or glass, or are made from degraded mudstone masquerading as dirt. One bowl contains remnant stone tools worn-out by years of wear. The audience is encouraged to carefully touch the exhibit in order to create a haptic relation with each earth object.

The format of the bowl prompts questions about the place of soils in modern society. Are they made from dirt to be cleaned away; or soil to be seen as a resource and provider for human wants; or are they ritual objects made from a sacred material, earth?

When placed in a gallery, or in the Canberra Hyatt, or on an altar in a church, *raw soil bowls speak of the relationship between food and earth, and the finger marks of the artist speak of the impact of humans on a shared resource, one we share with the many non-human others* (Pattenden, 2018: online) (Figs. 4.5 and 4.6).

Conclusion

As I reread the NSW Mining Industry factsheet about the rehabilitation of mines, quoted at the beginning of this chapter, the fundamental flaw in their logic comes into focus. Land is depicted as just a commodity. The land-people relationship is neither considered nor even acknowledged. Yet, ecological restoration specialists, Clewell and Aronson (2013), propose as fundamental, a holistic model of post-mining rehabilitation that includes both cultural and socio-economic values. They talk of ethics and the evolution of disturbed landscapes into something other than what has been planned. Their premise is that rehabilitation of lands must include members of the broader community. As part of the broader community, landscape artists are in a unique position to elucidate alternatives to the mechanistic view of rehabilitated post-mining land. Purdy, Wolseley and myself all make landscape art that asks the audience to become participators in the discussion, whether it be through looking or touching or thinking.

Art has the power to animate the human imagination to see Frankenstein lands as country to be cared for. Art makes our ears and hearts work, bypassing the intellectual ruminations of the mind. Our hearts are captured by bowls we can touch, or feathers embedded in painted stories of waterways, or the contemplative essence of landscape photograms. Art accesses the intricate deeply entangled relations between the human and non-human worlds and asks us to answer a question, where do we stand?

Collaborative landscape art, or eco-art, embodies the unspeakable connections between land and people and accesses our consciousness through our senses and our hearts. It is an instrument of reflection of our place as humans who create the new earth. Frankenstein lands need us to care and to attend our place in the sanctity of earth relationships. They need us to move beyond 'shadow places' to transformative points of animation in belonging-with in new acceptance of engineered land as country.

As I drive through the Hunter Valley, past the terraformed landscape, I recognise that I too am a witness to change.

Out there. Once. Out there once was a land known as the Tuscany of the southern hemisphere. And I still catch glimpses of it. Sun gold hills. Vines. Cattle. Mares with tags around their necks shepherding tiny foals. Green circles of Lucerne on river flats. But everywhere there are mine waste hills. White. Brown. Grey. They loom menacingly over public roads and sometimes devour them altogether. I take one last drive along Wallaby Scrub Road before it is closed for public use. The roadside paddocks are scrappy, covered in spindly undergrowth. Fences lean. Dams are empty. No-one has cared for this land for years knowing that the mine would eventually take the surrounds. It is another level of sadness for the local population and another reason to think about the future of Frankenstein lands in the Hunter Valley.

References

Bolt, B. (2008). A performative paradigm for the creative arts. *Working Papers in Art and Design, 5*. Retrieved March 10, 2019, from www.herts.ac.uk/__data/assets/pdf_file/0015/12417/WPIAAD_vol5_bolt.pdf

Clewell, A. F., & Aronson, J. (2013). *Ecological restoration: Principles, values, and structure of an emerging profession*. Washington, DC: Island Press.

Haraway, D. (2011). Speculative fabulations for technoculture's generations: Taking care of unexpected country. *Australian Humanities Review Issue, 50*(2011), 95–118.

Instone, L. (2015). Walking as respectful wayfinding. In D. B. Rose, K. Gibson, & R. Fincher (Eds.), *Manifesto for living in the anthropocene* (pp. 133–138). Brooklyn, NY: Punctum.

National Museum of Australia. (2016). *Evidence of first peoples: At least 52,000 years ago: Archaeological evidence of first peoples on the Australian continent*. Retrieved May 10, 2016, from http://www.nma.gov.au/online_features/defining_moments/featured/evidence_of_first_peoples

NSW Mining. (2018). Upper hunter mining dialogue. Retrieved 2018, from http://www.nswmining.com.au/environment/planning-land-use; http://www.nswmining.com.au/community/upper-hunter-mining-dialogue

NSW Mining Website: NSW Minerals Mine Rehabilitation Fact Sheet. (n.d.). Retrieved February 27, 2019, from http://www.nswmining.com.au/getattachment/Environment/Rehabilitation-mine-closure/NSW-Mining-Fact-Sheet_Rehabilitation.pdf

Orchard, C. (2017). Epistemology, aesthetics and mediation: Photographic reflections on landscape change in regional Australia. In N. Holmes & S. Taffel (Eds.), *Ecological entanglements in the anthropocene (ecocritical theory and practice)* (pp. 143–158). London: Lexington Books.

Pattenden, R. (2018, May–November). *Altar/d – Art installation series, Adamstown Uniting Church*. University of Newcastle Gallery. Retrieved from https://vimeo.com/299148159

Plumwood, V. (2008). Shadow places and the politics of dwelling. *Australian Humanities Review, 44*, 139–150.

Purdy, S. (2014). Susanpurdy.Net. Retrieved from http://www.susanpurdy.net/about/.01/02/16. Also see the Alternative Methods Photographic Seminar in Trentham Victoria run by Gold Street Studios.

Rose, D. B. (2015). The ecological humanities. In D. B. Rose, K. Gibson, & R. Fincher (Eds.), *Manifesto for living in the anthropocene* (pp. 1–6). Brooklyn, NY: Punctum.

Rose, D. B. (2016). Thinking like a Mantis? In *Love at the edge of extinction*, August 13. Retrieved from http://deborahbirdrose.com/

Wollstonecraft (Godwin) Shelley, M. (1818). *Frankenstein; or, the modern prometheus*, 1818. First Published 1823. Now Open Access.

Wylie, J. (2007). *Landscape*. London and New York: Routledge.

5

Kristin Linklater, Freeing the Natural Voice, in Orkney

Angela Campbell

Kristin

If place makes the poet,
she is Yesnaby and Skaill,
Merkister and Hundland,
on a hundred different days.
Every weather blows through her.
Light dances,
through a lifetime's work of effortlessness.
Obsession and delight,
make for such interesting bedfellows,
and still so much to discover.
Why is our span that bit less than the reach we aspire to?
There are galaxies in her fingertips,
and always the playfulness

A. Campbell (✉)
Federation University Australia, Ballarat, VIC, Australia
e-mail: a.campbell@federation.edu.au

A. Campbell et al. (eds.), *Located Research*,
https://doi.org/10.1007/978-981-32-9694-7_5

of

the

next

breath.

David McNeish
26 April 2019
Nearing the end of a week at the Voice Centre.

Attachment to place and creation of a sense of place have long fascinated artists and researchers in theatre and performance (Chaudhuri, 1995; Heddon & Turner, 2012; Kershaw, 2007; Mackey, 2016; McAuley, 2006; Pearson & Shanks, 2001; Tompkins, 2012, 2014). More recently, interest in the connections and modes of mobility between places have provided new paradigms for research and art practice (Heddon, Lavery, & Smith, 2009; Heddon & Turner, 2012; Sheller, 2017; Sheller & Urry, 2006; Wilkie, 2008, 2012). This focus on 'new mobilities' rests not so much on a singular place or instance of artistic practice but on the flow of ideas and practice between places, regions, and communities, back and forth. This interview with Kristin Linklater, master voice teacher, traces both a sense of place and the lines of flight that connect people and places, through breath and bone, across time and geography, from the centre to the periphery and back again.

The Kristin Linklater Voice Centre (KLVC) (KLVC: online) is perched some 16 kilometres off the northern coast of Scotland, on the top of a hill overlooking bright green fields, stone farmhouses, lochs, and coastline that cover much of the 523 square kilometres of the Mainland, the largest of the 20 inhabited Orkney Islands. From the Voice Centre, in a storm, the Atlantic can be heard crashing onto the pebbly shore of Skaill Bay at the bottom of the hill, even against the wind. Skaill House is close by, built in 1620. It was home to the 7th Laird of Breckness when, on one wild night in 1850, Skara Brae, Europe's most preserved Neolithic village, was unearthed from the dunes by wind and waves (Skaill House: online). The Voice Centre opened in 2014 and is a comfortable architecturally designed complex that includes a studio, kitchen, living space, and a sleeping 'Bothy'. Each bedroom in the Bothy is named after an Orcadian island. KLVC accommodates up to 14 students who travel to

Orkney from across the globe to work for weeks at a time on vocal production, language, and text with Kristin Linklater, the centre's founder.

When she retired from Columbia University in 2013 after working in America and internationally for over 50 years, Kristin returned to Orkney, the place of her birth, to build and then launch the KLVC. Her work in Voice is known worldwide through her teaching and workshops and also through her two books, both of which are seminal in the field (Smukler, 2009)—*Freeing the natural voice: Imagery and art in the practice of voice and language*, first published in 1976, revised and republished in 2006; and *Freeing Shakespeare's Voice: The Actor's Guide to Talking the Text* first published in 1992, then reprinted in 2009, and again in 2014. Kristin continues to write and to publish her findings (KLVC: online). Kristin began her career as an acting student at London Academy of Music and Dramatic Arts (LAMDA) in the 1950s. There, she came under the influence of Iris Warren whose approach to vocal training for actors aimed not just for a beautiful sound and clear diction, as was the fashion at the time in Britain (Eyre, 2000; Hodge, 2000), but sought emotional connection and truth, supported by strong technique. For Kristin, the 'constant emphasis was "I want to hear you, not your voice."' (Linklater, 2006: 6). This simple goal to free the 'natural voice' became her life's work.

After six years as voice teacher at LAMDA, Kristin took herself to the United States, arriving in 1963 at a time when truthfulness to emotion was the overwhelming aesthetic in both theatre and in film (Adler, 1988; Strasberg, 1988; Counsell, 1996; Malague, 2012). Her skills and desire 'to liberate the natural voice and thereby develop a vocal technique that serves the freedom of human expression' (Linklater, 2006: 7) were much in demand. She worked with acting companies such as Open Theatre under Joseph Chaikin, the Lincoln Centre Repertory Company under Harold Clurman and Elia Kazan, and the Tyrone Guthrie Theatre under Sir Tyrone Guthrie (Linklater, 2006: 2). Her coaching work with actors in theatre and film includes Bill Murray, Sigourney Weaver, Bernadette Peters, Mary Tyler Moore, Patrick Stewart, and Donald Sutherland. Kristin was a founding member of Shakespeare & Company. She held teaching positions at New York University Graduate School (now Tisch School of Arts), Emerson College, and Columbia University where she retired as Professor Emerita in 2013. Kristin has just celebrated her 83rd birthday.

I first encountered Kristin's work as a drama student in Melbourne, at the Victorian College of the Arts, University of Melbourne, in the late 1980s. At the time, I did not realise the exercises that occupied my first-year acting undergraduate mind and body for hours in Voice class might be attributed to Kristin Linklater. Thirty years later, my current role, teaching performing arts students at the Arts Academy, Federation University, a regional university in Australia, alerted me to Kristin's work at the KLVC and prompted me to sign up for a week-long residential 'Introduction to Shakespeare' workshop. In 2018, I found myself in Orkney, one of 14 students from the United States, United Kingdom, Australia, Italy, and Hong Kong studying in a unique learning environment. As I write in 2019, I am again at the Centre, in Orkney for two weeks this time, working on Advanced Voice, and Sound and Movement, with students from around the world, including a couple of Orcadians (local residents). Interestingly, this diversity is also reflected in the students' backgrounds; while mostly from film and theatre, my groups have included a yoga teacher, a poet, a documentary filmmaker, a Church of Scotland minister, a movement analyst, a lawyer, and a business person; all of us supported by Kristin's growing team of local and international colleagues.

The interview below captures, for me, two aspects of Kristin's work which relate directly to the importance of regional research, in this instance, regional practice-led research. Firstly, Kristin speaks of her reliance on what could be described as an emerging 'community of practice' based in Orkney but stretching far beyond. Lave and Wenger (1991) have argued extensively that 'learning does not rest with the individual but is a social process that is situated in a cultural and historical context' (Farnsworth, Kleanthous, & Wenger-Trayner, 2016: 140). The social, cultural, and historical context of the KLVC is situated geographically in Orkney but operates as the hub of a learning network that is reliant on and fed by a vibrant local and international community of artists, teachers, researchers, and interested bystanders. Kristin is conscious that her legacy is embedded not only in her publications but, importantly, is also embodied in breath, thought, and language, through and in the work of others. She runs the KLVC as an active laboratory where voice work and training with students is exploratory as well as iterative. She is constantly interrogating her own practice. There are growing numbers of Designated

Linklater Teachers (DLTs) who have undertaken rigorous training and accreditation. There is also a thriving Linklater Center in New York (The Linklater Centre: online). Consolidating her approach in the United Kingdom, a strong pedagogical focus is currently given to the Linklater Method at Royal Central School of Speech and Drama in London within the popular MA and MFA in Voice Studies. Kristin is keen to see the next stage of KLVC in Orkney flourish through a 'community of practice' that encourages ongoing engagement and professional development for local and international students, emerging DLTs and master teachers of Voice, as they work and study at the KLVC and around the world. A Designated Linklater Teacher (DLT) and fellow student from my first *Introduction to Shakespeare* workshop in 2018 is currently teaching Linklater work at the Shanghai Theatre Academy in China, for example.

Secondly, I wish to relate how an affinity with a particular regional landscape and its history can profoundly influence and communicate one's sense of what it is to be emotionally connected as an individual and truthful in art and in the world. In *Freeing Shakespeare's Voice*, Kristin reflects:

> As late as Shakespeare's day men and women spoke of the harmony of the spheres and the desire to be in harmony with them. The music in us still sings and so does Nature's music, but we seldom listen to ourselves or Nature with an ear tuned to such subtle vibrations. If we could actually experience that we live in a universe made of sound and light waves, there might be a reunion between the body and the brain and the world around us. (2014: 13)

She quotes Lewis Thomas from his book *Et Cetera, Et Cetera*:

> In order to get language to really work from the outset, as a means of human communication by speech… the words needed to express the feelings aroused by things, particularly the living things in the world…the sense of the roots must persist like genes in all the words to follow. (in Linklater, 2014: 12)

For Kristin it seems, the natural landscape is Orkney; sea cliffs, peat meadowland and wide sky, connected with a deep history which has marked, indeed 'performed', the landscape since the Neolithic (Marshall, 2017). The 'Heart of Neolithic Orkney' (UNESCO World Heritage List: online) was in fact designated in 1999 as a UNESCO World Heritage site. As Kristin says below, Orkney 'is in my DNA'.

AC: Why, what brought you back to Orkney; what was the impulse that drew you back from the United States to set up the KLVC?

KL: This question implies intention and that intention has not been the driving energy behind most of the things that I've done. Things happen in life in ways that appear on the surface intentional but in fact are usually are the result of a mix of different ingredients and impulses. I grew up in Orkney, so it is in my gene pool and in my DNA, it is somewhere in my flesh, and the first 12 years of my life were in Orkney. I've always been connected. My mother came back to Orkney after my father died in the 1970s, so I would visit Orkney from America and kept contact. It always seems by chance that I bought a small farmhouse here in 1999 when I was working in New York. I always thought of it as a place where I would get away from my work. I was very determined that it would not be a part of my work experience, but then when I decided that it was time to leave Colombia University and spend more time here, everything turned around because I'd been travelling a lot and doing workshops in many different parts of the world and thought I wanted to spend more time in Orkney and bring my work to me rather than travel to work all the time. So, it was kind of a series of happenstances that formed the decision to actually build the Voice Centre. I see it as all rather accidental.

AC: Accidental in the living of it but as you look back and see how those accidents played out, do you see any kind of pattern?

KL: Not a pattern because I've never done anything like this before but because it was clear that I'm going to go on working till the day I die because that is what I do, that is how I define myself, as a teacher. So, if I was living here, I would keep on teaching. So how does one do that? Well, one creates a space where one can do that, creates a space where people can come and work. As I say, once I had the idea, obviously I had to then think practically and seriously about how to do it. And I was very lucky, it all seems to me it was just a series of lucky incidents, but I'm sure they have a larger meaning. I spoke about it to a very generous friend who said he would give me a chunk of money to try and help do it. I've never had any money, so then I pulled a matching sum of money out of my pension fund. Then my teachers started to contribute as founding members, so the actual financing of the Voice Centre was very grass roots. And then I found a very clever local woman, Ruth Colston, who not only drew up a business plan but drew up the design for the building itself and project managed it to completion. So, it all started happening in a very bottom up way, not with a massive plan—but with Ruth.

AC: But nevertheless, it did happen in Orkney; it didn't happen in Massachusetts.

KL: It could have happened in Massachusetts, but I'd bought this little house in Orkney and that was it, I was stuck with Orkney. And it was romantic to come back here. When I first bought the house, whatever my subconscious was doing it was much smarter than my conscious mind because it was the right place to be, absolutely the right decision to make, although a lot of people thought it crazy and I thought it was crazy. It didn't make any kind of sense to make up a Voice Centre in this little island, where nobody knows where it is. But if I'd done it in Massachusetts, I would have had a preponderance of Americans whereas here I have people from all over the world.

AC: What has it meant for your practice, at this stage of your career, to teach from Orkney, has it changed anything?

KL: Enormously, because the most serious training experiences in my life have been under the auspices of academic training programmes. I've taught 17 years at Colombia University, I've taught 6 years at Emerson College, I've taught 12 years at Shakespeare and Company, not an academic institution, and before that another 12 years at NYU. As well as a lot of professional theatre work. I've always freelanced, doing workshops in different parts of the world, but workshop teaching is very different from organised teaching where you have a three-year programme with students where you see them through. Once I set the thing up here, I had to devise, to figure out how to teach in compressed intensive five-day chunks. Then what's been very interesting is tailoring for people who are not necessarily going to be actors. So, it is a much wider constituency.

AC: Were you and your work known to people in Orkney, were you a known figure?

KL: I was known in Orkney as my father's daughter and my mother's daughter. My father was a very well-known novelist, Eric Linklater. Linklater is an Orkney name, and my mother, when she came back here for the last 20 years of her life, she was a local activist, political activist, and so on, so that was really how I was known.

AC: You really come from a long line of storytellers, do you see yourself as part of that line?

KL: I don't recognise myself in that description. Why do you ask that, why do you emphasise that?

AC: I guess because of your love of language, your love of images, and the thrill of creating a place in the imagination for whatever is possible to emerge, the possibility of connecting with something that is mysterious. In your Voice work you offer connections between the psycho-physical-emotional spheres as some kind of explanation for that mystery, but really as I understand it, there is a deep mystery that you keep reaching for in your practice, reaching towards the unknown, what is unknowable in yourself, which to me is the thrill of life and art; to me, that is a storytelling impulse.

KL: I think if you take that as helping people to see the story of their lives and thereby shed light on their lives, help them to find the voice that can tell that story, that can express that, that might apply. I can see that. I must say, the older I get I have to curb my desire to tell anecdotes and stories from my life in the course of my teaching. I do like an audience. I'm sure that genetically the thing about words and language must have been influenced by my father and his stories. My paternal grandmother grew up as a small child with her father as a sailing ship captain in the Merchant Navy, so she spent many years as a child going around the world in sailing ships, great huge sailing ships, trading to Portland Oregon, Australia, and India and she would tell us those stories as a child. Yes, probably there is a storytelling background. That's why the voice is so important—so that we can tell our stories, whether in fragments or the whole thing. The words are there to explicate the otherwise unknown and unsayable parts of our lives. OK, I'll buy that, an inherited thing.

AC: In a way, you have woven yourself back into the story of Orkney.

KL: Well, I'm weaving myself. Moving between New York and Boston and Massachusetts, in my experience, it takes five or six years to feel like you are part of a community or an environment. Coming back to Orkney, because of the Voice Centre, I am now seen in a very different way. In the last five years, I'm now being seen in Orkney as my own person and not as the daughter of Eric and Marjorie and that is because of the Voice Centre and my visibility in that regard. The story is changing very much.

AC: What does the Voice Centre mean in Orkney? Was it something completely wild and unheard of before?

KL: The first reaction was, it must be a cult! What is going on out there, people coming in and staying for weeks and making weird noises. But luckily, I have this wonderful team, the people who work with me who are all local people and they make it quite clear that I'm not a weirdo. What has happened too is that I've worked with quite a lot of local amateur dramatic people. Orkney is extraordinary in the amount of theatre that they do. With a population of 20,000, it has 12 or 14 different amateur theatre groups and there are some very good actors, two or three really dynamic directors, all amateurs, they are enormously talented, hugely imaginative and they do some terrific work. I offer a free place in any of my workshops if I have room for one or two Orkney residents to apply for. I have had several people maybe 12–20 people who have come along for the full week workshop, so on an individual basis many people have had strong experiences and have taken the work out to the community to say this is good stuff, feet on the ground stuff, and enjoyable stuff too. I've worked with a group of young people who were chairing a science festival; they were going to be doing all the introductions to very eminent people who were coming to present talks on science. There is a vibrant science festival every year in Orkney. I worked with them for five days; there were all 20-year olds, 25-year olds, and they grew a lot through that experience. A choir came up and did a day-long workshop; a couple of theatre groups came up, so I've coached a lot of the local amateur groups and I myself directed an amateur production of Macbeth which was received with huge enthusiasm locally. But in terms of my input into the performance field, now there is a sense of the practicality of what it is that I'm offering here. The other thing quite simply is the contribution to Orkney, you have 12–14 people coming to a workshop, they go out and they shop, so I definitely contribute to the Orkney economy. Which I'm quite pleased about.

AC: And there are a number of people working with you at the Centre, doing the admin, the cooking…?

KL: Yes, I employ three wonderful people fairly full time and then two or three other people half time. So, from that point of view, that is good… everybody knows the Voice Centre.

AC: You are a Fellow of the University of the Highlands and Islands, is that right? Would you consider working with UHI in Orkney?

KL: I was very happy to have been made a Fellow. I have some thoughts about that, but I haven't at the moment any hands-on direct connection with them in terms of pedagogy or setting up programmes, but I hope it might happen.

AC: What is your marketing plan for the Centre?

KL: The internet, the website, word of mouth has been very, very, very good. Our first year was 2014, I always wanted to do a poetry workshop; the first year I had five people sign up, next year was seven, and I thought OK I'd better cancel and someone said, 'do do it again', then ten, this year I have a waiting list for my poetry workshop. We have to put in more and more Introductory Workshops every year. So, the demand is growing, and I think that is mostly word of mouth, the word is out. And it is very international, we have had people from over 35 different countries in the five years we have been going, so people from all around the world. We have a full group coming from Russia in May and a group of Japanese actors from Tokyo who will stay and train with me for two weeks in November. And I like that, the group says, we are coming and we'll bring our own interpreter.

AC: My experience, apart from the work that we did in the studio, was that I was enchanted by the history and landscape of Orkney and the wildlife. It is a magical place on many levels. What effect do you think that has on the students?

KL: Everybody thinks about it in different ways, but people, mostly the Americans are absolutely viscerally changed by going to Skara Brae and understanding that these people were living here 5000 years ago. A lot of people find the Neolithic experience a shock to the system somehow. As you know, looking out through the windows to the sea and the fields, according to what time of year you are here, of course, looking out on green fields does something very deep and has a calming effect. I think that green does something on a deep sensory level. And then the horizon widens the mental perspective. Mostly people who come here live in urban environments, so their view is always being cut off, but here the view goes on and on. That has to do something to the mind and therefore to the body. People describe it in different ways. They feel their minds have been liberated somehow by the impressions of nature and the natural clouds and different colours of the sky and sea, just even sheep and cows, talking to a sheep or cow! It is so fresh. It just refreshes people. People use words, like it's magical, it changes my imagination, I see things differently. Of course, if you are going to work on Shakespeare you've got to be able to see the natural world. And they get to exercise their ability to see the natural world. It's right there. And quite simply on the physical level, to get clean fresh air into the lungs for a lot of people it really is a physical rejuvenation, just in terms of the clean air that they can breathe in. Being in a community, feeling the sense … I don't know why that is so rare these days, but it seems to be. I think a lot of that is a waking-up. A deep waking-up. And maybe restoring some ancient primal relationship with nature.

AC: Do you see that at all as being political?

KL: Political? Everything is political. In what specific way?

AC: I remember you talking about your reading of the neolithic landscape of Orkney, and you gave a counter-reading against the interpretations that some of the archaeologists give of the deep history of the place. Your interpretation foregrounded birth and cycles of life, rather than death and burial ceremonies. This I'm imagining might encourage people to be awake to the power of the natural world, particularly at the moment when climate change is so frightening to so many people.

KL: Yes, I don't know any direct conversations I've had about that, except that I would suppose that when people read about climate change, they will maybe see things, because somehow if you are living in the middle of a city I don't think you see them as directly and as vividly as when someone says the sea is encroaching on the land. Well there it is a couple of fields away, so it may make people more appreciative of the loss that we are faced with. I haven't had many of those conversations with people. I think a lot of people get a big sense of community with the living situation at the Voice Centre and that I would consider a political contribution. A contribution to some kind of hope, from the point of view of civilisation. So many people feel that what is going on politically and globally is out of their control.

AC: I think the act of reconnecting with breath, with your body, with what you actually feel is a deeply political thing.

KL: I think so, I think the personal in that way is definitely political. And being encouraged to speak out, to talk, to say what you really think.

AC: You spent most of your career in the United States, and you have written 'I was lucky to come to the US when I did'. Fifty years ago, there was something the United States gave you …

KL: Wait a minute, this is about an hour's worth of talking if you want me to launch into that which I'm not going to do!

AC: What I'm assuming you were talking about there, is that you came to the United States at a time when the approaches to acting were opening out, working with impulses and emotional connections and psychological realism. Is that true?

KL: Oh, it was so much more than that! When I came to the United States, it was a time when they were building theatres right across the country, it was a political thing, it was a professional thing, it was a personal thing. There was hope. However, they had built all these theatres and art centres and they had employed the best actors and the best actors were all Actors Studio actors, but they couldn't be heard by the audiences because the Actors Studio had thrown out all the skills work of voice and body. This is a terrible oversimplification. My success in the late 60s was buoyed by the law of supply and demand. There was a huge demand for voice work. And in America, at the time, the profession of voice was the study of singing and/or diction. And I came in with the Iris Warren work. Which was emotional and psychological, and it kind of meshed with the Actors Studio terminology about emotional connection and truth. All those repertory companies wanted me to teach Voice and train their companies. That is the short version of this story.

AC: What is the future of the Voice Centre in Orkney, where do you see it going?

KL: Depends on how long I live. What I'm beginning to do is to get more of my senior teachers to come and teach with me. I hope that organically we will develop a group of teachers who will own the business and it will be more of a … it wouldn't be a Kristin Linklater Centre, it will be a Voice Centre that will house different teachers that are coming at it from different angles, based on my voice work. But the teaching won't depend on my name and my live presence all the time. We are beginning to do that. There are two workshops this year that will be taught by other people, and the more that that can happen, the better. I'm hoping that the Voice Centre will become independent of me, but I want to keep teaching till the day I die; otherwise, I won't know what to do with myself. Which is really why I created it. What am I going to do? I'm not going to go on cruises.

AC: How many workshops and students do you have going through the centre? If I look at the calendar, there is something on virtually all the time.

KL: I think on average it is about 25 weeks of the year, we will do 2 weeks in the month and then 2 weeks off. The Summer is very busy, the Spring and Fall are quieter. We have to follow the market, when people are available to come. We are tailoring it as much as possible. We have a couple of outside hires and I hope that will happen more, hiring out to people who I don't have to teach. I hope that will happen more.

AC: Is there anything you'd like to say?

KL: What I'm doing here, has this contributed at all to your thesis?

AC: Yes, it has because it is about the vitality of living and working and studying in regional areas. Actually, I have to really stop myself writing the story of you because the impulse is to say you spent the first 12 years of your life in Orkney and formed an imaginative world that you took elsewhere and that you then brought back.

KL: I think that is true, I think those first 12 years painted an inner landscape in me of cliffs, fields, sea, and sky, because all my teaching life in America, all my teaching life has been in big cities, but I've always called on that imagery. I didn't even consciously think of it as Orkney, but if I was teaching in New York, we'd be calling out to the sea, we'd be standing on a cliff and looking out to waves, we'd be standing in fields and looking out to the hills. As I said before, if you want to do Shakespeare or poetry, you have to be able to call up that rural landscape as well as the urban one. When I came back here, I realised this, and a lot of people who have worked with me in America in the past, when they come here, they say, I see where your imagery came from, it came from here. So, I think I did export the imagery of Orkney and somehow it has brought people back here, and they plug into it. Maybe that is a political contribution; that they get a very vivid plug into the world that we may be losing in the next 100 years. I know they carry this landscape with them when they go back.

Acknowledgements Many thanks to Kristin for her time and attention during this interview and in subsequent conversations. Thanks also to Fiona Gruber for her help in formulating my questions and also for her prompt to read Eric Linklater's children's classic from 1944, *The Wind on the Moon* (a book dedicated to a young Kristin and her sister), where happily I met the badly behaved siblings, Dinah and Dorinda. Thanks also to David McNeish for his poem, Kristin, written after our Sound and Movement workshop, as a gift to us all.

References

Adler, S. (1988). *The technique of acting*. New York: Bantam.

Chaudhuri, U. (1995). *Staging place: The geography of modern drama*. Ann Arbour: University of Michigan Press.

Counsell, C. (1996). *Signs of performance*. London: Routledge.

Eyre, R. (2000). *Changing stages: A view of British theatre in the 20th century*. London: Bloomsbury.

Farnsworth, V., Kleanthous, I., & Wenger-Trayner, E. (2016). Communities of practice as a social theory of learning: A conversation with Etienne Wenger. *British Journal of Educational Studies, 64*(2), 139–160.

Heddon, D., & Turner, C. (2012). Walking women: Shifting the tales and scales of mobility. *Contemporary Theatre Review, 22*(2), 224–236.

Heddon, D., Lavery, C., & Smith, P. (2009). *Walking, writing and performance: Autbiographical texts*. Bristol: Intellect Books.

Hodge, A. (2000). *Twentieth century actor training*. London and New York: Routledge.

Kershaw, B. (2007). *Theatre ecology: Environments and performance events*. Cambridge: Cambridge University Press.

Kristin Linklater Voice Centre. Retrieved May 3, 2019, from https://www.linklatervoice.com/

Lave, J., & Wenger, E. (1991). *Situated learning: Legitimate peripheral participation*. Cambridge: Cambridge University Press.

Linklater, K. (1992). *Freeing Shakespeare's voice: The actor's guide to talking the text*. London: Nick Hern Books.

Linklater, K. (2006). *Freeing the natural voice: Imagery and art in the practice of voice and language*. London: Nick Hern Books.

Mackey, S. (2016). Performing location: Place and applied theatre. In J. Hughes & H. Nicholson (Eds.), *Critical perspectives on applied theatre* (pp. 107–126). Cambridge: Cambridge University Press.

Malague, R. (2012). *An actress prepares: Women and the method*. London and New York: Routledge.

Marshall, J. (2017). Absence, presence, indexicality: The mise en scene of 'the Heart of Neolithic Orkney'. *Theatre Research International, 42*(1), 72–90.

McAuley, G. (2006). *Unstable ground: Performance and the politics of place*. Bruxelles: P.I.E. Peter Lang.

McNeish, D. (2019). *Kristin*. Unpublished Poem. Retrieved May 2, 2019, from stonesongmusic@gmail.com

Pearson, M., & Shanks, M. (2001). *Theatre/archaeology*. London: Routledge.

Sheller, M. (2017). From spatial turn to mobilities turn. *Current Sociology, 65*(4), 623–639.

Sheller, M., & Urry, J. (2006). The new mobilities paradigm. *Environment and Planning A: Economy and Space, 38*(2), 207–226.

Skaill House. Retrieved May 4, 2019, from https://skaillhouse.co.uk/

Smukler, D. (2009). Book review, freeing the natural voice: Imagery and art in the practice of voice and language by Kristin Linklater. *Voice and Speech Review, 6*(1), 465–466.

Strasberg, L. (1988). *A dream of passion: The development of the Method*. London: Bloomsbury.

The Linklater Center. Retrieved May 4, 2019, from https://www.thelinklater-center.com/

Tompkins, J. (2012). *Performing site-specific theatre: politics, place, practice* (Anna Birch and Joanne Tompkins, Eds.). Hampshire, UK: Palgrave Macmillan.

Tompkins, J. (2014). *Theatre's heterotopias: Performance and the cultural politics of space*. Hampshire, UK: Palgrave Macmillan.

UNESCO, World Heritage List. *Heart of Neolithic Orkney*. Retrieved August 5, 2019, from https://whc.unesco.org/en/list/514

Wilkie, F. (2008). The production of "Site": Site-specific theatre. In N. Holdsworth & M. Luckhurst (Eds.), *Concise companion to contemporary British and Irish theatre* (pp. 87–106). Oxford: Blackwell.

Wilkie, F. (2012). Site-specific performance and the mobility turn. *Contemporary Theatre Review, 22*(2), 203–212.

6

Local Innovation and Leadership from Regional Victoria: The City of Ballarat and UNESCO's Historic Urban Landscape Approach

Susan Fayad and Lisa Kendal

Introduction

A former work colleague once proposed that working in regional and rural local government is akin to being the country vet—a familiar and resourceful jack of all trades. Perhaps the country vet is not the best comparison for a highly bureaucratic political institution; however, it could be said that both the country vet and regional and rural local governments face similar circumstances—such as isolation, local familiarity and high community expectations, highly diverse and demanding roles and resulting self-reliance—which can often be incubators for localised ingenuity.

Unlike many of its metropolitan counterparts, the local government for the regional city of Ballarat in south-west Victoria, Australia, runs an airport, galleries, theatres and museums (and until very recently a livestock saleyard). It also manages a city, townships and rural areas alongside

S. Fayad (✉) • L. Kendal
City of Ballarat, Ballarat City, VIC, Australia
e-mail: susanfayad@ballarat.vic.gov.au; lisakendal@ballarat.vic.gov.au

A. Campbell et al. (eds.), *Located Research*,
https://doi.org/10.1007/978-981-32-9694-7_6

all the other roles required by a diverse and growing urban and rural community. Its activities are underpinned by approximately 70 Victorian state government legislated acts. It could be argued that the City of Ballarat is a large, well-resourced regional organisation compared to smaller regional and rural local governments (and indeed the country vet!). However, its isolation from the state's major population centres, increasing demands and areas of responsibility mean that it also needs to turn to innovation to address its local city management challenges. This can be difficult when the available frameworks have predominantly been developed with the larger population centres in mind.

In 2012, the City of Ballarat was propelled onto the global stage through its early adoption (Roders, 2019) and localised operationalisation of a new city management approach, becoming the first local government in the world to pilot the UNESCO *Recommendation on the Historic Urban Landscape* (UNESCO, 2011) or HUL approach, bypassing other Australian cities' approaches along the way (Buckley & Fayad, 2017).

This chapter explores Ballarat City's innovative application of the HUL approach to evolve the highly structured and legalised area of strategic land-use planning.

A Time of and for Change

The dynamic regional city of Ballarat has a population of over 107,000 people and is considered 'a regional capital for Victoria's Western District [providing] valuable retail, health and education services to more than 400,000 people' (City of Ballarat, 2017: online). It is one hour's drive away from the burgeoning state capital and one of Australia's biggest cities—Melbourne, which recently reached a population of five million people. In part, due to Melbourne's decreasing liveability but mostly due to an influx of people from the nearby regions, Ballarat is facing its own burgeoning population projections of a 60 per cent increase of the total population over a 25-year period (City of Ballarat, 2015). Together with climate change, economic restructuring, demographic change (i.e. ageing population) and trends in development to meet changing community needs, local government challenges have become increasingly complex and fast paced (City of Ballarat, 2017). Ballarat is renowned for its highly intact authentic historic urban and rural character and its internationally significant history, which includes a living Aboriginal cultural landscape and recognition of the city's pivotal role in the mid-nineteenth century global gold rushes.

The speed and types of change impacting historic Ballarat were among the key drivers for the city government to look for new approaches of management, especially in achieving locally distinct (contextual) development outcomes and integrating heritage conservation (Fayad, 2018). At this time, the local government's practitioners were becoming aware of shortcomings in the widely embedded doctrine and practice of city planning and heritage conservation, established over three decades ago (Fayad, 2019). As a regional local government, the City of Ballarat saw a need to drive innovative city management practice through the framework of UNESCO's HUL approach. Today, the City of Ballarat is recognised globally as a 'leading city' in the application of the HUL approach and also as an important contributor to '...the maturing of the [global] theoretical framework and ... the development of methods and tools to endorse the HUL Recommendation' (Roders, 2019: 34).

Recognising the Role of Culture, Local Government and Local Citizens in Achieving Global Sustainable Development

The HUL approach was developed by UNESCO to address emerging critical global issues, such as urbanisation, homogenisation of cities and climate change, and the impacts these were having on cities, their environs, local cultures and senses of identity (WHITRAP, City of Ballarat, 2016). Over the past two decades, recognition of the role of culture in facilitating sustainable development has been gaining momentum (Bandarin, 2019), leading to its inclusion in the United Nations' 2016 Sustainable Development Goals (SDGs), particularly SDG 11 which aims to '[m]ake cities and human settlements inclusive, safe, resilient and sustainable' (UNESCO, 2016: online).

What has also changed is the recognition of the roles of local governments and local communities in addressing these emerging complex city management challenges. Previously, sustainable development was predominately approached *top-down*, but now both the HUL approach and the SDGs acknowledge the need for local ownership and action. It is now understood that 'local governments are policy makers, catalysts of change and the level of government best-placed to link the [goals of sustainable development] with local communities' (UCLG, n.d.: online). It is perhaps then no surprise that much effort has gone into reaching out to local governments, such as the City of Ballarat,[1] to pilot, localise and shape new approaches to city management through localised practice-led action and applied research.

Ballarat's experience is demonstrating the success that can be achieved by regional cities and the increased capacity for innovative research and practice in the regions. Through Ballarat's participation in UNESCO's international HUL pilot program, the local government, researchers and local citizens are driving a bottom-up revolution to shape the city's future, something that has been proving more difficult to achieve in some other scenarios (Rey Pérez & González Martínez, 2018). Many local governments, both globally and in Australia, are excited by the possibilities they see in Ballarat, including several located in metropolitan areas in Australia. While waves of change that align with the goals of the HUL approach are being witnessed, some cities have reported difficulties in getting their own HUL programmes off the ground (Rey Pérez & González Martínez, 2018).

True to the depiction of regional cities as places that innovate, innovation is deeply embedded in the psyche of the community and the local government authority, City of Ballarat. Ballarat has transitioned to prosper as a niche manufacturing hub, with a growing knowledge sector and service industries, in particular in health and aged care, creative industries, digital innovation and entrepreneurship. Testament to its commitment to innovation, the City of Ballarat has been a leader in local

---

[1] The City of Ballarat is also pilot city for the United Nations SDGs, taking part in the United Nations Global Compact—Cities Programme (UNGCCP) Australian City partnership to apply the SDGs in practice, and is also the first Australian city to sign onto the Council of Europe's Intercultural Cities network, further highlighting the eagerness to drive change from the regions.

government over many years; for example, developing the first Regional Conservation Plan in 1991 and in 2019 completing a *Creative City Strategy* (aligned with the HUL approach) focused on stimulating cultural and creative industries to ensure the city's continuous innovation, economic growth and social sustainability (City of Ballarat, 2019).

Contemporary Challenges for Regional City Planning

Ballarat is currently experiencing its highest levels of growth and change since the gold rush occurred in the mid-1800s. There are several challenges for strategic city planning in this environment, to ensure that development results in positive and sustainable outcomes for the community, city and broader region. These include addressing levels of:

- poor agility and responsiveness in the strategic planning system which is slow, costly and highly legalised. It currently takes approximately 3–5 years and many tens of thousands of dollars for a local government authority to complete a strategic planning project and embed change in the relevant regulatory system.
- piecemeal and expert-driven planning policy which is often developed in isolation without prior citizen consultations.
- Regulatory updating to effectively manage population and jobs growth. For the City of Ballarat, this has been somewhat alleviated by an existing, approved residential growth on the city's western fringe that is absorbing (and creating) residential growth, however, there is a strong desire to retain a compact city with adequate infill development that relies on updated planning policy and development controls.
- While the planning system in Victoria is founded on sound principles and objectives, it is driven by State policies, which are not always tailored for local application. Substantial local work needs to be undertaken to justify introducing locally and/or contextually tailored policies. In some instances, there are policy gaps where crucial guidance is needed for effective local decision making, such as in regional natural resource management, landscape and environmental policies.

In many ways, rigid and legalised processes are among the most challenging, requiring very creative approaches to affect any meaningful *on-the-ground* change. The City of Ballarat expects that the HUL approach will ultimately help to mitigate these challenges, improving decision making and supporting new development that further strengthens the identity of the city.

Innovation in Practice

Operationalising UNESCO's HUL approach is different from *status quo* city management practice as it prioritises bottom-up participation, interdisciplinary practice and a deep understanding of place in order to set and guide change, as outlined in Fig. 6.1 below. This has helped the City of Ballarat and, increasingly other local actors, to ensure a better, local values-driven future than what might otherwise have been possible.

Historically there have been different trends in planning practice, such as blueprint, rational comprehensive and systems approaches (Townsend, 2015), but contemporary strategic planning theory aligns well with the objectives of HUL in that it requires an integrated approach and recognises the interrelationships between nature, place, economy, community and psyche as shown in Fig. 6.2 below (Townsend, 2015).

As outlined above, best strategic planning practice will tell the story of place (background and history, vision and aspirations for the future) and how the vision will be achieved (a change management plan). It will address national, state and regional issues, but in the context of a particular place and its stories. While the elements of HUL are not new concepts for contemporary strategic planning, HUL provides a methodological focus to ensure that planning facilitates progress while also conserving the things that are important to the identity of the city and community, notably in aiming to create places that are sustainable, liveable, prosperous, healthy and fulfilling (Townsend, 2015).

In applying HUL to the City of Ballarat's local area planning program, consideration has been given to understanding how its six critical steps relate to strategic planning processes. These include:

Fig. 6.1 UNESCO's HUL recommendation details the minimum layers of the historic urban landscape to be comprehensively surveyed and mapped (Fayad, 2019) (Image source: City of Ballarat, 2017)

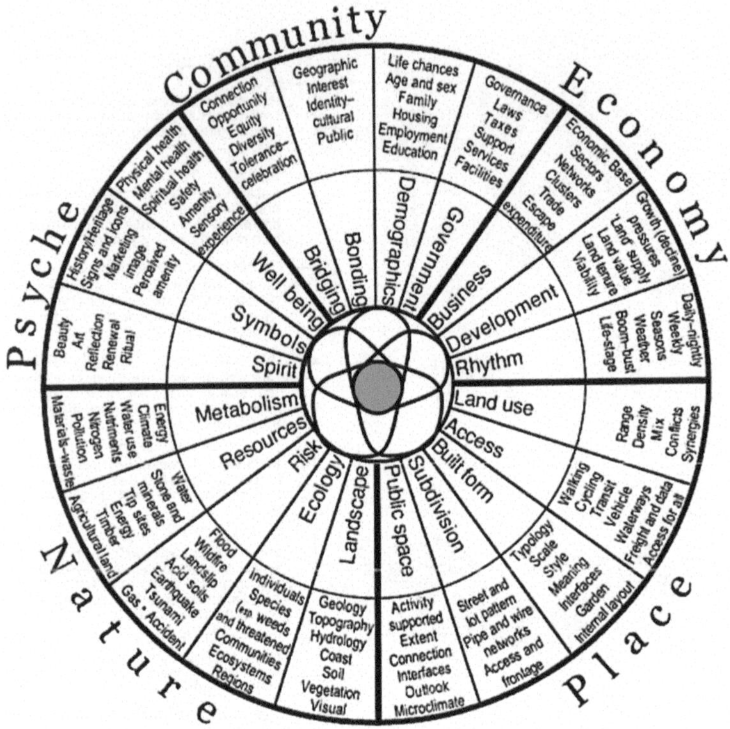

Fig. 6.2 Townsend, L 2015

1. Understanding place—surveying and mapping natural, cultural and human resources;
2. Participatory planning—working with local people to shape the future of their city to understand community values and determining their representative attributes;
3. Vulnerability assessment—potential negative impact on attributes, as associated values, as a result of socio-economic and climate stresses;
4. Dynamic place identity/heritage assessment and conservation—integrating urban values and their vulnerability status into a wider framework of city development, with careful attention to planning, design and implementation of development projects and areas of sensitivity;
5. Interdisciplinary action planning—prioritised action planning for conservation and development; and,

6. Partnerships for implementation—the establishment of partnerships and local management frameworks to guide implementation. (UNESCO, 2011)

In applying these to the City of Ballarat's local area planning processes, the focus has been on:

- Comprehensive research and review of technical studies;
- Participatory planning, in particular using a values-driven approach as established by *Ballarat Imagine* (see Box 6.1 below);
- Analysis of values, associated attributes and vulnerability to understand gaps and threats to the conservation of values;
- Application of landscape characterisation methodology to develop an integrated understanding of place; and,
- Development of a comprehensive, prioritised action toolkit to achieve a shared community vision.

Box 6.1 Ballarat Imagine

Ballarat Imagine was developed by City of Ballarat practitioners in 2013 to address the challenge of setting a community vision for the future of the city that would align with the HUL approach. Ordinarily, city visioning focusses on the *what* and *where* of change, however, the HUL approach requires cities use their valued characteristics and distinctive identity as a departure point for change.

Ballarat Imagine is used as the starting point for undertaking participatory engagement. It is set in the context of a changing city (the non-negotiable) and asks citizens and stakeholders:

- What do you *love* about Ballarat? (Values and attributes)
- What do you *imagine* for Ballarat? (Future scenario)
- What do you want to *retain* in Ballarat? (Limit of acceptable change)

Responses paint a distinctively localised picture of the city/neighbourhood and are used to set the collective vision and key priority areas for city strategies and local area planning projects.

This simple but innovative approach is now being used extensively by the City of Ballarat for many different types of community engagement programmes as well as by several cities around the world who are adopting the HUL approach (Fayad & Buckley, 2019).

To demonstrate, the following case study provides detail on how the HUL approach has been operationalised by the local government in applying it to a major strategic planning project in Ballarat.

The Ballarat East Local Area Plan

Following the adoption of the *Ballarat Strategy: Our Vision for 2040* (City of Ballarat, 2015) the City committed to a programme of integrated local area planning, to support growth and development while protecting local values by applying the HUL approach. Development of the Ballarat East Local Area Plan provided the local community and stakeholders with opportunities to contribute to a shared vision for their neighbourhood, ensuring that local values are conserved, enhanced and strengthened through good planning and ensuring that change and development are carefully managed to maintain liveability. This project has been undertaken over the past 2.5 years, and a final Local Area Plan will be prepared by mid-2019 for adoption by the Council, before it progresses to implementation into statutory land-use planning controls.

Methodology

Applying the HUL approach to Ballarat East involved developing a localised methodology to deliver an interrelated series of steps as shown below in Fig. 6.3.

At the beginning of this project, an HUL health-check assessment was undertaken to determine where there were opportunities to strengthen this approach. Additional work was then undertaken to:

- Comprehensively map assets and features;
- Understand community values, including preparation of a Social Significance Study which mapped values to tangible and intangible attributes; and,
- Understand place, including the preparation of a landscape characterisation study which captured the characteristics, community values, issues and threats and vulnerabilities.

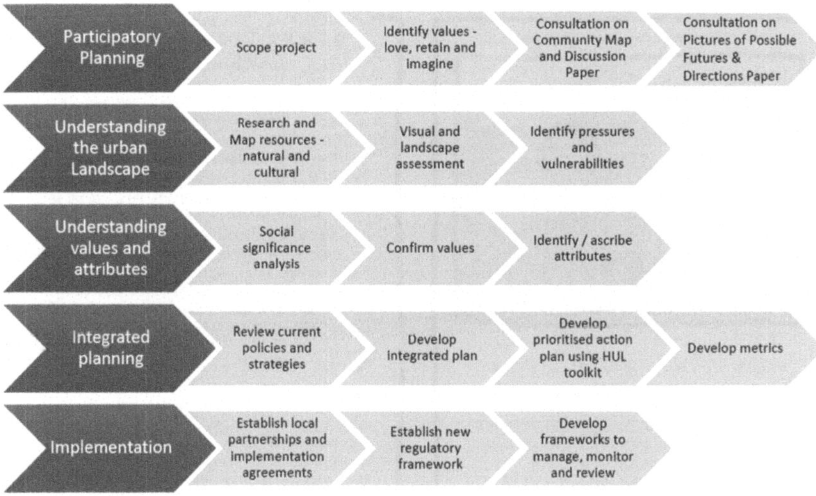

Fig. 6.3 HUL approach to integrated planning for Ballarat East (City of Ballarat, 2018)

Additional community consultations were undertaken to engage with key stakeholder groups, such as environmental advocacy groups, business, development industry, young people and the elderly. Throughout, creative engagement methods, such as Pictures of Possible Futures which used cartoon images and stories (see Fig. 6.4), were utilised to test preferences in relation to different future scenarios, including Historic East, Enlivened East, East in the Landscape, Landscape in the East, Eco Village East, Hidden gems of the east, Proximity East. A 'My Neighbourhood' community newsletter was also produced to provide updates on how the City of Ballarat was responding to the new opportunities identified through the process.

Benefits

This project involved extensive community engagement in developing all aspects of the plan, resulting in high levels of awareness and engagement throughout the community, business and development industries. Local knowledge has been integrated into the report to present a deep cultural understanding of place, alongside an action plan which addresses com-

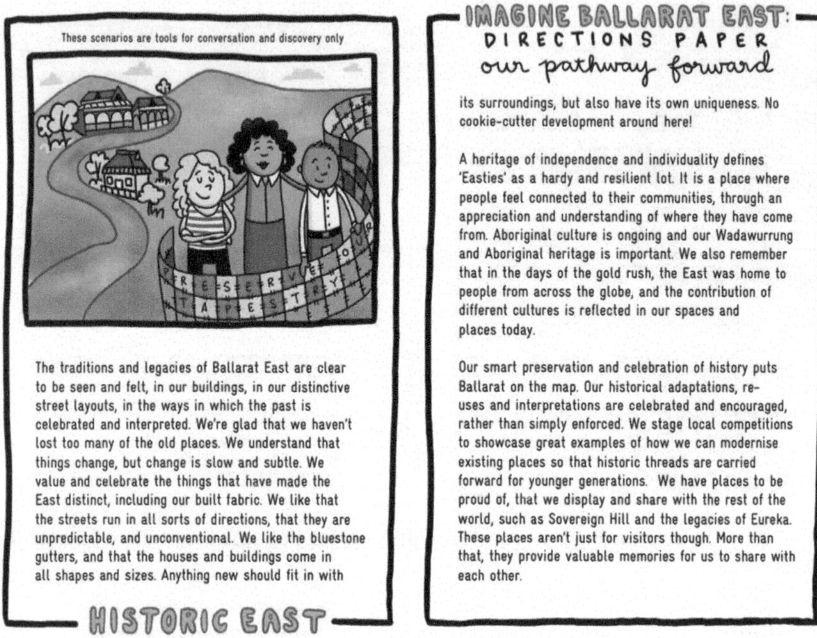

Fig. 6.4 Pictures of possible futures: Historic East (Foresite Lane in City of Ballarat, 2018)

munity priority issues. This process is expected to result in a clear vision for the future and certainty for community members and developers that will reduce conflict in decision making and achieve better outcomes for all.

Social Significance and Landscape Characterisation studies (City of Ballarat, 2018) have provided in-depth knowledge of tangible and intangible values of the East that will now be able to be considered for protection through regulatory systems. A number of specific actions have already been undertaken to respond to community priority interests, such as constructing sections of bicycle paths along waterways and preparing a conservation management plan. The latter can be delivered by the private sector in partnership with the local government and local community for a significant area of public open space—the original Ballarat East Town Hall Gardens. The large volume of knowledge and data acquired in this process will also be available for future planning projects through the HUL Ballarat website (City of Ballarat, online).

Learnings

A number of key challenges that might inform future strategic planning projects, in particular, scale, managing expectations and timeframes, emerged throughout the project. The scale and diversity of interests across the study area created challenges for adequately capturing, documenting and responding to needs through a single project, which ultimately affected the depth of research and timeframes for delivery. First, the Ballarat East study area, which is large and diverse in terms of community, urban form and environmental and cultural assets, was determined through a mapping exercise with the community. While the community associates the entire study area as being part of Ballarat East, it consists of a number of localities with extremely diverse values and issues ranging from an environmentally sensitive regional park interface along the eastern boundary to an urban environment with heritage significance and the major entrance to the City from the east. During the process, community representatives from one of the localities, Brown Hill, requested that their area be considered for separate local area planning due to the complexity and nature of the issues to be managed. A separate project—the Brown Hill Local Area Plan—is now also under way.

While an ideal process for working with multiple stakeholders to resolve complex issues, the high level of engagement required for participatory planning is very resource intensive, raises expectations from the community for ongoing engagement and is not easily applied to the more regulatory land-use planning processes (see Fig. 6.5 below). Ballarat City's commitment to public participation will continue to be refined, with consideration given to equitable engagement and the resources needed for maintaining high levels of engagement.

As identified earlier, it generally takes a long time to complete a strategic planning project and then additional time to see changes in the regulatory land-use planning scheme and real changes to development outcomes on the ground. The participatory approach required by the HUL approach further extends the time it takes to deliver strategic planning projects; however, it is recognised that it can also de-risk (and, as a result, quicken) actual implementation (Turner & Tomer, 2013).

Fig. 6.5 Land-use planning public participation spectrum (Kendal)

These long timeframes can be frustrating for community and other stakeholders, creating confusion and distrust regarding the process, which may ultimately risk the sustainability of this practice. These challenges may be somewhat alleviated and managed by reducing the scale and scope of projects, implementing short-term non-regulatory actions, as discussed above, building more effective decision-support tools and utilising the growing City's knowledge bank to create efficiencies for developers and decision makers.

Seeing the Results

While government is not often the first place many would look to find innovation and risk taking, application of the HUL approach in the regional city of Ballarat is notable for just that. Conversely, it is also because this is happening in a government setting that it is critical for early milestones and ongoing achievements to be clearly visible to ensure that the HUL approach is sustained, supported and seen as effective over the long term (Fayad & Buckley, 2019). While the HUL approach has transformed organisational practices in Ballarat City, for many its effects are not yet being felt *on the ground* (Fayad & Buckley, 2019).

At the time of writing, local planning outcomes that apply the HUL approach have not yet been embedded in the City of Ballarat's planning scheme, due to the timeframes required to amend land-use planning regulations. However, outside of the regulatory system, the HUL approach is playing a powerful role in managing change in the shorter term, for example, enabling:

- A more informed and engaged community, with an increased literacy in planning processes;
- Clarity of desired outcomes for the community, development industry and authorities;
- A growing knowledge bank of data and information about city values and assets to guide decision making;
- Proactively directing short-term investment for asset improvement aligned with community values; and,
- Advocacy for investment in community priorities.

Continued application of the HUL approach to strategic land-use planning and the progressive evolution of regulatory tools should soon start to address some of the contemporary challenges and detrimental impacts that the City of Ballarat faces as a regional city. As the City develops and implements its local area planning program, it will:

- Improve responsiveness—through easy access to knowledge and data to inform planning decisions and through a comprehensive programme of non-regulatory actions which contribute towards the common vision for the area;
- Manage growth—through improved locally relevant planning controls that reflect community values and achieve a common understanding of a desired identity and outcome for the city; and,
- Manage risks—ensure that risk and vulnerability are managed within the broader context of community values and vision for the city.

It is anticipated that the City of Ballarat will refine its application of the HUL approach in its future strategic planning projects. These are likely to include the development of protocols and a toolkit for use by the

City and for projects led by others in the City, for instance, State government investment. Continuing to build and ensure ready access to the City's knowledge bank will support clearer and faster decision making. Alongside these refinements, further attention is likely to be paid to the processes and requirements for projects of different scales, with attention given to site specific issues, such as precinct, locality and theme-based aspects. This is likely to include:

- focusing on progressing non-regulatory actions to maintain momentum while implementing regulatory controls for local area planning projects that have already been completed;
- planning for a changing climate and improving access to relevant information to inform decision making; and,
- Refining public participation processes to clarify the City's ongoing engagement and commitments in strategic land-use planning projects.

Conclusion

In applying UNESCO's HUL approach through its role in the international pilot program, the regional city of Ballarat has become recognised globally for its innovation and contribution to a new global sustainable development framework. As an early adopter of the approach, the city has adjusted implementation of the HUL approach to best suit its local context, working iteratively to contribute to the maturation of the HUL approach more broadly (Roders, 2019). Ballarat is showing the world that regional cities count as perfect incubators for this innovative approach to managing change for the future. Perhaps the final word is best left for Patricia O'Donnell, a global expert on HUL:

> We've known about the work in Ballarat since 2013. As a HUL expert, … I've been involved in the HUL work of building the [UNESCO] recommendation and in mainstreaming. In the group of cities across the world that's taken HUL to a mainstreaming position, Ballarat really stands out because they have embraced all of the ideas and concepts of HUL and then localised it in a very effective manner, which is quite unusual. Many cities

have done a component, have looked at perhaps one of the four tools, have tried to adapt their regulations to be more integrative of development and historic preservation, but in the work Ballarat has done … they really are standing out … (O'Donnell, 2019)

Acknowledgements The authors wish to acknowledge the Ballarat City Council who took a leap of faith and signed onto the global HUL pilot in 2013 and all the Council Officers, project partners and members of the local community who helped bring the programme to life and are taking it to new heights. Susan Fayad would also like to recognise Dr. Ron van Oers *(dec'd)* for the inspirational leadership and insight he brought from UNESCO to help guide Ballarat in the early days of the HUL pilot—he is greatly missed.

References

Bandarin, F. (2019). Reshaping urban conservation. In A. P. Roders & F. Bandarin (Eds.), *Reshaping urban conservation: The Historic Urban Landscape Approach in action* (pp. 3–20). Singapore: Springer.

Buckley, K., & Fayad, S. (2017). The HUL and the Australian Burra Charter – Some implications for local heritage practices. *Historic Environment, 29*(2), 46–57.

City of Ballarat. (2015). *Today, tomorrow, together: The Ballarat Strategy 2015–2040*. Ballarat: City of Ballarat. Retrieved May 7, 2019, from http://www.ballarat.vic.gov.au/city/city-strategy/ballarat-strategy-2040

City of Ballarat. (2017). *Our people, culture & place. A plan to sustain Ballarat's heritage 2017–2030*. Ballarat: City of Ballarat. Retrieved March 16, 2019, from http://www.hulballarat.org.au/cb_pages/ballarat_heritage_plan.php

City of Ballarat. (2018). *Draft Ballarat East local area plan*. Retrieved April 12, 2019, from https://mysay.ballarat.vic.gov.au/imagine-ballarat-east-local-area-plan/documents

City of Ballarat. (2019). *Historic urban landscape, Ballarat*. Retrieved March 16, 2019, from www.hulballarat.vic.gov.au

Fayad, S. (2018). The Historic Urban Landscape Approach: Finding a better way to manage change in the dynamic historic city of Ballarat. In J. Zhou & R. van Oers (Eds.), *Operationalising the Historic urban landscape – A practitioner's view* (pp. 39–63). Shanghai: WHITRAP.

Fayad, S. (2019). *The case of Ballarat, Australia*. Presentation, Al Ain (UAE).
Fayad, S., & Buckley, K. (2019). The transformational power of the HUL approach: Lessons from Ballarat, Australia 2012-2017. In A. P. Roders & F. Bandarin (Eds.), *Reshaping urban conservation: The Historic Urban Landscape Approach in action* (pp. 123–148). Singapore: Springer.
O'Donnell, P. (2019). Ballarat testimonial. Interview with Susan Fayad, City of Ballarat at Applying a HUL approach to the case of Al Ain city workshop, Al Ain, United Arab Emirates, 5 February, 2019.
Rey Pérez, J., & González Martínez, P. (2018). Lights and shadows over the recommendation on the historic urban landscape: "Managing change" in Ballarat and Cuenca through a radical approach focused on values and authenticity. *International Journal of Heritage Studies, 24*(1), 101–116.
Roders, A. P. (2019). The Historic Urban Landscape Approach in action: Eight years later. In A. P. Roders & F. Bandarin (Eds.), *Reshaping urban conservation: The Historic Urban Landscape Approach in action* (pp. 21–54). Singapore: Springer.
Townsend, L. (2015). Effective Strategic Planning, PLANET Training Notes, Planning Institute of Australia (unpublished).
Turner, M., & Tomer, T. (2013). Community participation and the tangible and intangible values of urban heritage. *Heritage & Society, 6*(2), 185–198.
UNESCO. (2011). *Recommendation on the historic urban landscape*. Paris: UNESCO. Retrieved March 3, 2019, from https://whc.unesco.org/en/hul/
UNESCO. (2016). *Culture: Urban future; global report on culture for sustainable development*. Paris: UNESCO. Retrieved September 19, 2019 from https://unesdoc.unesco.org/ark:/48223/pf0000245999
United Cities and Local Governments. (n.d.). *The sustainable development goals. What local governments need to know*. Retrieved March 16, 2019, from https://www.uclg.org/sites/default/files/the_sdgs_what_localgov_need_to_know_0.pdf
World Heritage Institute of Training and Research for the Asia and the Pacific Region [WHITRAP] and City of Ballarat. (2016). *The HUL guidebook: Managing heritage in dynamic and constantly changing urban environments. A practical guide to UNESCO's Recommendation on the Historic Urban Landscape*. Shanghai and Ballarat: WHITRAP.

7

The Role of Festival Networks in Regional Community Building

Judith Mair and Michelle Duffy

Introduction

Many regional and rural towns—both in Australia and elsewhere—face considerable challenges, including dwindling populations, limited employment opportunities and a decline in the traditional agricultural sector (Carson, Carson, Porter, Ahlin, & Sköld, 2016; Connell & Dufty-Jones, 2016). In the Australian context, these impacts have been exacerbated by changes in the mining and resources sector, as well as increased exposure to national, regional and global competitive forces (Tonts, Argent, & Plummer, 2012). In response to these dynamic and uneven processes, regional places and communities have had to transition from being resource-dependent to having a more diverse economic base, often

J. Mair (✉)
UQ Business School, The University of Queensland, Brisbane, QLD, Australia
e-mail: j.mair@business.uq.edu.au

M. Duffy
University of Newcastle, Callaghan, NSW, Australia
e-mail: Michelle.Duffy@newcastle.edu.au

A. Campbell et al. (eds.), *Located Research*,
https://doi.org/10.1007/978-981-32-9694-7_7

with a particular emphasis on tourism development. The research agenda of the 'new' rural geography explores the complexity of rural change and continuity as these places shift from a production-oriented focus to that of 'more complex and contested blends of production, consumption, and protection' (Tonts et al., 2012: 292).

Additional pressure is placed on the strategic, spatial, economic and environmental significance of rural and regional communities by the recent push behind the development of peri-urban regions (Buxton et al., 2006), as well as the impact of amenity-led migration (Drozdzewski, 2014). The rate of immigration to regional and rural Australia has increased substantially, particularly through the introduction of policies and schemes, such as the Regional Sponsored Migration Scheme, which seek to address a shortage of skilled workers in non-metropolitan areas and ease congestion in major cities (Kainth, 2018). One impact of the advancing urban frontier is that many rural communities have strong feelings of loss in terms of a sense of place and in what they perceive as the destruction of rural places in order to create so-called inauthentic and 'placeless' suburbs supposedly lacking any sense of belonging (Qviström, 2012).

In an effort to address the various challenges posed by these sorts of transitions, local councils and other stakeholders are increasingly using a festival structure to encourage economic diversification and tap into potential localised economic opportunities (Argent, 2002). As Gibson and Connell (2011: xvi) point out, in debates about economic restructuring, investment and government funding,

> festivals appear to be a positive, local and organically-generated activity, with the potential to contribute to regional development by stimulating both local society and the economy.

Thus, many stakeholders, especially in regional locations, have invested substantial resources with the expectation of positive social and economic outcomes for their local communities. Such desired positive social outcomes include increased social cohesion, community pride, sense of belonging and civic participation among others (Derrett, 2003; Duffy, Waitt, Gorman-Murray, & Gibson, 2011; Dunphy, 2009; Mulligan et al., 2006). Nonetheless, there have been criticisms of this instrumentalist use of festivals, with researchers suggesting that it leads to cultural commodification, a lack of authenticity and unproven benefits (Mair, 2019).

The underlying assumption that lies behind the staging of community festivals is that they function as a community building activity, creating or enhancing a sense of shared identity and belonging. The meaning of that community is understood as intimately linked to notions of place, a concept that suggests more than simply a location but instead something that is constituted out of the 'shared social experiences… constructed through shared practices and understandings' (Jarman, 2018: 336). While a festival is a temporary event, it offers a time, space and 'mechanism through which people continuously make and re-make collective identities and connections with place' (Quinn & Wilks, 2013: 17). This is especially significant for regional and rural places undergoing 'substantial transition', shifting from 'homogenous conservative agricultural communities… to diverse service oriented communities' where 'residents are having to deal with new challenges in their pursuit of an acceptable quality of life' (Derrett, 2005: 5).

Drawing on research carried out in the Hunter Valley of New South Wales, this chapter focuses on identifying the networks and ties between a range of actors and stakeholders in the community in the context of the Dungog Arts Festival. A recent festival, it originated as a response from members of the community to the demise of the very successful Dungog Film Festival, which had ceased in 2012 due to financial difficulties. Hosting such a festival was viewed by key stakeholders in the town as important because

> Dungog has gone through quite a number of hard issues [such as] the Tillegra Dam [which] was a huge issue that really divided the town. After that it was amalgamation which again divided the town, so [the festival] was something to bring the town together and something to bring people into the town… [as] an important part of the saving of Dungog Shire. (Interview 1, 2018)

Thus, participation in a community event like a festival offers a potent means of experiencing social connections and enhanced feelings of belonging—or conversely, feelings of alienation and exclusion. What these experiences reflect are the workings of social capital; that is, the social structures, both formal and informal, that encourage community residents to interact in positive ways and develop effective relationships (Putnam, 2000). As Quinn and Wilks (2013: 15) suggest, social capital 'offers significant opportunity to unravel the complexity of relationships and power dynamics that characterise communities'. However, while networks may facilitate the development of bridging social capital (where connections, or bridges, are built between diverse individuals and groups), networks are also responsible for bonding social capital (maintaining tightly knit networks that reinforce exclusive identities and homogeneity). Therefore, an investigation of the way that networks are constituted and maintained can inform research into community building.

Literature Review: Community, Festival and Networks

What is meant by *community* arises out of numerous competing narratives, ideologies and ideas that inform who (and what) constitutes a place or a community. Our ways of labelling a rural place invariably focus on notions of spatial location, of inhabiting particular bounded spaces separate from what we define as urban, and the imagery we draw on tends to focus on agricultural, traditional lifestyles and small populations (Bonifacio, 2011), or is a place of escape from the perceived ills of our modern city life (Short, 2006). Yet, this neglects the complex ways in which a rural place is experienced; rather than something separate and bounded, regional places are intimately connected with urban and other non-urban places through economic, political, social and cultural networks (Mair & Duffy, 2015). Massey's (1994) conceptualisation of place offers a means to better understand these processes, for, as she argues, the identity of a place or community is about 'the specificity of interaction with other places' (p. 12). Such a 'relational turn' offers ways to critically examine the ideas we hold about the rural as arising out of intertwined 'material and discursive phenomenon, processes and practices' (Heley & Jones, 2012: 9; see also Halfacree, 2007; Murdoch, 2003; Wood, 2011). It is such relational narratives that we create about place, identity, belonging, exclusion and so on that are activated within a community festival. Even so, there will be competing narratives, ideologies and ideas that inform who and what constitutes a place or community. Therefore, difference and diversity is not only something that is introduced *into* rural and regional places; rather, diversity already exists. Belonging is created through a sense of connection in particular ways or to particular collective identities, and these attachments and identities are embedded within the narratives that people tell about themselves as a community (Yuval-Davis, 2006).

Research examining festivals has focused on their role as a common framework for community celebration and reinvigorating notions of a shared community (Jepson & Clarke, 2015). Festivals are a popular means of creating and affirming a sense of community, often achieved through activities that enhance social connectedness and bring about a feeling of togetherness. The community festival framework uses community-oriented events as a way to draw attention to local communities situated in a particular place (De Bres & Davis, 2001; Duffy, 2000; Getz, 2007; Jaeger & Mykletun, 2013; van Winkle, Woosnam, & Mohammed, 2013). However, since the 1980s, governments around the world have increasingly turned to events like festivals as a means to facilitate development agendas in economically marginalised or depressed locations, often through a focus on culture, the arts, regeneration, education and tourism (Bowdin, Allen, O'Toole, Harris, & McDonnell, 2011). Yet, while a potentially valuable tourism product (Francesco & Oriol, 2009; Jago & Shaw, 1998), festivals are also instrumental in generating new forms of community identity. This role is an important process for regions undergoing transition, whether that be in terms of demographic change—such as in-migration through the settlement of culturally diverse communities or amenity-led migration—or through economic change brought about by transition from carbon-dependent industries (Duffy et al., 2011; Dunn, Thompson, Hanna, Murphy, & Burnley, 2001; Fincher & Iveson, 2008; Gorman-Murray, Waitt, & Gibson, 2012; Picard, 2015). Thus, festivals are not just about celebrating feelings of belonging and connection; they also play an important role in debates about how the idea of community is constituted and who does or does not belong (Duffy & Mair, 2018). This means that while a community festival may be successful because it activates opportunities for economic development or enhancing social connectedness through cultural activities, it is also potentially divisive because of the social changes wrought by in-migration populations and their associated networks (Mair & Duffy, 2018).

Rural festivals are often supported by local governments for their strategic importance in regional development, yet they can be at increased risk of failure, potentially because of the way they are managed or organised, which affects their leadership, planning and strategic decision-making (Getz, 2002). As Janiskee (1991) points out many rural festivals are run by volunteer-based organisations, usually made up of local residents, and as such, they are less likely to be involved in setting a strategic direction or addressing succession planning. Further, while involvement in a festival committee or organisation may bring long-term commitment from some, for others it will be a short-term contribution. This mix of motives and levels of commitment brings a range of challenges for the ongoing sustainability of a festival. Frost and Laing (2015) identify challenges for rural festival governance as the level of professional assistance required, concerns about the representativeness of organising committees, the potential for burnout, the lack of succession planning that is occurring and the limited resources that are available to volunteer organising committees.

A further issue for rural festivals is whether they can overcome any social divides or barriers within a community. Such divides may relate to the conservative nature of some rural communities with a concomitant mistrust of newcomers or may relate to existing and long-standing divisions. It could even be argued (e.g. Reid, 2007) that in some cases, committee members may prefer to keep control of the festival within a limited and select group—what Frost and Laing (2015) refer to as the inner sanctum. As Reid (2007: 95) points out, 'a neutral ground between these social groups may be achieved through organizing and planning rural events'; however, in practice, there may be concerns that the organising committees might be 'overly representative of one of these social groupings'. Distancing decision-making from a wide cross-section of the local community/ies has the potential for perceived or actual exclusion by some groups (Rogers & Anastasiadou, 2011).

Gibson and Stewart (2009), in their study of Australian rural festivals, concluded that the social and cultural concerns displayed and promoted in community festivals are closely linked to the 'pastimes, passions or pursuits of the individuals on organising committees' (p. 14). Thus, festival organisers play an important 'curatorial' role in the construction of these community events because they determine what particular sets of values underpin the event, they attach specific meanings to place and attempt, with varying degrees of success, to reproduce certain hegemonic meanings (Quinn, 2005). This means, as Clarke and Jepson (2011) point out, that the community festival is a demonstration of power in a community, and this power is embedded within notions of social capital.

Described by Serageldin (1996: 196) as 'the glue that holds society together', social capital refers to social structures and to the actions and interactions within these social structures in any given society (Coleman, 1988). Bourdieu's (1986) examination of capital situates social capital as one of three capitals operating within a society (the other two are economic and cultural capital), arguing that social capital is possessed and used by (mainly elite) individuals who use their membership of various networks and associations to benefit their social position. In contrast to this, Coleman (1988) and Putnam (1995, 2000) view social capital as something that can be accessible to all members of a community or society. Putnam suggests that social capital resides in 'those features of social organisation such as networks, norms and social trust that facilitate coordination and cooperation for mutual benefit' (1995: 67). Although social capital can refer to the notion of resources—those goods which are valued in society and embedded within an individual's network or associations (Lin, 1999)—the value of social capital lies in how it works through networks of relations, for, as Lee, Árnason, Nightingale, and Shucksmith (2005: 271) explain,

> it is meaningless to speak of social capital unless it is conceptualised as inhering in social relationships. While individuals, or groups, can hold 'slices' of social capital, it is only through social relationships, networks of social actors, that social capital can be mobilised and utilised.

Within these relationships, social capital operates through a generalised reciprocity that can transform individuals from self-seeking agents to members of a community with shared interests, a common identity and a commitment to the common good (Putnam, 2000). Despite the potential for positive outcomes from social capital, it may equally have negative consequences. Not all members of a community will hold equal levels of social capital (Portes, 2000), and, as a consequence, those who lack social capital may be excluded or isolated (Hawthorne, 2006), which in turn can reinforce social inequalities (Ooi, 2014). Therefore, the amount of social capital inherent in a community depends on the quality and quantity of interactions (Falk & Kilpatrick, 2000).

The importance of the lived, performed, felt and experienced modes of place and community made available through the festival means acknowledging the ways in which a festival can engage us within networks of emotion and affect that can heighten feelings of belonging or exclusion. As Waitt and Duffy (2010: 458) argue, 'festival spaces may create an affective ambience that encourages an openness to others, and sustain a social identification through the intangible feeling that encompasses an emotional space of belonging together'. Thus, a 'successful' festival in terms of social connectedness is one that brings about a feeling of togetherness, a collective joy that transforms individuals into a community (Ehrenreich, 2007). Such intense feelings can operate across different social structures including class, ethnicity or gender, and serve to re-affirm group identity and belonging (Duffy & Mair, 2018). Nevertheless, feelings of exclusion, disgust and even alienation can also be aroused and may instead close down processes of connectedness and social identification. These varying relationships between people, community and place troubles assumptions we may have about the community identities that are purposively constructed through events like festivals.

While there are a few studies that focus on the role of social capital in community festivals (see e.g. Misener, 2013; Quinn & Wilks, 2013; Wilks, 2011), Arcodia and Whitford (2006) propose such events can make important contributions to the development of networks, norms and resources. Festivals offer opportunities for developing new networks or revitalising existing ones via three main types of social capital: through bridging social capital (building new networks for individuals and organisations), through bonding social capital (strengthening existing networks for individuals and organisations) and through linking social capital (strengthening relationships between individuals, organisations and the public sector—with the result being improved structures for formal and informal community consultation, Mair & Duffy, 2014; Duffy & Mair, 2018). Regional development strategies often make use of the strong sense of local identity generated by these aspects of social capital, creating 'culture economies' that market 'the local' that also serves to deepen a sense of local identity (Lee et al., 2005; Ray, 1998). However, festivals, like other tourism events, can unsettle notions of rural and regional identity.

What is often neglected in such strategies is the significance of the often invisible networks that facilitate or inhibit social connections and therefore configures those who possess social capital and its attendant relations of power. As Quinn and Wilks (2013: 15) argue, we need to carefully examine the relationship between social capital and place because of the implications this has for 'the formation, nature and implications of social connections between various actors in festivals'. How a place is constituted is a political act (Pierce, Martin, & Murphy, 2011) because it is through social networks and their influence that a place is defined (Jarman, 2018). Nor is place simply that of location; it arises out of 'the specificity of interaction with other places' (Massey, 1994: 12) and an individual's 'conscious and unconscious interactions… that comprise their own relationship with a place' (Jarman, 2018: 337). The festival, then, is a complex activity *and* process. It can be understood as 'an iterative event support[ing] bonding social capital, regularly bringing communities together at a smaller scale and supporting local talent' (Jarman, 2018: 340). However, it is also part of economic development strategies to entice the '"outsider" and the tourist, who are seeking "authentic"' experiences (Fernandes, 2013). Thus, we find a tension in such strategies; festivals are used to initiate short-term results but are also framed as a means to establish long-term network relations (Klaster, Wilderom, & Muntslag, 2017).

Methodology

Case Study Context

Dungog is a small county town located in the Hunter Region of New South Wales, Australia. It is just over one-hour drive north of the city of Newcastle, and approximately three-hour drive north of Sydney. The Shire of Dungog covers an area of 2248 square kilometres and the main population centres are the towns of Dungog, Gresford, Paterson, Vacy and Clarence Town (Dungog Shire Council, 2019). Long occupied by Gringai Koori indigenous people (Williams, 2012), a clan of the Worimi people (Tindale, 1974), the area around Dungog was settled by Europeans in the early 1800s and has been an agricultural centre since then, with a focus on dairy farms and timber production. The Shire has a population of approximately 8500 people, of whom around 2000 live in the town of Dungog, making this the main settlement in the Shire (Dungog Shire Council, 2019).

One of the main historic buildings in the town of Dungog is the James Theatre, built in 1928 and known to be the oldest still operating purpose-built cinema in Australia (James Theatre, 2019). As a means to contribute towards the renovation and upkeep of this historic venue, a film festival was created in 2007. The Dungog Film Festival was very successful in terms of attracting famous names in Australian film, but after a period, the festival ceased to operate in 2012. A new festival was inaugurated in 2015, called simply the 'Dungog Festival', and run primarily by volunteers. The festival organisers sought to 'inject some much needed funds back into the local economy', following the devastation 'by a deadly super storm that hit the region in April [2015], destroying homes and businesses' (ABC News online, 2015). The then managing director, Tracy Norman, explained that what was needed was a 'showcasing [of] Dungog to the world as a wonderful tourism destination… letting the world know Dungog is back open for business' (quoted in ABC News online, 2015). Funding from Destination New South Wales was used to help attract tourists and increase multiple stays and year-round visitation to the area (interview 2, 2018).

In 2018, the Dungog Festival offered a wide range of experiences, including a street parade, a long table dinner, a sculpture exhibition (Sculpture on the Farm), a range of films at the James Theatre, a country garden ramble (where several local gardens were opened to the public), a street party with local and national talent, a vintage and agricultural vehicle show and, somewhat incongruously, daschund races! This new incarnation of the festival appears to be gaining traction in terms of reputation and visitor numbers. While the revised festival focuses on 'celebrat[ing] while sharing all those things that we love in Dungog and make our valley thrive by promoting the area' (Dungog Arts Foundation Director Michael Atkins, quoted in Dungog Chronicle online, 2017), there has arguably been a lack of clarity around the theme of the festival and its contents. This has caused some heated debate about the festival's role in and for the community both within the festival organisation itself and in the community more broadly.

Methods

Social Network Analysis (SNA) was chosen as a suitable method for investigating the structural properties of the networks and relationships in the context of the Dungog Festival. There are several levels of SNA including node-level analysis (at the level of any given individual), a dyadic network (between two people) and a network-level analysis (where an entire network is identified and analysed at the macro level) (Knoke & Yang, 2008). A core assumption of SNA is that the patterns of relationships identified have a significant effect on individual and group behaviour, including access (or otherwise) to resources and information (Steketee, Miyaoka, & Spiegelman, 2015). Thus, by analysing the networks in a given context, the researcher can assess how well-connected individuals are, and also how connected, or dense the overall network is. Broadly speaking, the more ties that individuals have within a network, the more cohesive the network can be considered to be. However, as Knoke and Yang (2008) point out, social relations are dynamic rather than static, and so an SNA is always only a snapshot of one point in time.

Some of the key measurements used in SNA include *Degree Centrality* (the quantity of ties that a node (an individual) has), *Betweenness Centrality* (which assesses how much influence a node has, with high betweenness centrality indicating the potential for a tie to act as a gatekeeper) and *Density*, defined as the proportion of actual ties that each member of a network has to the number of possible connections in the network (Jarman, 2018; Jarman, Theodoraki, Hall, & Ali-Knight, 2014; Steketee et al., 2015).

The sorts of questions that SNA can assist with include: identifying those individuals within a network that others consider to be most important (degree centrality); identifying those individuals who are most associated with the functioning of the network (betweenness centrality); and, identifying any sub-groups or cliques within a network (Jarman, 2016). In the context of the Dungog Festival, our research questions centred on the two initial questions suggested by Jarman (2016)—who is considered by those in the network to be most important to the Festival, and who is most associated with its functioning.

Data Collection and Analysis

Although there are several ways to generate the names required for an SNA, we opted to use name generators during interviews being carried out as part of a larger project on the Dungog Festival. Interviewees were selected initially as a result of a purposive strategy to identify those clearly associated with the Festival (i.e. the organiser, members of the organising committee) followed by a snowball sampling strategy where interviewees were asked to recommend others who may be able/willing to contribute to the project. A total of ten interviews were carried out, and each interviewee was asked to identify five individuals who they considered to be vital to the running of the Festival. Five was chosen as the required number of names following Jarman (2016), who argued that five names were sufficient for network-level analysis. This form of name generation is common in SNA but is subject to potential bias where informants forget individuals that are important, or mention names that are well-known to them for reasons other than the matter under discussion (in this case the festival; Knoke & Yang, 2008). However, where a broad consensus is reached by the majority of informants, it is likely that the influence of any bias will be limited. Basic data about each of the names generated was also collected at the time of name generation, including gender, workplace (where relevant) and role in relation to the festival.

The data was analysed using SocNetV-2.4, an online social network analysis software tool. Each informant was entered into the software as a node, and the five names that they identified were also entered as nodes with the connections (known as ties or an edges) between the informant and those they named being specified. This simply requires inputting source (the informant); target (the names they identified) and weight (this was automatically set at 1 as no information on the strength of the relationship was collected) (Jarman, 2016). Names were entered in no particular order. All participants were given codes (node numbers) to protect their identity.

Following data analysis, an initial sociogram (a visual representation of the network) was created. These sociograms are probably the most recognisable aspect of network analysis (Prell, 2012). An initial visualisation of the SNA for the Dungog Festival is shown in Fig. 7.1. The size of the

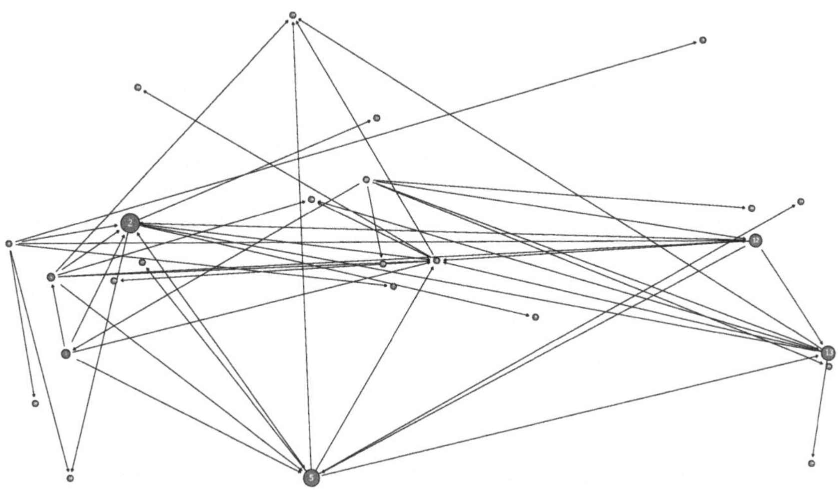

Fig. 7.1 Initial Visualisation of the festival social network graph

nodes represents the number of connections that any individual has, therefore, the larger the node, the more well connected that individual is.

Findings

Network-Level Values

Network-level measures relate to the structure of a network and are generally considered to relate to the cohesion of a network (Steketee et al., 2015). Network density relates to the level of inter-connectedness within a network and is calculated as measured by the proportion of actual ties to the number of possible ties. There are 25 nodes in this network which means that there are 300 possible ties. However, there are only 48 ties in this network, giving a density of 16% which can be considered to be fairly low.

Network distance is used to calculate how close or separated nodes generally are from each other. In this network, the average shortest path length (the distance on average from one individual to another) is 1.8 suggesting that most people within this network are not directly connected to each other.

Node-Level Values

Node-level values describe the characteristics of individuals within a network (Steketee et al., 2015). Node centrality is a key measure to identify which individuals are most central to a network. Degree centrality refers to the number of direct connections to a given node, while betweeness centrality relates to how often a node falls between other nodes (and thus is an important connector).

Nodes with particularly high betweenness centrality include Node 2 (38.2), Node 5 (22.56) and Node 3 (20.73). These nodes can, thus, be considered to be very important connectors in the network, acting either as bridges between many other nodes, or alternatively as gatekeepers. Within this network, Node 2 has a particularly high level of degree

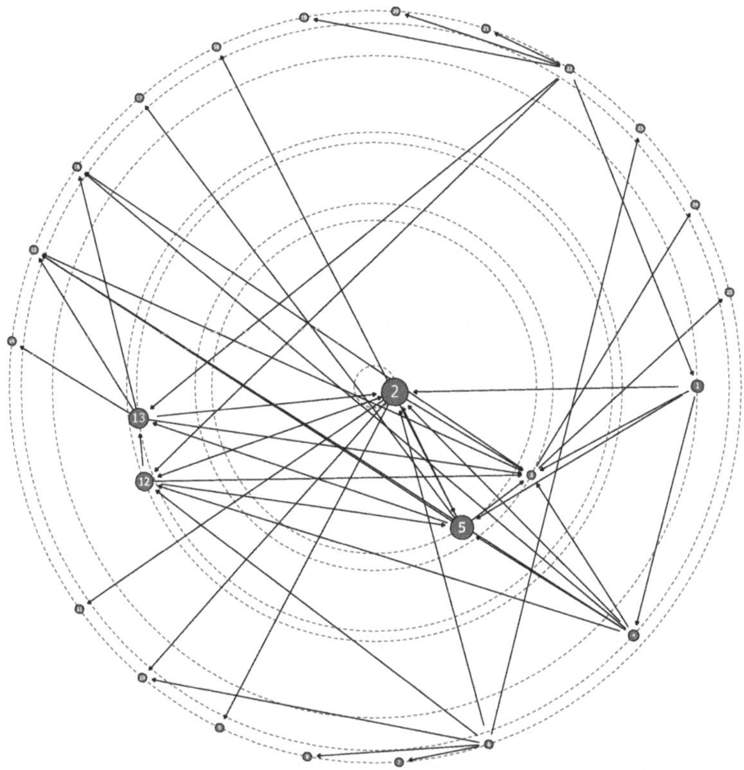

Fig. 7.2 Degree centrality shown as a radial

centrality (Node 2 = 0.250, compared with the mean of 0.080). However, other nearby nodes (particularly Node 5 = 0.245) are also considered by participants to be very important in the network (see Fig. 7.2).

Discussion

Perhaps unsurprisingly, given the population size of Dungog, our analysis demonstrated that there are specific individuals nominated as integral to the success of the Dungog festival. In addition, our interviews were conducted with individuals who held roles either in the community, local government or business and closely involved with the organisation and coordination of the festival. Nonetheless, of the 25 individuals nominated by our participants, five were nominated as critical to the festival, which can be explained as linked to the roles and responsibilities that they have in terms of the festival. In the context of the overall network associated with the festival, this is problematic. Table 7.1 indicates the roles held by each individual, although names are not provided to protect the anonymity of research participants. The network is not well connected and relies on a small number of individuals. In terms of the sustainability and longevitity of the festival, if any one of these individuals cease to be part of the festival network, this may in turn mean that those connected to that individual could also become disconnected from the festival

Table 7.1 Indicative role of individuals nominated in SNA

Node number	Indicative role in the festival/community or place of work
1	Tourist Information Centre
2	Councillor and local business operator (accommodation)
3	Festival Board of Directors
4	Member of festival organising committee
5	New resident and organiser of Sculpture Festival
6	Local business operator (IT)
7	Local resident and long-time volunteer
8	Local business owner (bar)
9	Dungog Shire Council
10	Volunteer Coordinator
11	Local resident and long-time volunteer
12	Chamber of Commerce
13	Festival organiser
14	Volunteer Coordinator
15	Dungog Arts Foundation board member and local business owner (accountancy)
16	Event Coordinator
17	James Theatre
18	Dungog Shire Council
19	Dungog Advertiser newspaper
20	President-Dungog Shire Events
21	Dungog Regional Tourism
22	Dungog Shire Council
23	James Theatre
24	Local resident and volunteer
25	Local resident

network. There is also potential for such key individuals to choose not to act as bridges or connectors within this network, thus turning into gatekeepers or barriers to the network.

In addition, almost all of the people identified as critical to the success of the festival in term of organisation, coordination and planning are volunteers apart from the festival director (although this role is part-time). As many of those we interviewed pointed out, community-based festivals are likely to be staffed by volunteers, generally who are local residents, and very much part of 'the life blood' of community events, and this is particularly so in rural and regional areas. As one stakeholder explained:

> I think to me one of the greatest parts of the festival is the 140 odd volunteers that we get and the experience that the volunteers have and the pride that they have about the festival. You know we all sit there the Monday morning after the festival or this time it will be Tuesday morning I guess, absolutely exhausted but having known that we'd put on this wonderful event. The friendships, I mean there's so many lasting friendships that have been forged by people volunteering together or meeting at the festival and, you know, that to me is a great outcome. (Interview 1, 2018)

Even so, the tightness of this festival's network, as well as the reliance on only a few people, leaves the festival open to failure (Frost & Laing, 2015; Janiskee, 1991). It is worth pointing out that the financial success of the festival has been strongly influenced by only a few important stakeholders, with one key backer standing out (not named to protect their anonymity). This is a concern in that should this one backer lose interest in the festival or find themselves unable to support the festival in future, the festival's viability could be placed in doubt. However, perhaps the most pressing concern for rural and regional events is volunteer burnout, most often the result of lack of resources, as demonstrated in a study by Stone and Millan (2011: 245) who recorded one respondent's concern: 'All of us who are running it at the moment are wearing too many hats and have too many responsibilities'. Burnout may lead to an untimely departure of volunteers that not only impacts on the available workforce for an event but also on organisational knowledge and appropriate training for new volunteers. While volunteers may have specialised skills and knowledge, volunteer-run events are often less 'engaged in elaborate goal setting or evaluation of effectiveness in reaching goals' (Getz & Frisby, 1988: 25). A lack of adequate human resources impacts on strategic planning, including planning for an orderly succession.

The challenges of human resources and strategic planning are important considerations, yet many of those involved in the current iteration of the Dungog festival were part of the older versions of the festival, including the successful film festival. Although their participation in this festival may differ, knowledge and experience of these previous events is important for the sustainability of this new festival, for, as Jarman (2018: 344) points out in his study of a Scottish festival's social network,

> longevity appears to engender centrality and importance, benefiting those willing and able to sustain a commitment to the festival. There are personal social capital rewards for this investment.

Nevertheless, there are opportunities to strengthen the festival network, which is an important consideration given the small number of individuals designated as critical to the festival. One avenue is to tap into networks that some individuals have outside that of the immediate Dungog festival network. For example, Node 5, Chair of one of the events that form part of the Dungog Festival has significant networks outside of Dungog that are not tapped into by the festival organisers at present. These networks offer strategic links between urban (especially Sydney's art scene) and regional locations that could increase tourist interest in Dungog. As Philippa Graham explained prior to the launch of this event in 2018, the hope is to 'attract Sydneysiders who are used to "the quirky, wonderful, challenging sculptures" of Bondi's Sculpture By The Sea' (quoted in Scully, 2018). This was a very popular event at the 2018 Dungog festival, and could be an event run independent of the festival as it was well-financed and made good use of the Chair's external networks (Getz, 2002). There were discussions around rescheduling this event as part of the 2018 seeDUNGOG Regional Arts and Culture trail, although at the time of writing this has not happened. Perhaps further complicating such tourism strategies in Dungog—again this is something common to regional places—is the fact that the Arts and Culture trail is coordinated by the Dungog Festival event manager, so there are competing interests in devising these different events.

Based on the networks we identified, the Dungog Arts Festival has varying success in creating and maintaining social networks and social capital. From the SNA and interviews, there is strong evidence of bonding social capital through specific key individuals who are well-known in Dungog. The involvement of these individuals has maintained existing social bonds within the community. The inclusion of the Sculpture on the Farm event offers strategic opportunities for Dungog; the Chair of this event is well connected to urban arts networks, and her knowledge and expertise can provide significant benefits to the community and its tourism industry. However, there may be some reluctance on the part of long-term residents to readily accept a newcomer's plans. Although not raised specifically in terms of the Sculpture on the Farm in interviews with stakeholders, Dungog's previous film festival was the initiative of in-migrants, and though very successful it ended after increasing financial concerns (Davis, 2014) and the departure of its directors from the area. In a similar vein, a few questioned how representative the Dungog festival is, as one participant stated

> I think there are too few people making the decisions and I don't think they're representative of—see, I could tell you the five people who are pivotal to this festival. But are they the right five people for the festival?

One way to address concerns about broader community representation and engagement has been to incorporate events at a number of towns in Dungog shire, such as the previously mentioned daschund race. Yet, as with strengthening networks beyond the immediate festival, the sustainability of this event requires critical evaluation. As Reid (Reid, 2007: 95) determined in her study of the social consequences of events on rural communities,

> The social divides or barriers that exist within rural communities are often deeply entrenched; in spite of this a neutral ground between these social groups may be achieved through organizing and planning rural events.

> Often this divide is based upon whether an individual is a 'local' (born within the community) or a 'nonlocal'. (Moved to the community)

This is not to suggest that all festivals will achieve goals of inclusivity and connectedness, but as suggested by the examination of the social networks in this example, there can be unexpected or unrecognised exclusionary practices in devising such events.

Conclusion

It is clear that the current iteration of the Dungog Festival has potential to become a long-standing part of the community events calendar. It appears well supported by both volunteers and those living locally and has certainly attracted significant interest across various sections of the community. However, the findings of this research suggest that there are areas which should be of concern to the festival organisers in relation to the ongoing success and viability of the festival.

The first issue of concern is the lack of a strategic approach to the festival organisation, with heavy reliance on volunteers and little evidence of any significant consideration given to forward thinking or succession planning. This may not become a problem in the short term, but could prove troubling in the medium to long term. Community-based and particularly regionally located festivals are more likely to be staffed by volunteers, generally who are local residents, and very much part of 'the life blood' of community events. This has important implications. The level of engagement in volunteering is significant not just in terms of the successful running of an event or group such activity supports, but is an indicator of social capital because this activity is understood to demonstrate the existence of community ties and neighbourliness (McAllum, 2017). Yet, the small population size of these regional communities means that festivals and other events are vulnerable because of volunteer burnout, most often the result of lack of resources (Stone & Millan, 2011).

Another issue which appears to be troubling is the dense and tight nature of the network of people associated with the festival. While around 25 people were nominated as part of the SNA as being of importance to the festival, it would seem that only around five people are really key to the success of the festival. With such a small number, volunteer burnout, or any of a range of other issues, could intervene and pose problems for the ongoing success of the festival.

The dense nature of the festival network also calls into question the extent to which the festival (and decisions made around who should be included, involved and consulted about the nature of the festival) can truly be said to be representative of the broader community and wider stakeholders. In many other cases, questions around 'whose festival is it' have led to disharmony and in some cases significant clashes between different stakeholder groups. This is never an optimal situation for a festival to find itself in and calls into question the impact that the festival is having on social capital, and indeed, whether it is helping to generate positive social capital or instead is leading to the potential exclusion of some.

Increasing tourism is one of the key reasons why communities and local governments support and stage festivals (Mayfield & Crompton, 1995). The ability of a festival to attract visitors to a host region and contribute to its economic and social wellbeing, especially in rural and regional locations where there is a transition away from more traditional rural industries, explains the significance afforded to festivals in many tourism policies and strategies (Mair & Whitford, 2013). The underlying assumption that lies behind the festival is that it functions as a community building activity, even though some may feel excluded from the public performances of identity displayed in a community festival. Yet, as a tourism event, festivals are also used to entice those from beyond the boundaries of a town and community in order to help generate economic as well as social benefits. Festivals are, therefore, complex sites for localness and belonging, celebrating localness and the connections beyond that confines of that locality. This means that the importance of a festival lies not simply in the additional visitors that are attracted to a destination; rather, increasing community investment in such events is a significant to achieving a community's economic and socio-cultural objectives (Ziakas, 2010). Community-based festivals are differentiated from those of a more commercial focus because the festival is generated from 'within the community in order to celebrate its unique identity and pride' (Rogers & Anastasiadou, 2011: 388). Even so, what is meant by 'the community' can prove challenging. This will, in turn, mean that there will be different perspectives as to what constitutes community involvement (Rogers & Anastasiadou, 2011).

These are important considerations for a town like Dungog. While there is strong evidence of bonding social capital through specific key individuals in Dungog, the small number of individuals designated as critical to the festival in this study is small and raises questions about its sustainability. There are a number of caveats to this analysis. First, we acknowledge that our study is limited as we only collected binary data; that is, our informants either identified someone or they did not. In future research, we could consider asking informants to rate the strength of their ties with other people on a scale, rather than simply identifying a connection, which would provide a more nuanced reading of Dungog's social relations. Second, this is a new festival and the festival committee and other associated stakeholders are still determining the focus and scale of the event. Given these factors, a critical exploration as to how social capital operates within regional places can offer important insights into the sustainability of festivals, especially in terms of addressing economic and social challenges.

Future research could examine in more detail the true nature of this festival in relation to how well, or otherwise, it functions as a community building process, and how those involved in the festival (and those who may feel excluded) feel about the different kinds of community and identity that are being performed at the festival. In addition, future research could usefully take a more quantitative approach to examine the way that the festival affects levels of individual and community social capital. As Klaster et al. (2017) point out, what is needed is a means to determine and measure the effectiveness of networks, yet this needs to be undertaken at a more nuanced level than the acknowledgement that predetermined goals have been met. While examples of effectiveness could be defined in terms of trust and shared learning, for example, this is most often 'mistakenly viewed as effective when participants are merely collaborating well' (Klaster et al., 2017: 677). In addition, the workings of such collaborative networks 'may be torn between obtaining short-term goals and establishing enduring network relations' (Klaster et al., 2017: 678). Determining the value, range and efficacy of networks in regional locations can help uncover the variety of networks and relations and how these may contribute to a festival's success as a community development tool, as well as the challenges faced by regional communities.

References

ABC News Australia. (2015). Dungog Festival aims to bring back tourists. Posted 22 October. Retrieved October 20, 2017, from www.abc.net.au/news/2015-10-22/dungog-festival-aims-to-bring-back-tourists/6875138

Arcodia, C., & Whitford, M. (2006). Festival attendance and the development of social capital. *Journal of Convention and Event Tourism, 8*(2), 1–18.

Argent, N. (2002). From pillar to post? In search of the post-productivist countryside in Australia. *Australian Geographer, 33*(1), 97–114.

Bonifacio, G. (2011). Introduction. In G. Bonifacio (Ed.), *Gender and rural migration* (pp. 1–19). New York and Oxon: Taylor and Francis.

Bourdieu, P. (1986). The forms of capital. In J. G. Richardson (Ed.), *Handbook of theory and research for the sociology of education* (pp. 241–253). Westport, CT: Greenwood Press.

Bowdin, G., Allen, J., O'Toole, W., Harris, R., & McDonnell, I. (2011). *Events management*. Oxford: Butterworth-Heinemann.

Buxton, M., Tienan, G., Bekessy, S., Budge, T., Mercer, D., Coote, M., et al. (2006). *Change and continuity in periurban Australia: Monograph 1 A review of the literature*. Melbourne: RMIT.

Carson, B., Carson, D., Porter, R., Ahlin, C. Y., & Sköld, P. (2016). Decline, adaptation or transformation: New perspectives on demographic change in resource peripheries in Australia and Sweden. *Comparative Population Studies, 41*(3/4), 379–406.

Clarke, A., & Jepson, A. (2011). Power and hegemony within a community festival. *International Journal of Event and Festival Management, 2*(1), 7–19.

Coleman, J. S. (1988). Social capital in the creation of human capital. *American Journal of Sociology, 94*, 95–120.

Connell, J., & Dufty-Jones, R. (2016). Introduction: Twenty-first century Rural Australia. In R. Dufty-Jones & J. Connell (Eds.), *Rural change in Australia: Population, economy, environment* (pp. 1–24). Farnham: Ashgate.

Davis, B.-J. (2014). Dungog Film Festival resurrected. *Newcastle Herald*. Posted 9 March. Retrieved May 31, 2019, from www.theherald.com.au/story/2137781/dungog-film-festival-resurrected/

De Bres, K., & Davis, J. (2001). Celebrating group and place identity: A case study of a new regional festival. *Tourism Geographies, 3*(3), 326–337.

Derrett, R. (2003). Making sense of how festivals demonstrate a community's sense of place. *Event Management, 8*(1), 49–58.

Derrett, R. (2005). Why do regional community cultural festivals survive? paper presented to The impacts of events: triple bottom line evaluation and event legacies. *Third international event management conference*, University of Technology Sydney, Sydney, NSW, 13–15 July.

Drozdzewski, D. (2014). 'They have no concept of what a farm is': Exploring rural change through tree change migration. In R. Dufty-Jones & J. Connell (Eds.), *Rural change in Australia* (pp. 83–101). Farnham, UK: Ashgate.

Duffy, M. (2000). Lines of drift: Festival participation and performing a sense of place. *Popular Music, 19*(1), 51–64.

Duffy, M., & Mair, J. (2018). *Festival encounters: Theoretical perspectives on festival events and social cohesion*. Abingdon and Oxon, UK: Routledge.

Duffy, M., Waitt, G., Gorman-Murray, A., & Gibson, C. (2011). Bodily rhythms: corporeal capacities to engage with festival spaces. *Emotion, Space and Society, 4*(1), 17–24.

Dungog Chronicle. (2017). Dungog Festival 2018 is shaping up to be bigger than ever. Posted 13 November. Retrieved November 20, 2017, from www.dungogchronicle.com.au/story/5051297/festival-revival-on-track/

Dungog Shire Council. (2019). About the Shire. Retrieved March 21, 2019, from https://www.dungog.nsw.gov.au/about-shire

Dunn, K., Thompson, S., Hanna, B., Murphy, P., & Burnley, I. (2001). Multicultural policy within local government in Australia. *Urban Studies, 38*(13), 2477–2494.

Dunphy, K. (2009). Developing and revitalizing rural communities through arts and creativity: Australia. Retrieved March 3, 2019, from Analysis and Policy Observatory Website: https://apo.org.au/node/17497

Ehrenreich, B. (2007). *Dancing in the streets: A history of collective joy*. London: Granta Books.

Falk, I., & Kilpatrick, S. (2000). What is social capital? A study of interaction in a rural community. *Sociologia Ruralis, 40*, 87–110.

Fernandes, C. (2013). The role of cultural events in building social capital and the implications for tourism development. In G. Richards & M. de Brito (Eds.), *Exploring the social impacts of events* (pp. 190–202). New York: Routledge.

Fincher, R., & Iveson, K. (2008). *Planning and diversity in the city: Redistribution, recognition and encounter*. Hampshire, UK: Palgrave Macmillan.

Francesco, G. R., & Oriol, M. I. (2009). Managing music festivals for tourism purposes in Catalonia (Spain). *Tourism Review, 64*(4), 53–65.

Frost, W., & Laing, J. (2015). Avoiding burnout: The succession planning, governance and resourcing of rural tourism festivals. *Journal of Sustainable Tourism, 23*(8–9), 1298–1317.

Getz, D. (2002). Why festivals fail. *Event Management, 7*, 209–219.

Getz, D. (2007). *Event studies: Theory, research and policy for planned events.* Oxford: Butterworth-Heinemann.

Getz, D., & Frisby, W. (1988). Evaluating management effectiveness in community-run festivals. *Journal of Travel Research, 27*(1), 22–27.

Gibson, C., & Connell, J. (2011). Preface. In C. Gibson & J. Connell (Eds.), *Festival places: Revitalising rural Australia* (pp. xv–xvii). Bristol, Buffalo, and Toronto: Channel View Publications.

Gibson, C. R., & Stewart, A. (2009). *Reinventing rural places: The extent and impact of festivals in rural and regional Australia.* Wollongong, NSW: University of Wollongong. Australian Centre for Cultural Environmental Research.

Gorman-Murray, A., Waitt, G., & Gibson, C. (2012). Chilling out in 'cosmopolitan country'? Interrogating the construction of Daylesford, Victoria, as a 'lesbian and gay rural idyll. *Journal of Rural Studies, 28*(1), 69–79.

Halfacree, K. (2007). Back-to-the-land in the twenty-first century: Making connections with rurality. *Tijdschrift Voor Economische En Sociale Geografie, 98*(1), 3–8.

Hawthorne, G. (2006). Measuring social isolation in older adults: Development and initial validation of the friendship scale. *Social Indicators Research, 77*(3), 521–548.

Heley, J., & Jones, L. (2012). Relational rurals: Some thoughts on relating things and theory in rural studies. *Journal of Rural Studies, 28*, 208–217.

Jaeger, K., & Mykletun, R. J. (2013). Festivals, identities, and belonging. *Event Management, 17*(3), 213–226.

Jago, L. K., & Shaw, R. N. (1998). Special events: A conceptual and definitional framework. *Festival Management and Event Tourism, 5*(1), 21–32.

James Theatre. (2019). History. Retrieved March 21, 2019, from https://www.jamestheatre.com.au/about/history/

Janiskee, R. (1991). Rural festivals in South Carolina. *Journal of Cultural Geography, 11*(2), 31–43.

Jarman, D. (2016). The strength of festival ties: Social network analysis and the 2014 Edinburgh International Science Festival. In I. Lamond & L. Platt (Eds.), *Critical event studies* (pp. 277–308). London: Palgrave Macmillan.

Jarman, D. (2018). Festival community networks and transformative place-making. *Journal of Place Management and Development, 11*(3), 335–349.

Jarman, D., Theodoraki, E., Hall, H., & Ali-Knight, J. (2014). Social network analysis and festival cities: An exploration of concepts, literature and methods. *International Journal of Event and Festival Management, 5*(3), 311–322.

Jepson, A., & Clarke, A. (2015). *Exploring community festivals and events*. Abingdon: Routledge.

Kainth, S. (2018). New plan to settle migrants in regional areas. SBS Punjabi. Posted 15 October 2018. Retrieved March 3, 2019, from www.sbs.com.au/language/english/new-plan-to-settle-migrants-in-regional-areas_1

Klaster, E., Wilderom, C., & Muntslag, D. (2017). Balancing relations and results in regional networks of public policy implementation. *Journal of Public Administration Research and Theory, 27*(4), 676–691.

Knoke, D., & Yang, S. (2008). *Social network analysis*. Thousand Oaks, CA: Sage.

Lee, J., Árnason, A., Nightingale, A., & Shucksmith, M. (2005). Networking: social capital and identities in European rural development. *Sociologia Ruralis, 45*(4), 269–283.

Lin, N. (1999). Building a network theory of social capital. *Connections, 22*(1), 28–51.

Mair, J. (2019). Introduction. In J. Mair (Ed.), *The Routledge handbook of festivals* (pp. 1–5). Abingdon: Routledge.

Mair, J., & Duffy, M. (2014). Festivals and sense of community: An Australia case study. In A. Jepson & A. Clarke (Eds.), *Exploring community festivals and events* (pp. 54–65). Oxon & New York: Routledge.

Mair, J., & Duffy, M. (2015). Community events and social justice in urban growth areas. *Journal of Policy Research in Tourism, Leisure and Events, 7*(3), 282–298.

Mair, J., & Duffy, M. (2018). The role of festivals in strengthening social capital in rural communities. *Event Management, 22*, 875–889.

Mair, J., & Whitford, M. (2013). An exploration of events research: Event topics, themes and emerging trends. *International Journal of Event and Festival Management, 4*(1), 6–30.

Massey, D. (1994). *Space, place and gender*. Cambridge: Polity Press.

Mayfield, T. L., & Crompton, J. L. (1995). Development of an instrument for identifying community reasons for staging a festival. *Journal of Travel Research, 33*(3), 37–44.

McAllum, K. (2017). Volunteers/volunteering. In C. R. Scott & L. Lewis (Eds.), *The international encyclopedia of organizational communication* (pp. 1–14). Malden, MA: John Wiley & Sons, Inc.

Misener, L. (2013). Events and social capital. In R. Finkel, D. McGillivray, D. McPherson, & P. Robinson (Eds.), *Research themes for events* (pp. 18–30). Wallingford, UK: CABI.

Mulligan, M., Humphrey, K., James, P., Scanlon, C., Smith, P., & Welch, N. (2006). *Creating community: Celebrations, arts and wellbeing within and across local communities*. Melbourne: Globalism Institute, RMIT Print Services.

Murdoch, J. (2003). Co-constructing the countryside: Hybrid networks and the extensive self. In P. Cloke (Ed.), *Country visions* (pp. 171–184). London: Pearson.

Ooi, N. (2014). *Socio-cultural sustainability of mountain resort tourism: A case study of Steamboat Springs*. Unpublished PhD thesis, The University of Queensland.

Picard, D. (2015). The festive frame: Festivals as mediators for social change. *Ethnos, 81*(4), 600–616.

Pierce, J., Martin, D., & Murphy, J. (2011). Relational place-making: The networked politics of place. *Transactions of the Institute of British Geographers, 36*, 54–70.

Portes, A. (2000). Social capital: Its origins and applications in modern sociology. In E. L. Lesser (Ed.), *Knowledge and social capital* (pp. 43–67). Boston: Butterworth-Heinemann.

Prell, C. (2012). *Social Network analysis: History, theory and methodology*. London: Sage.

Putnam, R. D. (2000). *Bowling alone: The collapse and revival of American community*. New York: Simon and Schuster.

Putnam, R. D. (1995). Bowling alone: America's declining social capital. *Journal of Democracy, 6*(1), 65–78.

Quinn, B. (2005). Changing festival places: Insights from Galway. *Social and Cultural Geography, 6*(2), 237–252.

Quinn, B., & Wilks, L. (2013). Festival connections: People, place and social capital. In G. Richards, M. P. de Brito, & L. Wilks (Eds.), *Exploring the social impacts of events* (pp. 15–30). Abingdon, UK: Routledge.

Qviström, M. (2012). Contested landscapes of Urban sprawl. *Landscape Research, 37*(4), 399–415.

Ray, C. (1998). Culture, intellectual property and territorial rural development. *Sociologia Ruralis, 38*, 3–19.

Reid, S. (2007). Identifying social consequences of rural events. Event Management, 11 (1/2): 89–98.

Rogers, P., & Anastasiadou, C. (2011). Community involvement in festivals: Exploring ways of increasing local participation. *Event Management, 15*, 387–399.

Scully, A. (2018). Dungog Festival's new $10k sculpture competition set to boost Hunter tourism. ABC News Online. Retrieved March 3, 2019, from http://www.abc.net.au/news/2018-03-28/dungog-festival-sculpture-exhibition-tourism-windfall/9596456

Serageldin, I. (1996). Sustainability as opportunity and the problem of social capital. *The Brown Journal of World Affairs, 3*(2), 187–203.

Short, B. (2006). Idyllic ruralities. In P. Cloke, T. Marsden, & P. Mooney (Eds.), *Handbook of rural studies* (pp. 133–148). London: SAGE.

Steketee, M., Miyaoka, A., & Spiegelman, M. (2015). Social network analysis. In N. J. Smelser & P. B. Baltes (Eds.), *International encyclopedia of the social and behavioral sciences* (Vol. 22, 2nd ed., pp. 461–467). Oxford: Elsevier.

Stone, C., & Millan, A. (2011). Empowering communities in the "big society": Voluntarism and event management issues at the Cheetham Hill Cross-Cultural Festival. *International Journal of Management Cases, 13*(3), 242–250.

Tindale, N. (1974). *Tribes of Australia: Their terrain, environmental controls, distribution, limits, and proper names.* Canberra: Australian National University Press.

Tonts, M., Argent, N., & Plummer, P. (2012). Evolutionary perspectives on rural Australia. *Geographical Research, 50*, 291–303.

Van Winkle, C., Woosnam, K., & Mohammed, A. (2013). Sense of community and festival attendance. *Event Management, 17*(2), 155–163.

Waitt, G., & Duffy, M. (2010). Listening and tourism studies. *Annals of Tourism Research, 37*(2), 457–477.

Wilks, L. (2011). Bridging and bonding: Social capital at music festivals. *Journal of Policy Research in Tourism, Leisure and Events, 3*(3), 281–297.

Williams, M. (2012). *A History in three rivers: Dungog Shire heritage study thematic history.* carste STUDIO. Retrieved May 28, 2019, from www.dungog.nsw.gov.au/sites/dungog/files/public/documents/Dungog%20Heritage%20Study%20Review%202014%20-%20Thematic%20History.pdf

Wood, M. (2011). *Rural.* Abingdon, UK: Routledge.

Yuval-Davis, N. (2006). Belonging and the politics of belonging. *Patterns of Prejudice, 40*(3), 197–214.

Ziakas, V. (2010). Understanding an event portfolio: The uncovering of interrelationships, synergies, and leveraging opportunities. *Journal of Policy Research in Tourism, Leisure & Events, 2*(2), 144–164.

8

Leadership: Untapping the Secret to Regional Wellbeing, Belonging and Resilience

Geraldine Jones, Kathleen A. Moore, and Damian Morgan

The Latrobe Valley is located 150 kilometres from Melbourne and has been home to the Victorian State Electricity generation and transmission since 1921. The closure of the Hazelwood Power Station in 2017 left not only workers but the region reeling. The area now finds itself simultaneously in transition and recovery as it seeks to rebuild regional wellbeing, the people's sense of belonging and resilience. The manner in which the closure occurred has drawn strong criticism from State and Federal government, community and businesses, both locally and from afar. Suffering of both workers and the region is palpable to such an extent that the Victorian Premier, Daniel Andrews, established the $266 million Latrobe Valley Authority in response and as part of his and the State Government's commitment to 'be with them every step of the way' in the rebuilding of 'pride of place' to enable a 'just transition' (ABC News Victoria, 2016; also Stanton, 2016). The closure placed leadership and leadership style firmly in the spotlight as an exemplar of 'how not to do it' but also,

G. Jones (✉) • K. A. Moore • D. Morgan
Federation University Australia, Ballarat, VIC, Australia
e-mail: g.jones@federation.edu.au; k.moore@federation.edu.au; damian.morgan@federation.edu.au

A. Campbell et al. (eds.), *Located Research*,
https://doi.org/10.1007/978-981-32-9694-7_8

importantly, the role that leadership in the workplace can have in influencing and fostering regional wellbeing, a sense of belonging and resilience.

Most researchers (Avolio, Walumbwa, & Weber, 2009; Bryman, 1996) agree that leadership involves the following three elements: influence, groups and goals. Influence concerns the ability to have an impact on and encourage others, either explicitly or implicitly, and to think or behave in certain ways. Groups refer to subordinates or followers of the leader who, generally within the workplace, are considered to be employees. Goals are the objectives or targets that the group needs to achieve. Leadership in the workplace can therefore be viewed as a fluid process that involves a dynamic interchange of explicit and implicit social and organisationally determined factors within the environment in which they occur (Gronn, 1999).

According to Anderson and Sun (2017) and Newstrom and Davis (1993), leadership style is the combined pattern of explicit and implicit attitudes, behaviours and actions that leaders hold or exhibit in the execution of leading. Hence, leaders can directly convey what efforts, relational considerations and approaches are needed on the part of employees and what employees can expect in return, including, for example, remuneration, recognition and the chance to access future opportunities (Maguire, 2002; Schein, 1980).

There have been four stages in the development of leadership theory and research over the last 70 years (Bryman, 1996): the trait (Cowley, 1931), behaviour (Pfeffer, 1977), contingency (Fiedler, 1967; Hersey, Blanchard, & Johnson, 1996) and new leadership (Anderson & Sun, 2017; Bryman, 1996) theories. Much of the research within these approaches has focused on defining leadership, leadership styles and the impact of leadership style in terms of followers and relationship exchange (Avolio et al., 2009). The primary outcomes of a better understanding of leadership have been to improve employees' performance, commitment, satisfaction wellbeing and resilience (Avolio et al., 2009).

With the onset of globalisation in the 1980s, combined with the ever-increasing pace of change and technology, the traditional workplace is becoming increasingly rare. In response, new leadership approaches such as transformational (Bass, 1985), visionary (Bennis & Nanus, 1985) and charismatic (House, 1977) have consequently emerged to address identified needs of promoting employee satisfaction, commitment, wellbeing and belonging in the workplace. In the absence of effective leadership styles, human resources practices, systems and processes tend to treat employees in a 'one size fits all' manner. The result has been termed a transactional leadership style. This style has proved ineffective in promoting employee's commitment and wellbeing (McDermott, Conway, Rousseau, & Flood, 2013). In contrast, a transformational style of leadership, where the employee is encouraged, motivated and supported, would appear to be optimal. However, research has suggested that this style is suboptimal (Bass, 1990; Sailor & Chelliah, 2014). So, while leadership styles, particularly those of frontline managers, can affect employees' work commitment, performance, engagement and satisfaction levels (Sahin, Cubuk, & Uslu, 2014), if neither a transformational nor transactional leadership approach is optimal, what style is effective?

Leadership can be viewed as a process that involves a dynamic exchange with employees of social, operationalised and reciprocated organisationally determined factors. As a result, leadership can be conceptualised as a fluid process (Anderson & Sun, 2017; Bryman, 1996). It may be that the missing element in leadership practice and, in turn employees' workplace wellbeing, sense of belonging and resilience, are related to what Argyris (1960) and Rousseau (1995) have called a *psychological contract*. Such a contract can be implicit or explicit to the leader-employee relationship.

The Psychological Contract

Argyris (1960) was one of the first to introduce the term psychological contract and likened it to an 'unwritten agreement' that exists when an employment relationship occurs. Levinson, Price, Munden, Mandl and Solley (1962: 21) extended this definition to a 'series of mutual expectations of which the parties to the relationship may not themselves be even dimly aware but [which] nevertheless governs their relationship with each other'.

In the 1990s, Rousseau extended Argyris (1960), Levinson et al. (1962) and Schein's (1980) work to include the notion of reciprocal exchange between the focal person and another party—in the organisational sense—between the leader and the employee. Thus, a psychological contract emerges when one party believes that a promise of future returns has been made, while the other knows that a contribution has been given, and an obligation has been created to provide future benefits (Rousseau, 1990, 1995).

The psychological contract is, therefore, a set of defined perceptions and beliefs held by leaders and subordinates linked to tacit and implicit determinants of employee commitment and performance and, in turn, organisational performance and its outcomes (Conway & Briner, 2002; Maguire, 2002; McDermott et al., 2013; Ng, Feldman, & Lam, 2010; Robinson & Rousseau, 1994; Salicru, 2015; Salicru & Chelliah, 2014). According to both Salicru and Chelliah (2014) and McDermott et al. (2013), the application of leadership agency in the psychological contracting process may provide the answer to both employee and organisational wellbeing, resilience and sense of belonging and, in turn, reduce workplace stress and anxiety among employees (Hakan & Jamel, 2016).

Model of the Psychological Contract

Researchers have argued that a psychological contract has three elements: transactional, relational and reciprocity (Conway & Briner, 2005; Coyle-Shapiro & Parzefall, 2008; Maguire, 2002; Morrison & Robinson, 1997; Ng & Feldman, 2009). We propose that these elements and their components, synthesised from extant literature, can be conceptualised in terms of ongoing exchanges within the employment relationship (Fig. 8.1).

The transactional component of the psychological contract revolves generally around the exchange of observable tangibles such as pay or other monetary benefits (Conway & Briner, 2002; Rousseau, 1990). The relational component involves a mix of socio-emotional or economic/non-economic interactions that carry with them an expectation of continuity of relations into the future (Morrison & Robinson, 1997; Rousseau, 1990). That is, the relational component is concerned with the exchange of intangibles such as loyalty and trust in exchange for ongoing

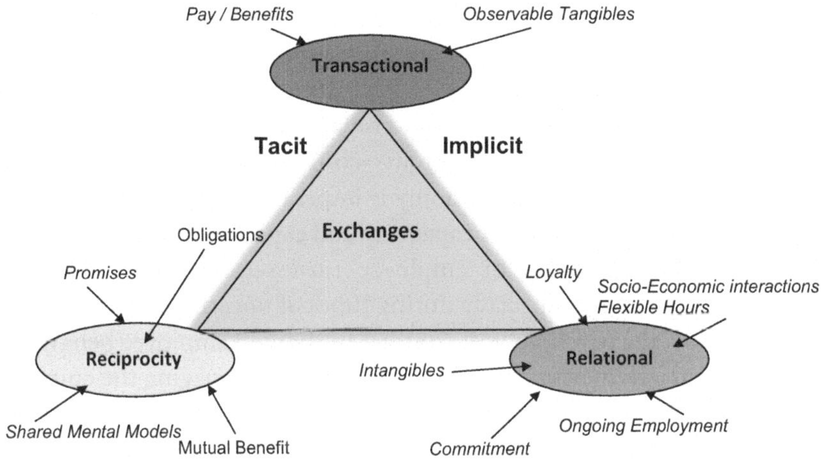

Fig. 8.1 Elements of the psychological contract

employment. This component deals with 'commitment loyalty and trust in management on behalf of employees in exchange for competent management, the opportunity for input and sense of belonging' (Maguire, 2002: 167). The third component of the psychological contract involves reciprocity. Reciprocity is about commitment to duty and mutual benefits with an emphasis on the promises and obligations made or inferred by both parties in the employment relationship (Robinson & Rousseau, 1994). The reciprocal items are at times explicit and other times implicit. Consequently, the parties share a mental model of expectations (Coyle-Shapiro & Parzefall, 2008; Robinson, 1996). Research has found that all three elements must exist at some level in order for a psychological contract to be perceived to exist (Conway & Briner, 2005; Maguire, 2002; Rousseau, 1990).

The psychological contract is an important construct as it provides a basis for capturing the complex basis of the relationship that can optimally exist between employees and employers. Its value lies in providing people with expectations of what they can receive and influence and 'what they can choose to carry out as part of this exchange' (Maguire, 2002: 167). Chang and Hsu (2009) argued that the psychological contract is one of the most important leadership factors to consider if an organisation wants to maintain strong organisational citizenship, productivity and organisational performance especially in the digital age. Horwitz, Heng, and Quazi (2003) have similarly made a connection between strong employee relations and organisational performance. They maintained that the importance of attracting, retaining and motivating employees in the knowledge economy is imperative. Organisations without employees with the correct capability and capacity fostered by leadership, or with high levels of employee turnover, lose their strategic competitive advantage especially during times of uncertainty and change. Consequently, the psychological contract in shaping employee behaviour can be used as a framework for understanding and managing the employment relationship due to its ability to exert either a negative or positive impact on employees' levels of engagement and commitment and, in turn, wellbeing, belonging and resilience (Coyle-Shapiro & Kessler, 2000; Guest, 1998). We suggest that a psychological contract is especially relevant in regional areas, where not only are people interlinked via non-work relationships and connections but where trust in the organisation to maintain constancy and longevity of employment in return for employee and community commitment is vital.

Does the Psychological Contract Exist in Reality?

From a series of interviews we conducted with organisational leaders in our region, we suggest not only that a psychological contract exists between leaders and employees but it is, moreover, imbedded within community. Participants from a range of public and private customer service orientated organisations indicated in our interviews that they thought about community as whole; workers are not just employees but also friends or acquaintances outside work. Workers and all aspects of self, both inside and outside of work, were considered by leaders in how they acted, behaved, responded and thought about employees. There was a general belief that a family-oriented culture within the workplace was important to generating wellbeing and building a sense of belonging.

One way workers fostered a sense of belonging was via their strong emphasis on collaborative and consultative behaviours and communication styles where they frequently used face-to-face communication, via meetings, one-on-one meetings or informal 'how are you going' enquires where possible rather than *just* email. A strong emphasis was placed on viewing and responding to mistakes or failure as a means of learning. Many interviewees considered this as essential for generating resilience and organisational wellbeing.

A dominant theme that emerged from the interviews was the leaders' belief and sense of duty in providing employees with a sense of meaning and purpose within the work environment. Further, these initiatives were linked to the vision of the organisation, the region and the community. Leaders also stated that everyone, including themselves and their employees, needed to feel a sense of purpose at work that connected to a bigger vision of service and civic social responsibility or contribution. Leaders felt a need to enable employees to derive some meaning from their work, and that, in their role, they have an obligation to create a place where people want to be at work. It was also noteworthy that most interviewees thought of their organisation as having a place in the region, inasmuch that it formed a part of the broader regional ecosystem. Leaders therefore considered that it was up to them, and hence their organisation, to demonstrate their commitment to both employees and the community. To this extent, many felt that they were personally accountable as leaders.

Clearly these leaders show high levels of interpersonal and organisational trust in employees. They provided encouragement and support which, in turn, fostered within workers a need to respond accordingly: that being reciprocity, a key component of the psychological contract. A strong focus on fostering trust was established through leading by example, preparing employees for career disruption, behavioural consistency, joint problem solving, the provision of autonomy and recognition of effort being key features. In these examples, the psychological contract became a form of effective leadership for creating wellbeing, belonging and resilience among employees.

Implications for Leadership Practice, Regional WellBeing, Belongingness and Resilience

The psychological contract and its maintenance have been linked to the tacit and implicit determinants of employee attraction, retention and commitment motivation and, in turn, organisational performance (McDermott et al., 2013; Ng et al., 2010). All these factors are vital to the creation of healthy workplace relationships, resilient organisations and overall regional wellbeing. Chang and Hsu (2009) argued that the psychological contract is one of the most important leadership factors to consider for an organisation hoping to maintain strong organisational citizenship, productivity and organisational performance in the digital age.

Horowitz, Heng and Quazi (2003) have found a connection between strong employee relations and organisational performance. They maintained the importance of attracting, retaining and motivating employees through their focus on the knowledge economy. They argued that organisations lacking employees with the correct capability and capacity, impaired by ineffective leadership and compounded by high levels of employee turnover, lose their strategic competitive advantage, especially during times of uncertainty and change. Change in response to external environments is a constant in the digital age for the modern organisation, and this highlights further the need for leadership practice that enables a positive psychological contract (Maguire, 2002).

As mentioned earlier, the psychological contract involves reciprocity. Therefore, exchanges within the carefully shaped psychological contract are able to provide a means for creating identity, trust, belonging, organisational self-esteem and organisational support—this underpins employee motivation and commitment and thus employee and organisation wellbeing. Therefore, the psychological contract is able to provide a mechanism/process for creating a positive sense of belonging, wellbeing, trust and commitment with employees, which underwrite motivation and performance. The psychological contract consequently produces an avenue for reducing organisational risk due to maximising employee attraction, retention and growth. These contracts also provide the organisation with a tool for mitigating employee cynicism, anxiety and stress. Breaches of a workplace psychological contract may have devastating effects for both parties, especially for workers in regional areas.

Work is an integral component of our modern lifestyle. The importance of this role is no less so for people living in regional areas where the workplace often reflects the social and economic trends occurring in the broader environment. These trends can include impact of technology, globalisation, economic and environmental changes and, more recently, the emerging impact of age upon the workforce, all of which can have a greater impact in regional communities. A further concern within regional areas is the closure of some industries and the exodus of some community members as well as youth seeking opportunities in 'greener pastures'. Such changes place considerable strain on regional communities and can lead to workplace stress, anxiety and even burnout among employees (Cartwright & Holmes, 2006).

Workplace anxiety and stress can result from the fundamentals of change and a lack of control over or input into this change by employees. Such reactions are especially problematic in circumstances where the economy or government policies force companies to downsize or relocate their operations in regional areas where alternative employment opportunities are often restricted. Frequently, a *misfit* can occur between the needs, expectations and capabilities of employees and what the organisation demands or offers them. This may cause disruption and anxiety for employees and their support networks alike (Cartwright & Cooper, 1997). The negative effects on employees' fitness, financial earnings and emotions can have a permanent effect, and it can take a long time for the individual, organisation and the broader economy to recover (Ng & Feldman, 2009). An example of the impact of change on a regional community can be viewed through the *Hazelwood* lens described below, but this change can also be viewed in terms of a breach of the psychological contract.

Breach of the Psychological Contract

On 3 November 2016, Engie, a French multinational electric utility company, announced that the entire Hazelwood plant located near Churchill in the Latrobe Valley would be closed at the end of March 2017. This closure affected some 500 staff directly employed by Engie and around 300 contractors, typically local, in a regional population of

125,000, of whom 32,560 people reported being in the workforce (Australian Bureau of Statistics, 2016). Workers and local communities were given only five months' notice of the closure. This period of notice contrasts with the two remaining coal-powered electricity stations in the area, which have signalled long-term that closures will not occur until at least 2032.

Many local people within the Latrobe and greater Gippsland region felt this closure was the death knell of both the region and local workforce, once considered the pinnacle of best practice and engineering ingenuity within the country and abroad. For many workers, the closure symbolised a betrayal by leaders and by the organisation of their trust and loyalty. Further, workers considered the company violated an explicit/implied agreement between workers and leaders within the workplace. Here, reciprocal offerings were considered to exist at a basic or transactional level where the transaction involved labour in exchange for wages and wages in exchange for labour.

These transactions or exchanges for many began with workers' grandfathers and fathers who worked their entire lives in the same plant. At a more abstract or, can it be said, human level, the idea of a shared vision was destroyed. For others in the community, this represented the continuing downward trend within this industry since the 1990s, when the effects of globalisation and the ever-increasing pressures and pace of technology and privatisation of electricity within Victoria, and in particular within the Latrobe Valley, began. But it is not only plant closures and job loss that have an impact on employees, their families and the community in general. Substantial impacts occur through the loss of history, identity and engagement, as well as the belief that leaders have let workers down (Insight SBS, 2017). In sum, every element of the psychological contract was violated bringing to an end the relationships built over three generations (ABC News Victoria, 2017).

Employee Disengagement

Gallup's (2017) *State of the Global Workplace Report* indicated that 76% of employees in Australia are disengaged from their work and the organisation; dissatisfaction with their direct leader being the major reason provided. Managers and leaders who were perceived as dishonest and did not provide strong relational support or recognition of employees were reported among the chief causes of this dissatisfaction. These latter effects are indicative of a lack of a psychological contract between the parties. Research by Firth, Mellor, Moore, and Loquet (2004) found that feeling unsupported by the organisation, exacerbated by poor communication, contributed to employees' feeling burnout, lacking in job satisfaction and intending to quit. These high levels of disengagement have implications for employees' job satisfaction, wellbeing, sense of belonging, resilience, job commitment and, not least of all, the productivity of the organisation and region.

In smaller regions, the leader or leaders are often your neighbour, the parent of your child's school friend, or a member of your club or church. How do such dual relationships affect individual and community wellbeing? In addition to the psychological contract or, indeed an essential unstated component of it, a requirement is good communication between parties.

Communication

While in terms of the *Hazelwood* closure, earlier communication to workers would not have stopped the closure, it may have provided staff with a better understanding of the rationale and enabled workers greater opportunities and time to prepare themselves psychologically. It might also have enabled more workers to find alternative employment if the closure had been staggered.

The importance of communication by leaders in this and other instances of major organisational change is paramount. As Moore (2001) found, it is two-way communication between management and staff that is important, not simply a 'top down' approach (e.g. management told us what was happening). Clearly communication is an essential component of a strong and healthy psychological contract between leaders and staff. But the ways in which employees themselves deal with life stressors have been attributed to their resilience levels and so it is important to consider this aspect.

Resilience

Resilience was described by Masten (2001) as an individual's ability to cope with stress and adversity and to positively adapt. Others have described it as 'a dynamic process of positive adaptation in the context of significant adversity' (Luthar & Cicchetti, 2000: 857) or, in colloquial terms, the ability to 'bounce back' from stressors. People who are considered resilient are said to have sufficient protective factors to cope with physical, psychological and socioecological stressors (Rutter, 2012). Clearly, being resilient contributes to feeling healthy and happy and enables people to successfully overcome difficulties and achieve what they want out of life, but we are not social islands.

People also need to feel secure, supported, accepted and included generally and notably in the workplace. It is the organisational leadership that can inspire these feelings of inclusion, which, in turn, contribute to individual wellbeing and job satisfaction. In a longitudinal study, Judge and Watanabe (1993) found that a pattern of mutual influence with people's job satisfaction had a positive impact on their life satisfaction. In the close and sometimes complex relationships in small regional areas, the interplay between leaders and their leadership style may hold the key to uncapping workplace and, in turn, regional wellbeing, belonging and resilience.

We suggest that a good leadership style includes the components of a psychological contract. The adoption and maintenance of such a contract is especially important in regional areas where interpersonal relationships are not confined to the workplace but are a reflection of the broader community: where community members themselves typically engage in transactions, relations and reciprocity—the foundations of a psychological contract.

References

ABC News Victoria (ABC1 Melbourne). (2016). Premier Daniel Andrews will personally oversee efforts to attract new industry to the Latrobe Valley in response to the impending closure of the Hazelwood power station. Broadcast: 31/10/2016.

ABC News Victoria (ABC1, Melbourne). (2017). Powering down: Hazelwood closure: After months of preparations, workers at the Hazelwood Power Station are tonight clocking off for the last time. Broadcast: 31/03/2017.

Anderson, M., & Sun, P. Y. T. (2017). Reviewing leadership styles: Overlaps and the need for a new 'full-range' theory. *International Journal of Management, 19*(1), 76–96.

Argyris, C. (1960). *Understanding organizational behavior.* Homewood, IL: Dorsey.

Australian Bureau of Statistics. (2016). Retrieved April 13, 2019, from https://quickstats.censusdata.abs.gov.au/census_services/getproduct/census/2016/quickstat/20504

Avolio, B. J., Walumbwa, F. O., & Weber, T. (2009). Leadership: Current theories, research, and future directions. *Annual Review of Psychology, 60*, 421–449.

Bass, B. M. (1985). *Leadership and performance beyond expectations.* New York, NY: Free Press.

Bass, B. M. (1990). *Bass & Stogdill's handbook of leadership* (3rd ed.). New York, NY: The Free Press.

Bennis, W., & Nanus, B. (1985). *Leaders: The strategies for taking charge.* New York, NY: Harper & Row.

Bryman, A. (1996). *Leadership in organizations in handbook of organization studies.* London: Sage Publications.

Cartwright, S., & Cooper, C. L. (1997). *Managing workplace stress.* London: Sage.

Cartwright, S., & Holmes, N. (2006). The meaning of work: The challenge of regaining employee engagement and reducing cynicism. *Human Resource Management Review, 16*, 199–208.

Chang, C. P., & Hsu, P. C. (2009). The psychological contract of the temporary employee in the public sector in Taiwan. *Social Behavior and Personality, 37*(6), 721–728.

Conway, N., & Briner, R. B. (2002). A daily diary study of affective responses to psychological contract breach and exceeded promises. *Journal of Organizational Behavior, 23*, 287–302.

Conway, N., & Briner, R. B. (2005). *Understanding psychological contracts at work. A critical evaluation of theory and research.* Oxford: Oxford University Press.

Cowley, W. H. (1931). The traits of face-to-face leaders. *The Journal of Abnormal and Social Psychology, 26*(3), 304–313.

Coyle-Shapiro, J., & Kessler, I. (2000). Consequences of the psychological contract for the employment relationship: A large scale survey. *Journal of Management Studies, 37*(7), 903–930.
Coyle-Shapiro, J., & Parzefall, M. (2008). Psychological contracts. In C. L. Cooper & J. Barling (Eds.), *The SAGE handbook of organizational behavior* (pp. 17–34). London: SAGE Publications.
Fiedler, F. E. (1967). *A theory of leadership effectiveness.* New York, NY: McGraw-Hill.
Firth, L., Mellor, D., Moore, K. A., & Loquet, C. (2004). How can managers reduce employee intention to quit? *Journal of Managerial Psychology, 19*(2), 170–187.
Gallup. (2017). *State of the global workplace.* Retrieved April 13, 2019, from http://news.gallup.com/reports/220313/state-global-workplace-2017
Gronn, P. (1999). *The making of educational leaders.* London: Cassell.
Guest, D. (1998). Is the psychological contract worth taking seriously. *Journal of Organizational Behaviour, 19*, 649–664.
Hakan, E., & Jamel, C. (2016). Benevolent leadership and psychological well-being: The moderating effects of psychological safety and psychological contract breach. *Leadership & Organization Development Journal, 37*(3), 369–386.
Hersey, P., Blanchard, K. H., & Johnson, D. E. (1996). *Management of organisational behaviour.* London: Prentice-Hall International.
Horwitz, F. M., Heng, C. T., & Quazi, H. A. (2003). Finders, keepers: Attracting, motivating and retaining knowledge workers. *Human Resource Management Journal, 13*(4), 23–44.
House, R. J. (1977). A 1976 theory of charismatic leadership. In J. G. Hunt & L. L. Larson (Eds.), *Leadership: The cutting edge* (pp. 189–204). Carbondale, IL: Southern Illinois University Press.
Insight (SBS Television). (2017). The Valley: We're about 150km east of Melbourne Broadcast: 21/02/2017.
Judge, T., & Watanabe, S. (1993). Another look at the job satisfaction-life satisfaction relationship. *Journal of Applied Psychology, 78*, 939–948.
Levinson, H., Price, C. R., Munden, K. J., Mandl, H. J., & Solley, C. M. (1962). *Men, management and mental health.* Cambridge, MA: Harvard University Press.
Luthar, S. S., & Cicchetti, D. (2000). The construct of resilience: Implications for intervention and social policy. *Developmental Psychopathology, 12*, 857–885.
Maguire, H. (2002). Psychological contracts: Are they still relevant? *Career Development International, 7*(3), 167–180.

Masten, A. S. (2001). Ordinary magic: Resilience processes in development. *American Psychologist, 56*, 227–238.

McDermott, A. M., Conway, E., Rousseau, D. M., & Flood, P. C. (2013). Promoting effective psychological contracts through leadership: The missing link between HR strategy and performance. *Human Resource Management, 52*(2), 289–310.

Moore, K. A. (2001). Hospital restructuring: Impact on nurses mediated by social support and a perception of challenge. *Journal of Health and Human Services Administration, 23*, 490–517.

Morrison, E. W., & Robinson, S. L. (1997). When employees feel betrayed: A model of how psychological contract violation develops. *Academy of Management Review, 22*, 226–256.

Newstrom, J. W., & Davis, K. (1993). *Organizational behavior: Human behavior at work*. New York, NY: McGraw-Hill.

Ng, T. W. H., & Feldman, D. C. (2009). Age, work experience and the psychological contract. *Journal of Organisational Behaviour, 30*, 1053–1073.

Ng, T. W. H., Feldman, D. C., & Lam, S. S. K. (2010). Psychological contract breaches, organizational commitment, and innovation-related behaviors: A latent growth modeling approach. *Journal of Applied Psychology, 95*(4), 744–751.

Pfeffer, J. (1977). The ambiguity of leadership. *Academy of Management Review, 2*(1), 104–114.

Robinson, S. L. (1996). Trust and breach of the Psychological Contract. *Administrative Science Quarterly, 41*, 574–599.

Robinson, S. L., & Rousseau, D. M. (1994). Violating the psychological contract: Not the exception but the norm. *Journal of Organizational Behavior, 15*, 245–259.

Rousseau, D. M. (1990). New hire perceptions of their own and their employer's obligations: A study of psychological contracts. *Journal of Organizational Behavior, 11*, 389–400.

Rousseau, D. M. (1995). *Psychological contracts in organisations: Understanding written and unwritten agreements*. Thousand Oaks, CA: SAGE Publications.

Rutter, M. (2012). Resilience as a dynamic concept. *Development and Psychopathology, 24*, 335–344.

Sahin, D. R., Cubuk, D., & Uslu, T. (2014). The effect of organizational support, transformational leadership, personnel empowerment, work engagement, performance, and demographical variables on the factors of psychological capital. *Emerging Markets Journal, 3*, 1–17.

Salicru, S. (2015). The global leadership psychological contract model – Actionable to shape the future to 2050. In M. Sowcik, A. A. Andenoro, M. McNutt, & S. E. Murphy (Eds.), *Leadership 2050: Critical challenges, key contexts, and emerging trends* (pp. 159–173). Bingley, UK: Emerald Group Publishing Limited.

Salicru, S., & Chelliah, J. (2014). Messing with corporate heads? Psychological contracts and leadership integrity. *Journal of Business Strategy, 35*(3), 38–46.

Schein, E. H. (1980). *Organizational psychology*. Englewood Cliffs, NJ: Prentice-Hall.

Stanton, K. Old King coal a dying dynasty. *The Citizen*, Melbourne University, 28th June 2016. Retrieved May 28, 2019, from https://www.thecitizen.org.au/articles/old-king-coal-dying-dynasty

The Valley: We're about 150km east of Melbourne [online]. *Insight (SBS TELEVISION);* Time: 20:30; Broadcast Date: Tuesday, 21st February 2017; Duration: 59 min., 1 sec.

Part II

Transition and Resilience

9

Mental Health and Mining: Research Challenges and Influences

Carole L. James, Jane Louise Rich, and Brian Kelly

Importance of Working with Industry Partners and Communities

Researching the issues of mental health for the workplace, in this case, mining, provides the opportunity to address industry-identified needs and to contribute to an overall aim held by Universities Australia, to ensure that research conducted by academics is relevant to 'real world' issues, that research is sustainable and builds capacity both within and beyond the university sector by engaging with industry partners and translating research into practice (Universities Australia, 2013). This is an important overarching aim for researchers as working with industry

C. L. James (✉) • J. L. Rich • B. Kelly
Centre for Resources Health and Safety, Faculty of Health and Medicine, University of Newcastle (UON), Callaghan, NSW, Australia
e-mail: Carole.James@newcastle.edu.au; Jane.Rich@newcastle.edu.au; Brian.Kelly@newcastle.edu.au

A. Campbell et al. (eds.), *Located Research*,
https://doi.org/10.1007/978-981-32-9694-7_9

challenges us to produce research and conduct interventions in a timely, rigorous and sustainable fashion—meeting national, industry and academic outcomes.

Vital to this agenda is 'research translation' and how researchers engage with end users to translate research into practice and to achieve impact to the economy, society, environment and/or culture as a result of research outcomes (Australian Research Council, 2018). Our experience in working with industry has demonstrated that effective translation of research including methodology, process, intervention and outcomes has benefits not only for the application of any outcomes, which may lead to increased national competitiveness of industry, but also importantly for the relationship between industry and researchers and for capacity building at the organisational or individual level as a result of research intervention.

The Australian mining sector is an integral part of Australia's economy. The mining workforce is characterised by a predominantly male workforce (85%), who are increasingly exposed to high market volatility, as characterised by the economic boom and bust cycle (Australia Mining, 2018), and subsequent job insecurity that accompanies such economic circumstances (Masige, 2016). The Australian mining industry is considered a world-leader in the area of occupational health and safety (OHS) (Poplin et al., 2008). Although addressing mental health in the mining workforce has not traditionally been a focus of the industry's health and safety priorities, it is increasingly being recognised within occupational health safety (OHS) frameworks and governing bodies (Bloom et al., 2011; Safe Work Australia, 2019). This focus reflects growing community recognition of the significance of unaddressed mental health needs and particularly the public attention to concerns about suicide among males in this industry (Milner, 2016).

Higher suicide rates have been reported among 'blue-collar' workers, however, the evidence in specific industry sectors is lacking (Milner, Spittal, Pirkis, & LaMontagne, 2013). Data from our group (Kelly, 2015) suggests the prevalence of psychological distress in mining employees is at moderate (31%) and high levels (13%) (Kessler et al., 2002), which is higher than community levels (20% and 7%, respectively) (Australian Bureau of Statistics, 2008).

Furthermore, growing evidence regarding the cost to the workplace of unaddressed mental health problems has also provided a powerful impetus of such a focus. Approximately 35–45% of workplace absenteeism is attributed to mental health problems; presenteeism (less than optimal productivity) is associated with moderate levels of psychological distress and up to 18 lost work days per year (Hilton et al., 2008). Mental health problems are also associated with greater rates of workplace injuries, and ultimately lower productivity. Evidence also indicates the beneficial return on investment for workplaces that instigate workplace mental health programs (Coopers, 2014). Further research noted that industry productivity losses from mental illness may range between $AU288 and $AU429 million per year, and in coal mining, it is estimated at $5.0 million ($AUS) per year (Ling et al., 2016).

An Interdisciplinary Approach

The University of Newcastle's Centre for Resources Health and Safety was established to facilitate transdisciplinary research to address emerging issues in these industries and related communities. This was facilitated through funding from Aspen Medical, a private company of healthcare and medical services globally, whose support enabled a Chair in Resources, Health and Safety to be established and core Centre members to lead this initiative. The Centre sits within the University of Newcastle's Institute for Energy and Resources (NIER) (University of Newcastle, 2019). A research agenda was developed in consultation with key regional industry stakeholders in the mining and resources sector and within the local health services along with the engagement of researchers in public health and health sciences, psychology, environmental sciences, social science disciplines such as geography, health anthropology and sociology, and engineering. The four theme areas emerging from such planning were health services and workforce; environmental exposures and health; 'working well' (workplace and health); and adapting to workplace risks and hazards. Such diverse expertise, from across the university, enables researchers to fully address industry needs and to deliver valid research outcomes and interventions that can be implemented appropriately within the mining sector that operates in regional, rural and very remote environments. With an interdisciplinary team, a genuine understanding of feasibility and acceptability can be understood when designing, implementing and evaluating interventions, including financial effectiveness, health outcomes, organisational change along with structural barriers. Sitting within NIER enabled access to the established industry links, industry engagement process and existing cross-disciplinary programmes of this diverse and successful established research network. Consequently, the CRHS researchers developed Working Groups with interested experts from across the university and industry, across all disciplines and faculties. Members of the CRHS bring in the relevant expertise from across the working groups to design research plans that meet industry-identified problems, develop grant proposals in conjunction with industry partners and conduct research and deliver findings. The core researchers at the CRHS (authors of this chapter) include a psychiatrist, occupational therapist and human geographer, all with strong research records in their respective fields, yet able to access the expertise of engineers, scientists, population health researchers, nurses, social scientists, statisticians and importantly industry representatives regularly to ensure that research is rigorous, appropriate and 'real world' to meet the needs and problems identified by the industry. This skill mix also enables research to be innovative and fresh—drawing on the various methodologies, theoretical frameworks and lenses that each discipline brings.

Importance of Face-to-Face Engagement with Companies

Research Process

The research to date highlights the gains and challenges for health and social science researchers in working collaboratively with industry to address industry problems and needs. To be collaborative partners, both sides need to share in the creation of research questions, methods and interventions and identify challenges and risks in implementing the research. Through this process of co-design or 'end-user' engagement, genuine trust is built between researchers and industry, and the likelihood of effective dissemination and implementation of research findings are strengthened (Wolfenden, Williams, Wiggers, Nathan, & Yoong, 2016). Maintaining and developing such industry relationships, ensures feasibility and acceptability of research—and the translation, implementation and sustainability of research outcomes into industry practice.

One of the challenges of working with an industry such as mining and the resources sector is the ebb and flow of their economic state. Much of our research has been conducted throughout times of significant economic change in the mining industry, which, as noted previously, is characterised by an economic boom and bust cycle. Engaging with employees and staff who have experienced changes to rosters, staffing levels and concerns about redundancy and job security impact upon engagement in any research in, the first instance, and also create important contextual factors, such as those relating to the psychological and social wellbeing of employees that need to be identified, documented and synthesised in any analysis of findings. While a challenge, this has also provided us the opportunity to investigate the impact of such volatility on the workforce and the place for interventions to mitigate the adverse personal impacts. The latter is relevant to a project currently in development that aims to address related skills in financial management and planning within a health promotion programme in this industry.

Industrial relations are an important consideration when working with the mining and resources sector, which are traditionally unionised workplaces. This is especially pertinent in times of economic and subsequent workplace change. Early engagement with employee representatives and ongoing consultation throughout projects to promote confidence that the research or its findings would not inadvertently have a prejudicial impact on employees, for example, identifying those with current mental health problems, has been essential for the research process to proceed. The essential process of university institutional ethics reviews, assurances and robust processes regarding protection of confidentiality and ongoing communication about findings are key elements that assist in building such confidence and trust in the research process at each individual and organisational level. Strong involvement and engagement from management and employee representatives at the worksite has assisted us to achieve high levels of recruitment and retainment in our studies, as shown in the 'Working Well – Mental Health and Mining' research where recruitment was between 30% and 72% (Considine et al., 2017). This highlights the importance of engaging face-to-face with workplace site-level management, local management, supervisors and employee representatives to discuss the research and the research process.

In addition, we have found that internal workplace champions, often individuals with a personal interest in mental health, took it upon themselves to be key supporters and campaigners of the mental health research, promoting the research at the workplace site level, which encouraged local participation. It is vital to the success of the research for workplace champions to be identified to assist with the promotion and acceptance of mental health within the workplace.

Dedication from both parties, researchers and industry, provides opportunities to gain 'shared learnings'. Industry can move beyond the systematic approach of performance research from a quality assurance viewpoint, to understand and gain insight into valid research methods and rigorous processes. These include human ethics reviews and appreciating the benefits of doing research in a highly systematic manner. While for researchers, this shared learning provides the opportunity to deliver and evaluate interventions in a real-life setting; understand industry stressors and constraints; and present valid, applicable and reliable outcomes that can be translated directly into workplace practice.

Other elements of the research process that facilitate a stronger relationship with industry include, willingness to learn about the 'real-world' of the industry partner—workplace site visits, firsthand experience of the settings in which the industry is conducted, and research staff who can present a credible openness to learn from the partners and flexibility to adapt to the demands of the work setting (e.g. rosters, locations for worker engagement such as tea rooms, mine sites themselves or mine site training days).

A further element of such engagement is commitment to a multilevel communication plan to provide reports on research progress and findings, at regular intervals, at the individual site level, including brief reports and presentations to management and occupational health and safety committees, and flyers or posters to be displayed widely around the individual workplace. This involves de-identification of sites (an assurance necessary in the research implementation) while providing each site with its profile against the collective industry data to inform its own health and safety programmes. In addition, reports on the collective industry data need to be made available to peak industry bodies and agencies to disseminate the findings from the research.

A critical element to such industry engagement is the role of industry-appointed 'monitors'. This is a process formalised by the Australian Coal Association Research Program (ACARP) funding body that entails ACARP representatives meeting with researchers throughout the life of the project to assist with engagement of workplace sites and provide advice on the design of research data collection methodology. This formal process of consultation and translation ensures relevance and appropriateness of the research as it assists with information about emerging contextual factors that affect the industry over the life of a longitudinal study and aids interpretation of findings. Underpinning this relationship is the explicit acknowledgement by the industry research funders and monitors of the need for academic independence in data reporting; while advice and feedback is regularly sought, the researchers' maintain independence in the publication and dissemination of findings. Using 'research monitors' is an element we have extended beyond ACARP-funded projects, having found that industry representatives complement industry engagement with research projects more broadly.

Methodological Challenges of Research in Remote Locations/Worksites

Completing research within an applied and real-life industry setting naturally encounters challenges related to the competing interests of all involved. These competing interests range from workplace production demands, workplace health and safety demands, employee interest and expectations and the requirements from an academic view with completing milestones in a research project. There is an expected tension related to time allocated to participate in a research project, with both data collection and with interventions and with productivity commitments. As such, access to workers for research data collection (Kelly et al., 2016) and to attend interventions requires all parties (management, supervisors, researchers and employees) to be dedicated, committed, flexible and responsive to the constantly changing workplace environment. In our research, we have found training days, which regularly occur at some mine sites, are an ideal time to approach participants for components of the study. Employees routinely attend periodic training days at the workplace, throughout the year, which take them out of production for that shift where they undergo a range of continuing professional type development/education.

The significance of staffing changes at the workplace must be considered when conducting research at the workplace. Economic constraints in the mining industry, with staffing changes as a result of redundancy and roster changes, may impact on any results and on motivation of employees to become involved in any research studies (Kelly et al., 2016).

A challenge we have encountered with our workplace research using a pre-post study design was the lack of employees who included an identifying code to enable longitudinal data analysis. Despite attempting to match participants, an identifying code known only to the individual was not successful, as participants chose not to complete this component of the survey (Sayers, Rich, Rahman, Kelly, & James, 2019). Reasons for this are unknown; however, it may be associated with participants being concerned information would not remain unidentifiable and confidential, or may also be related to the workplace culture/stigma; employee and management divide and a lack of trust between employees and management, particularly during times of industrial tension between unions and management. Therefore, cross-sectional analysis of data was completed with the limitation that this does not allow causal inferences to be made. Overcoming this challenge of clear identification of employees over time is a significant consideration for future workplace research (Sayers et al., 2019).

The process of data collection needs careful consideration when involving workplaces, particularly those in rural or remote locations where access to technology can be challenging. Not all employees at workplaces have access to computers and/or the internet onsite and good access to the internet at some remote workplaces is limited. As a result, to collect data we have found to encourage engagement in the research, presentations by a researcher to each of the crews at the worksite to provide information of the research and provide details of requirements to potential participants and distributing paper-based surveys rather than using online survey forums is more successful for recruitment and participation. This does however add an element of cost to the research with printing, postage and data entry of paper-based surveys. In addition, multiple visits to the workplace by the researchers to capture the employees working on different shifts and rosters also has a cost associated with travel, accommodation and researcher time, often amplified by the rural and remote locations of mining operations. Costs also need to be factored into research for workplace visits to provide face-to-face presentations and reports back to the individual workplace site

As noted above, for any work-based research, the presence of internal champions at the workplace, who have an interest in and who can promote participation in the research, is essential to success. Such champions need to come from a mixture of management, supervisors and employees to support the research, encourage ongoing visibility of the research and to inspire involvement in any research interventions, enhancing uptake among employees (Tynan et al., 2018).

When researching mental health in the workplace, researchers need to be aware of mental health issues more broadly in the community and of any media reactions to suicide and drug and alcohol problems both at a national and at a local level respective to the workplace. Such media publicity may impact upon problems being stereotyped or normalised and deflect from the purpose of the research or intervention being presented. This is also important for researchers when they are at the workplace to be aware of any mental health issues broadly so enabling appropriate discussion with workers, for example if there had been a recent workplace injury or death.

When completing mental health and health-related research at the workplace, there is little relevant data nationally that can be used to compare the mining and resource industry to other industries within Australia. This is important to consider when developing survey tools to ensure, wherever possible, compatible information is being collected that will enable appropriate comparisons of data. It is also important for researchers to lobby regulatory agencies, such as SafeWork Australia and the respective state equivalent agencies, to collect and provide access to appropriate data to facilitate comparative analysis. When studying mental health in the mining industry, we have used data from the ANSMHWB (Australian Bureau of Statistics, 2008), which was collected in 2007 as no more recent data has been available. Factors which impact on the health of the community may have changed in this time period, however, the ANSMHWB is in line with international evidence and was the best available data at the time (László et al., 2010; Shandro et al., 2011).

Methodological Challenges of Intervention Options at Workplaces

The workplace provides a unique opportunity to both recognise and promote effective response to health problems; however, in the mining industry it has been identified that in many instances health promotion/programmes are provided on an adhoc basis rather than as an integrated, system-based and sustainable approach, with limited evaluation of the outcomes (Chapman, Rich, & James, 2018). There is also a lack of research on the effectiveness of health programmes in the workplace environment to inform appropriate workplace decisions. Some of the challenges in researching and then providing interventions within a workplace include the location of the workplace, ease of access to employees for participation, and the overall acceptability and/or restrictions placed on research that may limit intervention and evaluation. For example, there is great benefit in conducting a small pilot study to ensure acceptability and feasibility.

The availability of appropriate health interventions in rural and remote locations impacts upon the ability to develop evidence relative to the workplace. This is common with other health services. In relation to mental health, treatments for mental health problems are effective, but access to treatment is poor in Australia, especially in the rural and remote areas in which mines commonly operate (Meadows, Enticott, Inder, Russell, & Gurr, 2015). General practitioners (GPs) are usually the first point of healthcare access for people experiencing symptoms of a mental health problem (Parslow, Lewis, & Marsh, 2011); however, although workers may see a GP at home, access to and continuity of care between mines sites and community for FIFO workers in Western Australia has been identified as an impediment to care (Henry, Hamilton, Watson, & MacDonald, 2013). Our research identified the importance of a comprehensive and holistic approach in investigating and supporting the mental health of mine employees; however, it also identified that there was a gap between those with mental health problems and professional help-seeking behaviour, despite the presence of Employee Assistance Programs and/or onsite health services at many mine sites (James et al., 2017; Kelly et al., 2015).

This is an important consideration for workplace interventions, not only for those addressing mental health, as many places do not have access to health professionals to provide appropriate services or interventions, or due to limited resources, the timeliness to access services is compromised (James et al., 2017). Implementing interventions, and research of such, at the workplace can be cost prohibitive. Costs associated with taking services to rural and remote workplaces include direct costs of travel and accommodation, costs of the intervention itself plus the indirect costs associated with disruptions to productivity commitments at the worksite itself. This presents a tension between productivity and the intervention or research.

This tension may present constraints as to when a workplace intervention is offered to minimise disruption at the workplace—can it be offered during a worker's shift, at the beginning or end, after a shift or on a training day? What are the best methods to access the largest number of employees at a workplace? In the pre-post intervention data collection during the 'Working Well – Mental Health and Mining' study at coal mining sites, only 51% of the respondents of the phase 2 survey reported that they had completed any training. This may be the result of challenges in accessing all employees at sites and/or changes in staffing with participants completing the phase 2 survey who had not completed the phase 1 survey. It may also be as a result of this being self-reported, with employees not being aware of the specific name of the training they had received, which is a potential confounder to these results (Kelly et al., 2016). These factors need to be discussed and determined in conjunction with the workplace to provide the most appropriate and least disruptive timing.

The appropriateness of interventions for the particular workplace and the particular audience is another consideration to ensure the appropriateness of such service delivery which is vital for effectiveness and sustainability (Kelly et al., 2016). In our research, the presence of internal mining champions to promote participation in the programme at a site level was vital to the success of the intervention. More broadly, such champions are seen as important to assist acceptance of mental health programmes within the workplace (Tynan et al., 2018) and should be considered integral for any intervention offered at the workplace to enhance promotion and uptake by employees.

Our research in the mining industry has identified a clear preference for face-to-face format of delivery, closely followed by online resources. There was a clear effect of age, with younger participants preferring the use of online resources. This suggests that the peer-based model and face-to-face format of the multicomponent intervention (Mates in Mining) was appropriate and well received by participants. It is suggested that in future there be a combination of face-to-face mental health support complemented by online components of service delivery (Kelly et al., 2016). However, within rural and remote parts of Australia, including at some mining workplaces, access to the internet is disjointed or limited and this is an important consideration for type of service delivery (Christensen, Griffiths, & Evans, 2002). Choice of access to different methods of service delivery is encouraged to meet the needs and preferences of those seeking help (Kelly et al., 2016).

Outcomes of Our Mental Health Research with the Mining Industry

We have investigated the prevalence of mental health and psychological distress in mining in coal and metalliferous mines across Australia with a representative sample of men and women from a broad range of employment categories. We have implemented and evaluated 'Working Well –

Mental Health and Mining'—a multicomponent programme targeting mental health in mining which included the following: 1. Mates in Mining (MIM)—Adapted from the Mates in Construction (MIC) peer-assisted, evidence-based model of mental health awareness and support with the incorporation of wider mental health awareness education to complement its focus on suicide prevention. Mates In Mining includes several components: General Awareness Training (GAT), Connectors; Applied Suicide Intervention Skills Training (ASIST), Case management and field officers to support the participating sites external to the employer; 24/7 Help Line; 2. Supervisor training providing an introduction and overview of the issues relevant to managing staff experiencing mental health problems and/or suicidal behaviour, and to build skills in responding to these issues and 3. An organisational Policy review to identify how mental health is understood and addressed by the organisation to identify areas of policy that consider mental health, the impacts and how to respond to them. (James et al., 2017; Kelly et al., 2015).

Psychological Distress

Psychological distress levels of miners within our sample from coal and metalliferous mines across Australia were significantly higher in comparison with a community sample of employed Australians, with 45% of employees self-reporting moderate-to-high or very high levels of psychological distress (James et al., 2018). The following factors contributed significantly to levels of psychological distress: lower social networks; a past history of depression, anxiety or drug/alcohol problems; high recent alcohol use; work role (managers) and a set of work characteristics (level of satisfaction with work, financial factors and job insecurity); perception of lower workplace support for people with mental health problems (Considine et al., 2017). There was also a significant relationship between the perception of stigma and psychological distress, with those who reported higher psychological distress more likely to feel that someone with a mental health problem would be treated poorly in the workplace (Considine et al., 2017; James et al., 2017; James et al., 2018).

Alcohol Use

Risky alcohol use has been identified as being at considerably higher levels in Australian miners (coal and metalliferous) than both national and international averages. More than half of the male (53.7%) and almost one-third of the female (29.3%) participants consumed alcohol above the threshold considered risky or hazardous. There was a significant contribution of many individual level factors associated with AUDIT scores: younger age, male, current smoking status; illicit substance use; previous alcohol and other drug use (AOD) problems; and higher psychological distress (Tynan et al., 2017).

The evaluation of the 'Working Well – Mental Health and Mining' project found the programme was well received and identified positive changes in knowledge, attitudes and help-seeking behaviours in workers. Workers who completed training showed significant improvements in confidence in identifying workmates with mental health problems and support available, and willingness to start a conversation with a workmate about their mental health. In addition, those supervisors who completed the supervisor training were subsequently more confident that they could identify someone experiencing mental ill-health in the workplace; identify and recommend support services to a person experiencing mental ill-health; and have an effective conversation about performance issues that may be due to mental ill-health (Kelly et al., 2016; Tynan et al., 2018).

In our sample, following the 'Working Well – Mental Health and Mining' project, help-seeking behaviours increased, with participants' gender, age, relationship status, shift type and psychological distress significantly associated with likelihood of seeking help. When considering help-seeking behaviours for mental health, participants were more likely to report contact with non-professional sources of support (e.g. friend or family member; 46.8%) when compared to professional sources of support (28.6%). There was also a significant contribution of workplace variables, with job security and satisfaction with work significantly associated with help-seeking behaviour (Sayers et al., 2019; Tynan et al., 2016).

Our research supports the importance of a focus on mental health for the mining industry and of the important role peers play in addressing mental health problems in the workplace. The success of the 'Working Well – Mental Health and Mining' project shows that evidence-based initiatives embedded within workplace policy and undertaken with a robust organisation-wide strategic approach can achieve positive health and safety outcomes for workers.

Summary

This chapter discusses research challenges associated with working with industry in real-world environments particularly in regional, rural and remote locations. It identifies the importance of working with industry partners and on industry-identified problems (mental health) and the need and benefits of collaboration between researchers and industry stakeholders. An interdisciplinary approach to research is required to broaden the scope, capacity and translation of outcomes. The challenges associated with research methodology and providing interventions at the workplace, especially in rural and remote locations, needs consideration; however, our research supports the importance of developing evidence into the effectiveness of health programmes/interventions in the workplace environment to inform appropriate workplace decisions.

References

Australia Mining. (2018). Has mining learnt from the boom and bust cycle? Retrived March 2, 2019, from https://www.australianmining.com.au/news/has-mining-learnt-from-the-boom-and-bust-cycle/

Australian Bureau of Statistics. (2008). *National survey of mental health and wellbeing 2007: Summary of results*. Canberra, ACT: ABS.

Australian Research Council. (2018). *Engagement and impact assessment 2018–19 national report*. Canberra, ACT: ARC.

Bloom, D. E., Cafiero, E. T., Jané-Llopis, E., Abrahams-Gessel, S., Bloom, L. R., Fathima, S., et al. (2011). *The global economic burden of noncommunicable diseases*. Retrieved June 3, 2019, from http://www3.weforum.org/docs/WEF_Harvard_HE_GlobalEconomicBurdenNonCommunicableDiseases_2011.pdf

Chapman, P., Rich, J., & James, C. (2018). *Obesity and NSW coal mining – Final report*. Retrieved April 23, 2019, from http://www.nswmining.com.au/NSWMining/media/NSW-Mining/Obesity-Blueprint.pdf.

Christensen, H., Griffiths, K. M., & Evans, K. (2002). *e-Mental health in Australia: Implications of the internet and related technologies for policy*. Retrieved April, 23, 2019, from https://pdfs.semanticscholar.org/891b/2e4683f4ffa4c45e5025a4f566d9d9860f6b.pdf

Considine, R., Tynan, R., James, C., Wiggers, J., Lewin, T., Inder, K., et al. (2017). The contribution of individual, social and work characteristics to employee mental health in a coal mining industry population. *PLOS ONE, 12*(1), e0168445.

Coopers, P. W. (2014). *Creating a mentally healthy workplace: Return on investment analysis*. Retrieved April 24, 2019, from https://www.headsup.org.au/docs/default-source/resources/bl1269-brochure%2D%2D-pwc-roi-analysis.pdf?sfvrsn=6

Henry, P., Hamilton, K., Watson, S., & MacDonald, N. (2013). *FIFO/DIDO mental health research report 2013*. Retrieved April 23, 2019, from http://www.workplacehealth.org.au/_literature_175869/FIFO_DIDO_Mental_Health_Research_Report

Hilton, M. F., Whiteford, H. A., Sheridan, J. S., Cleary, C. M., Chant, D. C., Wang, P. S., et al. (2008). The prevalence of psychological distress in employees and associated occupational risk factors. *Journal of Occupational and Environmental Medicine, 50*(7), 746–757.

James, C., Tynan, R., Roach, D., Leigh, L., Oldmeadow, C., & Kelly, B. (2017). *Mental health in the australian metalliferous mining sector; psychological distress and associated factors in metalliferous mine employees*. Retrieved from https://journals.plos.org/plosone/article?id=10.1371/journal.pone.0209377

James, C., Tynan, R., Roach, D., Leigh, L., Oldmeadow, C., Rahman, M., et al. (2018). Correlates of psychological distress among workers in the mining industry in remote Australia: Evidence from a multi-site cross-sectional survey. *PLOS ONE, 13*(12), e0209377.

Kelly, B., Considine, R., Tynan, R., Lewin, T., James, C. L., & Wiggers, J. (2015). *Mental Health of employees in the New South Wales and Queensland Mining Industry: Baseline Findings - Report 1*. Australian Coal Industry's Research Program: Brisbane. Retrieved April 23, 2019, from http://www.qldminingsafety.org.au/wp-content/uploads/2015/09/Kelly_The-Extent-of-Mental-Health-Problems.pdf

Kelly, B., Considine, R., Rich, J., Tynan, R., Skehan, J., Perkins, D., et al. (2016). *Feasibility and Acceptability of strategies to address mental health in coal mining in*

New South Wales and Queensland – Report 2. Retrieved May 4, 2019, from https://www.researchgate.net/publication/328674232_Feasibility_and_acceptability_of_strategies_to_address_mental_health_and_mental_ill-health_in_the_Australian_coal_mining_industry

Kessler, R. C., Andrews, G., Colpe, L. J., Hiripi, E., Mroczek, D. K., Normand, S. L., et al. (2002). Short screening scales to monitor population prevalences and trends in non-specific psychological distress. *Psychological Medicine, 32*(6), 959–976.

László, K. D., Pikhart, H., Kopp, M. S., Bobak, M., Pajak, A., Malyutina, S., et al. (2010). Job insecurity and health: A study of 16 European countries. *Social Science & Medicine (1982), 70*(6), 867–874.

Ling, R., Kelly, B., Considine, R., Tynan, R., Searles, A., & Doran, C. M. (2016). The economic impact of psychological distress in the Australian coal mining industry. *Journal of Occupational and Environmental Medicine, 58*(5), 171–176.

Masige, S. (2016). More than a fifth of coal jobs lost: Coal services. Retrieved April 4, 2019, from https://www.australianmining.com.au/news/fifth-coal-jobs-lost-coal-services/

Meadows, G., Enticott, J., Inder, B., Russell, G. M., & Gurr, R. (2015). Better access to mental health care and the failure of the Medicare principle of universality. *The Medical Journal of Australia, 202*(4), 190–194.

Milner, A. (2016). *Suicide in the construction industry*. Retrieved March 2, 2019, from http://micaus.bpndw46jvgfycmdxu.maxcdn-edge.com/wp-content/uploads/2015/11/MIC-Annual-suicide-report-MIC-and-Deakin-University.pdf

Milner, A., Spittal, M. J., Pirkis, J., & LaMontagne, A. D. (2013). *Suicide by occupation: Systematic review and meta-analysis. British Journal of Psychiatry, 203*(6), 409–416.

Parslow, R., Lewis, V., & Marsh, G. (2011). The general practitioner's role in providing mental health services to Australians, 1997 and 2007: Findings from the national surveys of mental health and wellbeing. *The Medical Journal of Australia, 195*(4), 205–209.

Poplin, G. S., Miller, H. B., Ranger-Moore, J., Bofinger, C. M., Kurzius-Spencer, M., Harris, R. B., et al. (2008). International evaluation of injury rates in coal mining: A comparison of risk and compliance-based regulatory approaches. *Safety Science, 46*(8), 1196–1204.

Safe Work Australia. (2019). *Work-related psychological health and safety – A systematic approach to meeting your duties*. Retrieved March 2, 2019, from https://www.safeworkaustralia.gov.au/system/files/documents/1901/work-related_psychological_health_and_safety_guide.pdf

Sayers, E., Rich, J., Rahman, M. M., Kelly, B., & James, C. (2019). Does help seeking behavior change over time following a workplace mental health intervention in the coal mining industry? *Journal of Occupational and Environmental Medicine*. Publish Ahead of Print. https://doi.org/10.1097/jom.0000000000001605

Shandro, J., Koehoorn, M., Scoble, M., Ostry, A., Gibson, N., & Veiga, M. (2011). Mental health, cardiovascular disease and declining economies in British Columbia mining communities. *Minerals, 1*(1), 30–48.

Tynan, R., Considine, R., Rich, J. L., Skehan, J., Wiggers, J., Lewin, T. J., et al. (2016). Help-seeking for mental health problems by employees in the Australian Mining Industry. *BMC Health Services Research, 16*(1), 498.

Tynan, R., Considine, R., Wiggers, J., Lewin, T. J., James, C., Inder, K., et al. (2017). Alcohol consumption in the Australian coal mining industry. *Occupational and Environmental Medicine, 74*(4), 259–267. https://doi.org/10.1136/oemed-2016-103602

Tynan, R., James, C., Considine, R., Skehan, J., Gullestrup, J., Lewin, T., et al. (2018). Feasibility and acceptability of strategies to address mental health and mental ill-health in the Australian coal mining industry. *International Journal of Mental Health Systems, 12*(1), 66. https://doi.org/10.1186/s13033-018-0245-8

Universities Australia. (2013). The voice of Australia's Universities. Retrieved May 3, 2019, from https://www.universitiesaustralia.edu.au/research-innovation-workforce

University of Newcastle. (2019). Centre for resources health and safety. Retrieved May 3, 2019, from https://www.newcastle.edu.au/research-and-innovation/centre/resources-health-and-safety/about-us

Wolfenden, L., Williams, C. M., Wiggers, J., Nathan, N., & Yoong, S. L. (2016). Improving the translation of health promotion interventions using effectiveness-implementation hybrid designs in program evaluations. *Health Promotion Journal of Australia, 27*(3), 204–207.

10

A Coalition of Hope! A Regional Governance Approach to Indigenous Australian Cultural Wellbeing

Anne Poelina

Introduction

The Rainbow Snake is a universal Indigenous Australian living creature responsible for the creation and protection of waterways. Countless generations of traditional owners have cared for the sacred ancestral being; it is a spiritual guardian. The ancient creation songs and stories shared across the continent create meaning and purpose for our collective responsibility in managing the health and survival of the rivers, wetlands, springs, billabongs, floodplains and soaks. This chapter presents an *insider* view about how to promote remote Aboriginal people's wellbeing through a co-operative regional earth-centred governance model. The story reveals a powerful policy and investment approach to the planning and development of regional governance and showcases the unique cultural and environmental values of the Fitzroy River, and its Indigenous

A. Poelina (✉)
Nulungu Research Institute, University of Notre Dame, Broome, WA, Australia
e-mail: majala@wn.com.au

A. Campbell et al. (eds.), *Located Research*,
https://doi.org/10.1007/978-981-32-9694-7_10

people as having local, national and international significance. The central theme is the responsibility of Indigenous leaders to facilitate knowledge building, which requires sharing a deeper understanding of continuing colonisation and the collective responsibility, as Australians, for managing water.

The chapter is structured as follows. First, I introduce my position as an Indigenous person and our experience as colonised Indigenous Australians. The reader is introduced to the Fitzroy River and the Rainbow Snake as a universal Indigenous Australian representative at all our meetings of water management. The sacred ancestral being is a central theme that links people and places throughout the region and to the Australian continent. The methodology captures the importance of stories as an Indigenous research method for the insider researcher. The intention is to showcase the lived experiences of Aboriginal people and links the cultural authority of Indigenous regional governance in a context for how Indigenous people are strengthening their capacity in these modern times.

The theoretical framework below identifies the characteristics of the pedagogy of oppression through the work of Paulo Freire and makes the shift towards a pedagogy of hope (1970). These characteristics are explored through the journey of colonisation and oppression for Kimberley Indigenous people to illuminate resilience in transition. In the section on the Fitzroy River Water Planning and Management, we learn how these Indigenous people are impacted by global climate change and how they are learning from the experiences of fellow Indigenous Australians confronted with the national disaster of the Murray Darling Basin. The author makes the case for decolonising through an Earth-Centred Regional Governance approach. This approach is focused on cultural wellbeing, resilience and what is required to make the transition to justice, freedom and hope.

Ngayoo Yimardoowarra marnin in my Indigenous language Nyikina means *I am a woman who belongs to the Mardoowarra*. This centres my responsibility as a guardian of the Martuwarra in the West Kimberley region in remote north-western Australia. Martuwarra (or Mardoowarra in Nyikina) is the generic name for the Fitzroy River in several of the

Indigenous languages in the region (Pannell, 2009). The words 'Aboriginal' and 'Indigenous' are used interchangeably here to describe the original people of the Australian mainland. Throughout this chapter, I highlight the ongoing colonisation of the original Australians in north-western Australia. The colonial story of invasion and occupation of the Fitzroy River 'country and peoples were violent and brutal. It resulted in the subjugation and slavery of Kimberley Aboriginal people. Invasion was defined as development: the colonial states were established to create wealth for private and foreign interests at the expense of Indigenous people, our lands and waters' (Poelina, 2016). Despite the racist laws, policies and continuing violence perpetrated against the original Australians, we have survived.

The Kimberley is a large remote region with a unique mix of people, histories, landscape and development located in the north-western corner of Western Australia (Department of Regional Development, 2014; Kimberley Development Commission, 2018; Pepper & Scott, 2014.) Water management is the central theme that has always linked people and places throughout the region and with other regions. The Rainbow Snake is a universal Indigenous Australian representation of water. Stories from all over Australia tell of how the Rainbow Snake went up in the sky and down in the ground to carve river valleys and gorges, billabongs, soaks, springs, floodplains and aquifers (Poelina & McDuffie, 2017). As an insider, I know that these ancient stories and songs explain the interrelationship between surface and groundwater and how water is connected within each region and through a network of catchments and basins on a continental scale.

The unique cultural and environmental values of the river are of local, national and international significance. The five distinct representations of the Rainbow Serpent are listed in both the National Heritage and Aboriginal Heritage Lists (Department of Environment, 2011; Pannell, 2009; Western Australian Aboriginal Heritage Act, 1972). In listing the Fitzroy River for National Heritage status, traditional owners documented their cultural values and beliefs particularly regarding the river's *right to life*. The evidence demonstrates why their sacred ancestral being, the Martuwarra, must be protected (Pannell, 2009).

Right now, it is the guardianship responsibility of the Traditional Owners, as it has been for countless generations, to protect the survival of the Fitzroy River for current and future generations (Lim, Poelina, & Bagnall, 2017). As a sacred living entity, the Fitzroy River can teach others about reciprocal bonds between human and non-human life forces. These values and ethics can promote health in its holistic context by fostering a deep relationship between human and non-human beings (Lim et al., 2017; Poelina & Nordensvard, 2018; Taylor, Moggridge, & Poelina, 2016). In a time when climate change is quickly spiralling into *climate chaos*, we need to revalue water as a precious and rare commodity. There is an urgent need to change from the current development models if we are to give humanity, non-human beings and our planet earth a *climate chance* (Borland & Lindgreen, 2013). An earth-centred governance prioritises the health of the land and water above all and sets cautious limits to extraction and invasive development (Graham & Maloney, 2019).

The Martuwarra Fitzroy River Council (MFRC) is the contemporary regional model for consolidating the independent Indigenous nations' shared values and joint responsibility for managing the national heritage values of the river. It has the potential to create a transition to diverse economies, sustainable communities and culturally enriching livelihoods (CENRM, 2010; Pascoe, 2014). At the same time, it is necessary to keep an eye on potential cumulative impacts from extractive development to ensure they don't impact negatively on the environmental and human rights of current and future citizens (Brueckner, 2014; Burdon, 2014; Watson, 2018). Managing this requires careful dialogue. This global perspective illuminates the experience of invasion and colonisation that Indigenous Australians share with other Indigenous people across the world. There are similarities in the colonial processes used; however, the outcomes of these experiences are unique to each location. My research seeks to reframe the dominant colonial experience in Australia from a colonised insider position. It focuses on the Kimberley region of West Australia as an example of the ongoing colonisation of original Australians (Poelina, 2016).

The process and outcome of creating dialogue for building hope and the capacity to reach our full potential as Indigenous peoples is a process and outcome of decolonising (Smith, 1999). Decolonising requires dialogic action based on mutual respect involving open and honest dialogue (Freire, 1970); mutual respect that includes open and honest dialogue, action and critical reflection (Poelina, A., 2009, 2017; Poelina, Taylor, & Perdrisat, 2019).

In the Kimberley, as elsewhere in Australia, Indigenous people's identity is grounded in their cultural relationship with their environment and with each other. Aboriginal people in the Kimberley generally identify with one or more of four landscapes; the sea, river, desert and hill country (Kimberley Aboriginal Law and Culture Centre (KALACC), 2017). This knowledge building and adaptation is promoted through an ethic of caring for diverse cultural landscapes through people who have a deep respect for the earth. This Indigenous Australian story presents a new way of approaching knowledge-building and adaptive water management to promote Aboriginal people's wellbeing through a co-operative regional earth-centred governance model. Through this learning with others, we are building knowledge to reform legal practice and this requires new ways of thinking and critical reflection (Graham & Maloney, 2019).

Method: Indigenous Insider Researcher

All About Stories

Martin and Mirraboopa emphasise if research is to serve Aboriginal people, it must centralise the core structures of Aboriginal ontology (Martin & Mirraboopa, 2009). Through collaboration with Indigenous members of the MFRC, Kovach claims that the conversational method fits in with Indigenous worldviews and involves dialogic participation in which people share their stories with the purpose of helping others—a deeply relational process. Oral stories, used to transmit knowledge, enable people to observe and maintain community traditions while contributing to innovative research (Kovach, 2010)

Wilson, in crafting the definition of an Indigenous methodology, emphasises the concept of relational accountability, in which researchers consider all relationships while doing research, with the obligation to fulfil responsibilities to the world around us (Wilson, 2001). Similarly, Lambert argues that Indigenous methodologies are steeped in the relationships between the researcher, community members and the evidence being gathered, making our collaboration as researchers accountable to both the community and the information we build and gather together. This must do justice to, and be respectful of, all involved in sharing and analysing knowledge since as insiders we have been entrusted with it (Lambert, 2011). Important to generating new knowledge is the question of how the information informs policy and then, in turn, influences investment to transform the lives of Indigenous Australians.

Inside the Kimberley

As already noted, the Kimberley is a vast remote wilderness region in the remote north-western corner of Australia. Within this region, the culture, society and politics of historical groups are formed around four distinct interrelated landscapes from which diverse cultural groups share kinship, ceremonies, knowledge-making and practice. My identity is formed by the ancient and contemporary stories of my family's relationship with the Fitzroy River.

Following an academic career in capital cities and regional centres in Indigenous health, education and development, I returned to focus on my homelands in the Kimberley. Throughout the past 20 years, my role as an Indigenous leader in the Kimberley region has focused on promoting Indigenous human rights and environmental justice. My capacity as an Indigenous community leader, academic researcher and cultural and environmental advocate provides me with a unique *insider* perspective. As Deputy Chair of the Walalakoo Registered Native Title Body Corporate, I am jointly responsible, in collaboration with other Nyikina and Mangala leaders, for managing 27,000 square kilometres of our people's lands and waters (Walalakoo Aboriginal Corporation (WAC), 2016; Watson et al., 2011).

My view is further informed through my role as Chair of the MFRC. The MFRC is a federation of six Registered Native Title Body Corporations and registered claimants with traditional land tenure and title along the river. The WAC and the MFRC have mandates for traditional custodianship of the Fitzroy River. The WAC represents Nyikina and Mangala Native Title traditional owners who have responsibility for 40% of the river (Poelina et al., 2019).

The Fitzroy River Declaration and the MFRC are specific examples of the earth-centred approach. The approach values the river as the connection between the health and wellbeing of its diverse Indigenous peoples. Through our continuing guardianship and cultural governance of the river, we are strengthening earth-centred policy-making towards a decolonising future as an example of better practice for the rest of the country and globally. First Law promotes an earth-centred Indigenous governance approach to Environmental Law and the Rights of Nature (Lim et al., 2017). The Declaration is a regional earth-centred governance approach to decolonising Native Title lands and waters (Graham & Maloney, 2019; Kimberley Aboriginal Law and Culture Centre (KALACC), 2017).

I wholly concur with our leading organisation, the Kimberley Aboriginal Law and Culture Centre (KALACC) research, that increasing active participation in cultural practices for people in riverside communities improves their health and wellbeing. Conversely, the same research identified where cultural practices diminish, health and wellbeing follow (KALACC, 2017). My values and ethics, the way I see and act in the world, have been shaped by Indigenous experience of a deep connection and relationship with the natural world. Indigenous community responses contribute to building an evidence base that links cultural health to wellbeing for both individuals and communities (Hemming & Rigney, 2008; Poelina et al., 2019).

Along with Graham and Maloney (2019), I believe that improving Indigenous Australian wellbeing can be achieved through better management of Australia's biodiversity. This requires a multidisciplinary approach (Dockery, 2009, 2010; Preece, 2017; Pyke, 2017). My efforts include collaborating with critical friends, non-government and government agencies as well as developing a network of academic and community advocates from a disparate range of intersecting disciplines nationally and from around the globe (Lim et al., 2017; Poelina & Nordensvard, 2018; Taylor et al., 2016).

Aboriginal Voices

Emerging Aboriginal voices are being recognised as integral to providing an alternative narrative to Australian history (Altman & Kerins, 2012; Marshall, 1989; Pascoe, 2014) not just to reset perspectives but to consider a range of options and strategies to orient Aboriginal determined actions for change (Griffiths & Kinnane, 2011). In telling this story, I focus on the current socio-political context for Aboriginal people as an insider, and in doing so, I remain conscious of my responsibilities and obligations to our sacred river, family, community and the broader Australian community.

When Aboriginal people champion their rights as free people, it is referred to as *decolonising* (Smith, 1999). The process and outcome of decolonising require us to explore and analyse within an Indigenous context, and importantly to honour First Law, customary law (legal systems prior to British annexation), Indigenous science and technology (knowledge-making, adaptation, application and culture from the beginning of time to now) (Black, 2011; Graham & Maloney, 2019; Pascoe, 2014). This contrasts with 230 years of Australian Western knowledge-making which has typically focused on single issue approaches which do not reflect the interconnected fabric of Indigenous lives (Boutlon, edited 2016; Hemming & Rigney, 2008).

Fitzroy River Declaration

Since the beginning of time, traditional owners in this region continued to agree that the Fitzroy River is a continuous living entity that cannot be divided into a series of autonomous sections. The Mardoowarra is a living water system and therefore must be treated with the dignity and respect of a sacred ancestral being with its own right to life. The Fitzroy River Declaration (The Declaration) expresses the ethics and collective position of Native Title Traditional Owners in relation to maintaining the spiritual, cultural and environmental health of the river (Kimberley Land Council (KLC), 2016; Lim et al., 2017). Grace (2016) validates the actions of the Aboriginal nations who came together for The Declaration.

Through this process, traditional owners identified knowledge-gathering and sharing as necessary ingredients for building capacity for sustainable regional development (Lim et al., 2017).

This socio-political and cultural regional governance approach to Indigenous land and water management principles prioritises the health of the ecosystem above all other interests and has wider application internationally (Jackson, S., 2015; Jackson, Pusey, & Douglas, 2011; Toussant, 2008; Toussant, Sullivan, Yu, & Mulardy, 2001; Vasseur et al., 2017).

Martuwarra Fitzroy River Council

In June 2018, the Kimberley Land Council (KLC) in partnership with the Pew Charitable Trust brought together Native Title Traditional Owners from the region to establish the Martuwarra Fitzroy River Council (MFRC). The MFRC is an earth-centred regional governance cultural authority founded on the principles of the Fitzroy River Declaration. The MFRC is a federation of six Indigenous nations with custodianship for managing 733 kilometres of our sacred Fitzroy River (KLC, 2016).

Theoretical Framework

Anti-dialogic Action

Freire's *Pedagogy of the oppressed* (1970) describes the instruments of anti-dialogic action enacted through, 'conflict, invasion, manipulation, divide and conquer,' which leads to chaos, uncertainty, despair, disease and hopelessness (Freire, 1970: 119–147; Poelina, 2009). Aspects of anti-dialogic action can be seen in the 'deficit discourse' within which Aboriginal identity is bound to narratives of 'negativity, deficiency and disempowerment' (Fforde, Bamblett, Lovett, Gorringe, & Fogarty, 2013).

Overseas and in Australia deficit discourses are linked with negative outcomes, such as stereotyping and perpetuating the notion of lack of agency (Fforde et al., 2013: 166). An anti-dialogic framing of the so-called

'Aboriginal problem' would interpret Aboriginal people as being the problem, as opposed to an Aboriginal dialogic perspective that sees the dominant Western model of governance as systemic racism and therefore problematic. The consequences of colonisation from legislated inequality are termed 'structural violence' and are evident in three forms: (1) physical violence manifested in mortality and morbidity rates; (2) psychological violence manifested in poor mental health and high levels of substance misuse; and (3) individual and community violence manifested in family and community breakdown (Altman & Kerins, 2012; Boulton, edited, 2016; Galtung, 1970, 1990).

The anti-dialogic lived experiences of Indigenous Australians are direct outcomes of the continuing 'invasion, manipulation and divide and conquer' tactics by governments, which deny both the history of colonisation and the contemporary Australian context of racism and structural violence (Poelina, 2009, 2017). Inevitably, this systemic frustration reduces aspirations and inhibits the ability of an Indigenous person to reach their full potential as a human being. There needs to be thorough consideration of the social determinants of health and wellbeing in order to better understand the extent of oppression Aboriginal people experience in the Kimberley (Boutlon, edited, 2016). This anti-dialogic approach is opposite to the earth-centred, decolonising approach, the focus of which is on the wellbeing of the Fitzroy River, and through the River, its people. The river is the central life force of these diverse cultural landscapes and provides food, medicine, recreational and ceremonial activities which promote and support the health and wellbeing of the people. In turn, the people reciprocate as guardians of the Fitzroy River through an ethic of care.

Social Determinants of Indigenous Health and Wellbeing

The colonial states of Australia were established to create wealth for private and foreign interests, principally the British Crown at the expense of Indigenous people, our lands and waters, and natural and cultural resources (Poelina, 2009, 2017). For many years, the pastoral industry was propped up by Aboriginal slaves:

> For 100 years Aboriginal people were the back bone of a vast [sheep and] cattle industry in the Kimberley, based on their land and their slave labour. But when equal pay was introduced in the mid-1960s, most of the black stock workers lost their jobs, and they and their families were kicked off the stations to gravitate to the fringe camps of the white townships. (Mayfan & Australian Film Finance Corporation, 1992)

Indigenous Australians continue to live as colonialised peoples (Watson, 2018). Brown identifies that social determinants of health in Indigenous culture which impact on wellbeing include 'oppression, racism, environmental circumstances, economic factors, stress, trauma, grief [and] cultural genocide' (Brown, 2001). Twenty years ago, the World Health Organization (WHO) expressed concern regarding the lack of appropriate care, especially to vulnerable groups such as Indigenous peoples (World Health Organisation (WHO), 1999). At that time, the WHO published an international overview of the world's Indigenous populations:

> The needs and rights of Indigenous peoples have been of little concern to those larger and powerful nations. [Furthermore] it is remarkable that during the same period of colonialism there has been no lack of knowledge of the brutalities to which the Indigenous peoples of the world have been and continue to be subjected. (WHO, 1999: 7)

Over time, I have come to realise the Australian Indigenous colonial experience is shared with other indigenous people internationally (Césaire, 2010; Poelina, 2009, 2017). Forty years ago, a historically significant event occurred on my grandmother's country at the remote pastoral station, Noonkanbah. Film Australia produced the movie *On Sacred Ground* which showcased Aboriginal people from across the Kimberley and beyond gathered to protest against oil drilling exploration by Amax Oil and Gas Company on our sacred ancestral lands (Hughes & Howes, 1981). The voices of the wise old people from

Noonkanbah echo in my head. Their words and actions to protect our sacred sites resonated around the globe:

> We hope one day that the government might understand that we are a human being, [who] hardly can read and write, but he has protection over his land, what he can see, what he can hold on to, they just don't realise we don't want the mining company to ruin up our sacred places those are the things we see. They just can't see the point we are trying to get at, how it hurt our memories. (Paddy Moolambin in Hughes and Howes, 1981)

Another senior leader added:

> All we are saying is that we want land rights and human rights and a treaty to be signed by the Australian government so that Aboriginal people can be recognised as an owner [of our land]. (Jimmy Beindurry in Hughes and Howes, 1981)

Paddy Moolambin and Jimmy Beindurry were men of high degree, senior elders in First Law who stood to protect their land, waters and sacred sites from mining. As leaders, they sought rights and recognition from government 40 years ago.

Kimberley Aboriginal people continue to experience imposed invasive development because the *Native Title Act (1993)* does not allow traditional owners the right to veto extractive and invasive industries on their land (Blowes, 1993; Hepburn, 2005; Triggs, 1999). The colonial footprint has expanded in multiple forms with differing impacts at different times and in different locations (Poelina, 2009, 2017). Fresh corporate interests are poised to implement government policies that promote invasive development such as mining, fracking and irrigated agriculture throughout the region (Poelina et al., 2019). These examples of invasive development are also instances of structural violence that represent an ongoing process of colonisation.

Given the structural violence inherent in oppressive colonial systems, it is not surprising that Aboriginal people in the West Kimberley experience high levels of community and family breakdown, loss of cultural identity, racial discrimination, poor education and low standards of living, which contribute to high rates of poor health and wellbeing, perpetuating a cycle of disadvantage (Boulton, edited, 2016; Poelina, 2009, 2017). The *Overcoming Indigenous Disadvantage Report* (2016) identified overwhelming evidence that Indigenous life expectancy remains approximately 20 years lower than other Australians (Steering Committee for the Review of Government Service Provision (SGRGSP), 2016). Every health indicator for Indigenous people in the West Kimberley is almost two to three times worse than non-Indigenous people's health (Department of Indigenous Affairs (DIA), 2005). Many of our young people in the Kimberley live in despair and poverty. They have the highest suicide rates in the world, as evidenced in the recent Coronial Inquest (Aboriginal Policy and Coordination Unit, 2018).

While being Indigenous is not itself a risk factor, Indigenous individuals are more likely than non-Indigenous Australians to experience poor mental and physical health, imprisonment, shortened life expectancy, poverty, unemployment, homelessness, poor education, drug and alcohol misuse and violence. The Australian Bureau of Statistics (ABS) Aboriginal Health and Wellbeing of Aboriginal Australians report, 2017 confirms very little has changed for Aboriginal Australians (ABS, 2017). This was validated in a recent stocktake of similar evidence in the *Rising Inequality Report* by the Productivity Commission (2018). These reports speak significantly of a crisis that must be addressed, effectively, through a paradigm shift to holistic policy, investment and political action.

Indigenous Cultural Health and Wellbeing

The Indigenous concept of health is holistic and includes the social, cultural and environmental factors that impact on the status of individual, family and community wellbeing. Importantly, it affirms the need to connect and learn from the past to act in the present for forging a sustainable future.

Indigenous peoples' concept of health and survival is both a collective and an individual intergenerational continuum encompassing holistic perspectives incorporating four distinct shared dimensions of life. These dimensions are the spiritual, intellectual, physical and emotion. Linking these four dimensions, health and survival manifest itself on multiple levels where the past, present and future co-exist simultaneously (WHO, 1999: 3).

The historical connection of Indigenous Australian people to their country, culture and kin remains integral to their contemporary experience of wellbeing (Altman & Kerins, 2012; Burgess, Johnstone, Bowman, & Whitehead, 2005).

Culture

Culture is defined as a shared system of beliefs, knowledge, values, symbols, stories and ways of life in a group passed on through generations. Chu (1998: 126) highlights that, 'from infancy to old age, culture mediates one's experience and guides one's perceptions, interpretations, and behaviour.' Indigenous researcher Stephen Kinnane states:

> 'Culture' has been described as the glue that holds the diverse people of the Kimberley together. But it is a glue that has had to weather policies of removal, enforced concentration on the outskirts of towns, the impacts of grog and substance abuse and increased migration of kartiya [white people] as the potential development of natural resources is realised. (Kinnane, 2003)

Kinnane validates the notion of decolonising by maintaining cultural practices through resistance and reconciliation, which serves to maintain Indigenous identities and sustain resilience. He makes the point that this will be tested as developers come to exploit the natural assets of the region. Importantly, part of decolonising for the River's Traditional Owners is to unite and become organised as people who share culture and hold the same conceptual map; that is, they view the world similarly because of their common history and shared lived experiences (Hall, 1997). Similarly, Stockton (1995) identifies the relevance of culture for improving wellbeing. Traditional owners in the Kimberley seek to improve wellbeing by passing culture on to young people through community participation in cultural activities (KALACC, 2017). This requires educating the wider community to accept that colonisation has continuing influences on Indigenous health and wellbeing (Dockery, 2009, 2010; Ganesharajah, 2009). Collaborations between diverse groups promote the concept that sustainable lives and sustainable development are vital ingredients to shifting from being welfare dependent to creating wealth, equity and hope.

Pedagogy of Hope

Despite wellbeing becoming a key criterion in policy development and evaluation within the community sector, the original people living on their traditional lands along the Martuwarra must reframe their lives in the pursuit of justice and freedom through a post Native Title determined lens. For the first time in 40 years since the Noonkanbah dispute, Indigenous leaders are standing united towards building a pedagogy of hope. Their collective dream for the future now is to build trust and capacity to face the challenges from government policy objectives that conflict with maintaining their wellbeing and the health and sustainability of the Fitzroy River (KLC, 2016; Lim et al., 2017; Poelina et al., 2019).

Indigenous Concept of Health

In order to act in good faith and nurture an ethic of care for both human and non-human beings, we need to free ourselves from the paradigm of *me* (Western individualism) with man dominating nature (Burdon, 2014; Graham & Maloney, 2019) to *we* (Indigenous community-ism). We need to turn our *we* dialogue into action by starting with the Indigenous Australian concept of health:

> holistic, encompassing mental health and physical, cultural and spiritual health. Land is central to well-being. This holistic concept does not merely refer to the 'whole body' but in fact is steeped in the harmonised interrelations, which constitute cultural well-being. These inter-relating factors can be categorised largely as spiritual, environmental, ideological, political, social, economic, mental and physical. Crucially, it must be understood

that when the harmony of these inter-relations is disrupted, Aboriginal ill health will persist. (Swan & Raphael, 1995: 14)

Dialogue is a process action we need to engage in with other interests in the river to promote and maintain our human right to a safe environment when living on our traditional lands (Rimmer, 2018).

Dialogic Action

Paulo Freire (1970) proposed that development needs to incorporate dialogue to action. The concept of cultural synthesis identifies the need to generate dialogue to promote harmony (Freire, 1970). Freire challenged us to understand the tasks of knowledge building to unravel the increasing complex problems such as water management by first understanding the social and cultural contexts of oppressed people's lives.

The formation of the MFRC has provided a regional governance model for traditional owners who have shared values and beliefs with an opportunity to participate in regional planning and management. Through this collaboration with a wide range of partners, we are building new knowledge to inform policy and investment for Indigenous-led workforce development and business (Lim et al., 2017; Poelina et al., 2019). By promoting our values and beliefs of the Fitzroy River as a major spiritual, cultural, social, environmental and economic asset, we signal to potential partners our intention to contribute to regional investment, planning, management and assessment.

Fitzroy River Water Planning and Management Through an Earth-Centred Regional Governance

We live at the beginning of the Anthropocene where the attempts of humans to dominate and control the natural environment through technological and economic advancement have altered earth systems. There is a strong international consciousness that human activity has induced climate change which is quickly spiralling into *climate chaos* (Borland & Lindgreen, 2013). In 2019, the world earth systems are facing unprecedented risks that are pushing the planetary boundaries towards ultimate collapse (Gallagher, 2012). The Intergovernmental Partnership on Climate Change (IPCC, 2018) report contextualises the transformative changes needed to overcome the societal challenges and biodiversity threats associated with climate change.

Among developed countries, Australia is one of the most vulnerable to the impacts of climate change (Climate Council, 2015). Australia ranks as the second worst in the world on biodiversity conservation, corresponding to a total loss of 5–10% (Preece, 2017). Expansion of agriculture and aquaculture is a direct cause of loss of biodiversity (Preece, 2017). The current government management approach planned for the Fitzroy River is likely to increase the number of rare, endangered and extinct species.

Government attempts to initiate an unprecedented amount of forced invasive exploitation along the river demonstrates continuing colonial development for the benefit of established power and wealth (Borland & Lindgreen, 2013). In these, governments and investors fail to see and value the multiple and diverse new economies of culture, science, conservation and tourism. Indigenous people in the Kimberley refer to these multiple and complimentary industries that contribute towards the balance of life as the 'forever industries'—productive activities that support sustainable life and sustainable livelihoods forever (Poelina & Nordensvard, 2018).

The first step towards creating sustainable regional communities supported by 'forever industries' is recognising that the planet has limits to industrial growth and that human wellbeing is crucially dependent on rich and healthy ecosystems (Alexander, 2016; Assadourian, Prugh, Starke, & Springer Link, 2013). Rather than positioning social and natural systems as conditional subsystems to the economic system (anthropocentrism), the natural system must be at the centre, balancing environmental, human and subsequently economic activities (ecocentrism) (Graham & Maloney, 2019; Watson, 2018).

To establish a harmonious outcome, it is necessary to promote an inclusive co-operative process rather than a conflictual approach. It is also important to base reasoning and action around an examination of processes, which facilitates development of economic interests from within the region. The push for development creates direct pressure on environmental and cultural values but also on 'the capacity of existing state-based legal frameworks to protect these values' (Lim et al., 2017: 21). Building on the momentum garnered to secure Indigenous rights and the legal and ethical protection for the river, I have advocated for planning and development to be transparent and inclusive of all stakeholders, particularly those who live in the region and frequent the river for cultural affirmation, food, recreation and work, seeking to realise the pre-election promises of the current state government to develop a Fitzroy River Management Plan.

Fitzroy River Management Plan

The Draft Fitzroy River Management and Draft Fitzroy Water Allocation Plans were presented at a meeting involving Indigenous community and government representatives in Broome on 21 February 2019. They offer opportunities for national and state governments to invest into a structure for integrated and adaptive water management (Western Australian Government, 2018). This would contribute to better water planning, which is a priority for improving the health of water places in northern Australia (Douglas et al., 2011).

The MFRC is currently of the view that there should be a moratorium on issuing water licences. The Draft Plan requires broader scientific and industry knowledge as well as independent peer review of the government science. Most important, it is vital to include Traditional Owners in the co-design of any planning process because they have intergenerational knowledge spanning tens of thousands of years of knowledge-making, adaptation and mitigation (Graham & Maloney, 2019; Watson, 2018). These processes would facilitate opportunities for disparate interest groups to collaborate in order to sustain life and livelihoods in the region.

This inclusive and consultative approach to managing limits to water is supported by Dale (2014) who indicates that northern development has focused on progressing irrigated agricultural and mining demands, particularly gas extraction. Dale also cautions against development at all cost, pointing out the futility of seeking perpetual growth on an earth with finite resources. Additionally, Dale (2014: 143) posits co-operation to prevent further ecological decline as an alternative to current 'conflict based' 'competitive' approaches.

Murray Darling Basin

The Report of the South Australian Murray Darling Basin Royal Commission (SAMDBRC) in 2019 detailed the failure of the management of the Murray Darling Basin (MDB). Commissioner Bret Walker described the massive fish kills, polluted river ways, dry river beds, water theft as gross neglect, collusion, corruption as a direct consequence of government maladministration. The lessons learnt from the lack of water management planning that was responsible for the MDB disaster signal the need for sound policy investment in order to prevent a similar disaster occurring in the Martuwarra. Critical to the River's survival is the need to understand the potential cumulative impacts from development currently being promoted in the region (Poelina et al., 2019). The development of a Fitzroy River Management Plan has the benefit of being informed by the South Australian Royal Commission into management of the MDB. The MDB example demonstrates that Indigenous Australian voices are necessary for delivering justice in water and land management governance (SAMDBRC, 2019).

The Commissioner acknowledged that Australian laws and institutions continue to disadvantage Indigenous Australians. The Commissioner's findings state 'it appeared unconscionable that cultural flows have been put at the bottom of the pile' (SAMDBRC, 2019: 448–452). The Commissioner's findings alerted the MFRC to proceed with caution on any negotiations with government development proposals for the Fitzroy River, as justice may not be delivered.

The MDB Royal Commission shed a light on the true value of Indigenous knowledge for responsible water and land management practices. These cultural landscapes are not a blank canvas (SAMDBRC, 2019: 500–501). Australian rivers and waterways are already being 'developed' by Aboriginal people who continue to manage these natural assets for sustaining life and livelihoods (Gammage, 2011; Pascoe, 2014; Watson et al., 2011). The body of knowledge and practices for responsible river management has been handed down from thousands of generations of traditional owners for future generations.

Multi-stakeholder Partnerships

There are many government agencies and departments with a single objective from their portfolio's perspective. Despite wide acknowledgement of the need for intergovernmental trans-departmental community partnerships in the literature and policy, there continues to be a predominantly siloed approach to Indigenous investment and development and the ongoing failure to close the gap on Indigenous disadvantage (Brann & Brann, 2017; Commonwealth of Australia, Department of Prime Minister and Cabinet, 2017).

Negotiating a multi-stakeholder partnership is an important dialogic process for the disparate interests within MFRC to collaborate to achieve shared interests in adaptive water management and governance. Some describe this working together as the revolution required to create a sustainable world (Brouwer, Woodhill, Hemmati, Verhoosel, & van Vugt, 2016; Senge, Smith, Kruschwitz, Laur, & Schley, 2008). Aboriginal people are developing collaborative partnerships with government and industry to explore alternative low-impact development futures involving science, culture and conservation economies (Hemming & Rigney, 2008; Hill, 2008).

Conservation management of terrestrial landscapes, such as the Martuwarra, needs to incorporate Aboriginal cultural, legal, political, social, governance and environmental management knowledge (Poelina et al., 2019). The partnership approach endeavours to strengthen Aboriginal peoples' capacity by building their social, cultural, environmental and economic capital. Through an Indigenous holistic lens, all forms of capital are seen as necessary in order to shift from poverty to sustainable wealth creation. By drawing on a range of disciplines, disparate sectors can contribute to creating legal and policy instruments which can provide robust and equitable water governance for the Fitzroy River (Lim et al., 2017). The MFRC's perspective is informed by Lim's 'principle of fundamental truth or laws on which to base reasoning or action' (Lim, 2014: 98). The MFRC endorses these principles will create a better understanding of the cumulative impacts of development across the region.

Summary

The MFRC is a federation of registered Native Title bodies with traditional custodian responsibility for the Martuwarra Fitzroy River catchment. There are multiple economies in culture, science, heritage and conservation that are grounded in our deep relationship with the river and our connection to each other. Guided by First Law, our *living water* systems are our life force, connecting surface to groundwater, uniting the diverse cultural landscape and people throughout the Kimberley.

In these modern times, Aboriginal people living along the Fitzroy River and throughout the region continue to frame our health and wellbeing in terms of First Law ethics, values and governance. 'In accord with Aboriginal law we continue to have the authority and the responsibility to care for and ensure our territories are kept alive and well for future generations' (Watson, 2018: 2). First Laws are the natural *laws of the land* which frame the principles of adaptive water management through regional-scale earth-centred governance. This contemporary story champions decolonising invasive development through an Indigenous Australian wellbeing approach to regional governance.

Conclusion

This story is grounded in a regional governance approach to Indigenous Australian cultural wellbeing. The approach is less than a year in its development. It is at the early stages of engaging in dialogue to generate action and is demonstrating a useful approach to influencing policy and investment. As Indigenous people living in this region of Australia, we are reframing development policy towards an earth-centred regional governance approach. We are exploring new ways of doing business by asserting our Indigenous and human rights and by engaging with governments and other stakeholders as equal partners. We are advocating that sustainable life of the river requires respect for the cultural and environmental values of the diverse Indigenous peoples along the river. We believe sustainability of the River and the Traditional Owners transition to resilience and cultural wellbeing requires investment to strengthen their capacity to shift from a pedagogy of oppression towards a pedagogy of hope and freedom.

Acknowledgement This research is supported by an Australian Government Research Training Program (RTP) Scholarship.

References

Aboriginal Policy and Coordination Unit. (2018). *Government response to learning from the message stick: The report of the inquiry into aboriginal youth suicide in remote areas*. Perth, WA: Department of Premier and Cabinet.

Alexander, S. (2016). *Limits to growth: Policies to steer the economy away from disaster*. Retrieved from The Conversation: https://theconversation.com/limits-to-growth-policies-to-steer-the-economy-away-from-disaster-57721

Altman, J., & Kerins, S. (2012). *People on country; Vital landscapes, indigenous futures*. Sydney: The Federation Press.

Assadourian, E., Prugh, T., Starke, L., & Link, S. (2013). *State of the world 2013: Is sustainability still possible?* Washington, DC: Island Press.

Australian Bureau of Statistics (ABS). (2017). *Census of population and housing – Counts of aboriginal and Torres Strait Islander disability 2012*. ABS. Retrieved October 15, 2017, from http://www.abs.gov.au/Aboriginal-and-Torres-Strait-Islander-Peoples

Black, C. (2011). *The land is the source of the law: A dialogic encounter with indigenous jurisprudence*. New York, NY: Routledge.

Blowes, R. (1993). Governments: Can you trust them with your Traditional Title: Mabo and Fiduciary Obligations of Governments. *Sydney Law Review, 15*, 254–267.

Borland, H., & Lindgreen, A. (2013). Sustainability, epistemology, ecocentric business, and marketing strategy: Ideology, reality, and vision. *Journal of Business Ethics, 117*(1), 173–187.

Boutlon, J. (2016). *Aboriginal children, history and health: Beyond social determination*. New York, NY: Routledge.

Brann, O. G., & Brann, M. (2017). Environmentalists launch campaign against Gogo irrigation plans. ABC Rural. Retrieved September 26, 2017, from http://www.abc.net.au/news/rural/2017-09-25/green-groups-launch-campaign-against-gogo-irrigation-plans/8980912

Brouwer, H., Woodhill, J., Hemmati, M., Verhoosel, K., & van Vugt, S. (2016). *The MSP guide: How to design and facilitate multi-stakeholder partnerships*. Practical Action Publishing. Retrieved August 31, 2016, from http://www.developmentbookshelf.com/doi/book/10.3362/9781780446691

Brown, R. (2001). Australian indigenous mental health. *Australian and New Zealand Journal of Mental Health Nursing, 10*, 33–41.

Brueckner, M. (2014). On the social sustainability of development in Western Australia: A community perspective. In M. Brueckner, A. Durey, R. Mayes, & C. Pforr (Eds.), *Resource curse or cure? On the sustainability of development in Western Australia* (pp. 239–255). Berlin; Heidelberg: Springer-Verlag.

Burdon, P. (2014). *Earth jurisprudence: Private property and the environment* (1st ed.). Abingdon, Oxon, UK: Routledge.

Burgess, P., Johnstone, D. M., Bowman, J. S., & Whitehead, P. J. (2005). Healthy country: Healthy people? Exploring the health benefits of Indigenous natural resource management. *Australian and New Zealand Journal of Public Health, 29*(2), 117.

CENRM. (2010). *Fitzroy catchment management plan*. University of Western Australia. Centre of Excellence in Natural Resource Management. Retrieved November 10, 2016, from http://www.environskimberley.org.au/wp-content/uploads/2014/11/Fitzroy-Catchment-Action-Management-Plan.pdf

Césaire, A. (2010). *Discourse on colonialism* (J. Pinkham, Trans.). Delhi, India: Aakar Books.

Chu, C. (1998). Cross-cultural health issues in contemporary Australia. *Ethnicity and Health, 3*(1/2), 125–130.

Climate Council. (2015). Thirsty country: Climate change and drought in Australia. Retrieved March 2, 2019, from http://climatecouncil.org.au/uploads/37d4a0d2a372656332d75d0163d9e8b8.pdf

Commonwealth of Australia, Department of Prime Minister and Cabinet. (2017). *Closing the gap Prime Minister's report*. Canberra, ACT: Australian Government Publishing Service.

Dale, A. (2014). *Beyond the North South culture wars; Reconciling Northern Australia's recent past with its future*. Cairns, QLD, Australia: The Cairns Institute, James Cook University.
Department Environment and Energy. (2011). National Heritage Place – West Kimberley. Retrieved November 14, 2019, from http://environment.gov.au/heritage/places/national/west-kimberley/
Department of Indigenous Affairs (DIA). (2005). *Overcoming indigenous disadvantage in Western Australia*. Perth, WA: Department of Indigenous Affairs.
Department of Regional Development. (2014). *Kimberley: A region in profile 2014*. Retrieved October 14, 2016, from http://www.drd.wa.gov.au/Publications/Documents/A_region_in_profile_2014_Kimberley.pdf
Dockery, A. (2009). Culture and wellbeing: A case for Indigenous Australians. *CLMR discussion paper series 09/01*. Perth: Centre for Labour Market Research and School of Finance, Curtin University of Technology.
Dockery, A. (2010). Culture and wellbeing: The case of Indigenous Australians. *Social Indicators Research, 99*, 315–332.
Douglas, M., Jackson, S., Setterfield, S., Pusey, B., Davies, P., Kennard, M., et al. (2011). Northern futures: Threats and opportunities for freshwater ecosystems. In B. Pusey (Ed.), *Aquatic biodiversity of the wet-dry tropics of Northern Australia: Patterns, threats and future*. Darwin, NT: Charles Darwin University Press.
Fforde, C., Bamblett, L., Lovett, R., Gorringe, S., & Fogarty, B. (2013). Discourse, deficit and identity: Aboriginality, the race paradigm and the language of representation in contemporary Australia. *Media International Australia, 149*(1), 162–173.
Freire, P. (1970). *Pedagogy of the oppressed* (M. Ramos, Trans.). The Continuum Publishing Company: Penguin Books.
Gallagher, D. R. (2012). *Environmental leadership: A reference handbook*. Thousand Oaks, CA: SAGE Reference.
Galtung, J. (1970). Feudal systems, structural violence, and the structural theory of revolution'. *Revista Latinamericana de Ciencia Politica, 1*(1), 25–79.
Galtung, J. (1990). Cultural violence. *Journal of Peace Research, 27*(3), 291–305.
Gammage, B. (2011). *The biggest estate on earth: How Aborigines made Australia*. Crowsnest, NSW, Australia: Allen and Unwin.
Ganesharajah, C. (2009). *Culture and wellbeing: The importance of country. Native Title research report 1/2009*. Canberra: Australian Institute of Aboriginal Studies (AIATSIS).
Grace, P. (2016). Indigenous-Led Conservation: Experiences from the Kimberley. *Australian Environmental Law Digest, 3*(3), 21–26.

Graham, M., & Maloney, M. (2019). Caring for country and rights of nature: A conversation between Earth jurisprudence and Aboriginal law and ethics. In C. La Follette & C. Maser (Eds.), *Sustainability and the rights of nature in practice*. Boca Raton: CRC Press.

Griffiths, S., & Kinnane, S. (2011). *Kimberley Aboriginal caring for country plan*. Broome, WA: Nulungu Centre for Indigenous Studies.

Hall, S. (1997). *Representation and signifying practices*. London: Sage Publications.

Hemming, S., & Rigney, D. (2008). Unsettling sustainability: Ngarrindjeri political literacies, strategies of engagement and transformation. *Continuum: Journal of Media and Cultural Studies, 22*(6), 757–775.

Hepburn, S. (2005). Disinterested truth: Legitimation of the Doctrine of Tenure Post-Mabo. *Melbourne University Law Review, 29*(1), 1–38.

Hill, R., Land and Water Australia, & Australian Conservation Foundation. (2008). *A cultural and conservation economy for Northern Australia: A proof-of-concept study*. Canberra, ACT: Land and Water Australia.

Hughes, O., & Howes, R. (Directors). (1981). *On Sacred Ground* [Motion Picture]. Lindfield, NSW, Australia: Film Australia.

IPCC. (2018). *Global warming of 1.5°C; Summary for Policymakers*. Retrieved March 20, 2019, from http://report.ipcc.ch/sr15/pdf/sr15_spm_final.pdf

Jackson, S. (2015). *Indigenous social and cultural values relating to water in the Fitzroy Valley, Kimberley (WA): Information availability, knowledge gaps and research needs*. Northern Australia Environmental Science Programme, Australia. Retrieved May 3, 2019, from https://www.agric.wa.gov.au/sites/gateway/files/indigenous%20social%20and%20cultural%20values%20relating%20to%20water%20in%20the%20fitzroy%20valley.pdf

Jackson, S., Pusey, B., & Douglas, M. (2011). Indigenous community-based management of tropical freshwater systems. In B. Pusey (Ed.), *Aquatic biodiversity in Northern Australia: Patterns, threats and future*. Darwin, NT: Charles Darwin University Press.

Kimberley Aboriginal Law and Culture Centre (KALACC). (2017). *Cultural solutions: Shared pathways for engagement in the Kimberley*. Fitzroy Crossing, WA: Kimberley Law and Culture Centre.

Kimberley Development Commission. (2018). *Annual report 2017–2018*. Kununurra, WA: Kimberley Development Commission.

Kimberley Land Council (KLC). (2016). *Native title at Heart of Fitzroy River Declaration*. Retrieved February 1, 2018, from https://www.klc.org.au/native-title-at-heart-of-fitzroy-river-declaration/

Kinnane, S. (2003). *Shadow lines*. Fremantle, WA: Fremantle Art Centre Press.

Kovach, M. (2010). Conversational method in indigenous research. *First Peoples Child and Family Review, 5*(1), 40–48.

Lambert, L. (2011). *Eberhard Wenzel Oration: Two eyed seeing: Indigenous research methodologies.* Retrieved March 23, 2019, from https://www.healthpromotion.org.au/images/eberhard_oration_by_lori_lambert.doc

Lim, M. (2014). Is water different from biodiversity? Governance criteria for the effective management of transboundary resources. *Review of European, Comparative & International Environmental Law, 23*(1), 96–110.

Lim, M., Poelina, A., & Bagnall, D. (2017). Can the Fitzroy River Declaration ensure the realisation of the first law of the river and secure sustainable and equitable futures for the West Kimberley. *Australian Environment Review, 32*(1), 18–24.

Marshall, P. (1989). *Raparapa: Stories from the Fitzroy River Drovers*. Broome, WA: Magabala Books.

Martin, K., & Mirraboopa, B. (2009). Ways of knowing, being and doing: A theoretical framework and methods for indigenous and indigenist re-search. *Journal of Australian Studies, 27*(206), 203–214.

Mayfan, P., Australian Film Finance Corporation (Producers), Chase, G., & Isaac, G. (Directors). (1992). *The Kimberley Mob* [Motion Picture]. Ronin Films.

Pannell, S. (2009). *Martuwarra/Marduwarra ('River Country'): A report on the Indigenous heritage values of the Fitzroy River drainage system. Report to the Kimberley land Council for the National Heritage assessment of the West Kimberley*. Broome, WA: Kimberley Land Council.

Pascoe, B. (2014). *Dark Emu, black seeds; Agriculture or accident?* Broome, WA: Magabala Books.

Pepper, M., & Scott, M. (2014). Biogeography of the Kimberley, Western Australia: A review of landscape evolution and biotic response in an ancient refugium. *Biogeography, 41*(8), 1449.

Poelina, A. (2009). *Action research to build the capacity of Nyikina Indigenous Australians*. Unpublished doctoral dissertation. Armidale, NSW: University of New England.

Poelina, A. (2016). A lucky country and a fair go. In I. Tardos (Ed.), *This is my country* (pp. 18–23). New York: Foto Evidence Press.

Poelina, A. (2017). Protecting the river of life. In K. Aigner (Ed.), *Australia Vatican Museums ethnological collection*. Canberra: Aboriginal Studies Press.

Poelina, A., & McDuffie, M. (Directors). (2017). *Mardoowarra's right to life* [Motion Picture]. Retrieved June 18, 2018, from, vimeo: https://vimeo.com/205996720. Password: Kimberley.

Poelina, A., & Nordensvard, J. (2018). Sustainable Luxury tourism, indigenous communities and governance. In M. Gardetti & S. S. Muthu (Eds.), *Sustainable luxury, entrepreneurship, and innovation* (pp. 147–166). Singapore: Springer Singapore.

Poelina, A., Taylor, K., & Perdrisat, I. (2019). Martuwarra Fitzroy River Council; an Indigenous cultural approach to collaborative water governance. (Taylor, & Francis, Eds.) *Environmental review, Aboriginal water management, special edition.*

Preece, N. (2017). Australia among the world's worst on biodiversity conservation. Retrieved March 6, 2019, from https://theconversation.com/australia-among-the-worlds-worst-on-biodiversity-conversation-86685

Productivity Commission. (2018). *Rising inequality? A stocktake of the evidence, Commission Research Paper*. Canberra, ACT.

Pyke, M. L. (2017). *Cultural systems; science and natural resource management Aboriginal management of wetlands in the west Kimberley, Australia*. PhD Thesis. Perth, WA: University of Western Australia.

Rimmer, M. (2018). Northern exposure: Alaska, climate change, indigenous rights, and atmospheric trust litigation. *Intellectual Property and Clean Energy*. Retrieved May 3, 2019, from https://doi.org/10.1007/978-981-13-2155-9_23

Senge, P. M., Smith, B., Kruschwitz, N., Laur, J., & Schley, S. (2008). *The necessary revolution: How individuals and organizations are working together to create a sustainable world.* London: Nicholas Brealey.

Smith, L. (1999). *Decolonizing methodologies: Research and indigenous peoples.* Dunedin, NZ: Zed Books, University of Otago Press.

South Australian Murray Darling Basin Royal Commission (SAMDBRC). (2019). *Report*. Adelaide: Government of South Australia.

Steering Committee for the Review of Government Service Provision (SGRGSP). (2016). *Overcoming Indigenous disadvantage: Key Indicators*. Canberra: Productivity Commission.

Stockton. (1995). *The Aboriginal gift – Spirituality for a nation*. Alexandria, NSW: Millennium Books.

Swan, P., & Raphael, B. (1995). *Ways forward: National consultancy report on Aboriginal and Torres Strait islander Health*. Aboriginal and Torres Strait Islander Health. Canberra, ACT: AGPS.

Taylor, K., Moggridge, B. J., & Poelina, A. (2016). Australian Indigenous Water Policy and the impacts of the ever-changing political cycle. *Australasian Journal of Water Resources, 20*(2), 132–147.

Toussant, S. (2008). Kimberley friction: Complex attachments to water-places in northern Australia. *Oceania, 78*, 46–61.
Toussant, S., Sullivan, P., Yu, S., & Mulardy, M. (2001). Fitzroy Valley indigenous cultural values study. *Report for the water and rivers commission*. University of Western Australia. Retrieved March 4, 2019, from https://www.water.wa.gov.au/___data/assets/pdf__file/0018/4842/46326.pdf
Triggs, G. (1999). Australia's indigenous peoples and international law: Validity of the Native Title Amendment Act 1998 (Clth). *Melbourne University Law Review, 23*(2), 372.
Vasseur, L., Horning, D., Thornbush, M., Cohe-Shaacham, E., Andrade, A., Barrow, E., et al. (2017). Complex problems and unchallenged solutions: Bringing ecosystem governance to the forefront of the UN sustainable development goals. *Ambio, 46*(7), 731–742.
Walalakoo Aboriginal Corporation (WAC) RNTBC. (2016). *Walalakoo Health Country Plan 2017–2027*.
Watson, I. (2018). Aboriginal relationships to the natural world: Colonial 'protection' of human rights and the environment. *Journal of Human Rights and the Environment, 9*(2), 119–140.
Watson, J., Watson, A., Poelina, A., Poelina, N., Watson, W., Cameilleri, J., et al. (2011). *Nyikina and Mangala Mardoowarra Wila Booroo: Natural and cultural heritage plan. Providing information to plan for and manage activities on Nyikina and Mangala country while protecting culture and the environment*. Nyikina-Mangala Aboriginal Corporation and World Wide Fund for Nature, Australia.
Western Australian Aboriginal Heritage Act. (1972). State Law Publisher. Retrieved from https://www.slp.wa.gov.au/PDF
Western Australian Government. (2018). Media statements – Consultation begins on visionary plan for Fitzroy River. Retrieved March 10, 2018, from https://www.mediastatements.wa.gov.au/Pages/McGowan/2018/03/Consultation-begins-on-visionary-plan-for-Fitzroy-River.aspx
Wilson, S. (2001). What is an indigenous research methodology? *Canadian Journal of Native Education, 25*(2), 175–179.
World Health Organisation (WHO). (1999). *The mental health of indigenous peoples: An international overview* (E. Cohan, Ed.). Geneva: Department of Mental Health, WHO.

11

Real and Virtual Communities of Practice: A Case Study from the Outer Hebrides

Simon Bradley and Anna Wendy Stevenson

This case study focuses on the impact of, and opportunities arising from, the presence of the University of the Highlands and Islands (UHI) on the remote, rural Scottish islands known as the Outer Hebrides. UHI was created in 2011 and comprises 13 independent colleges and research institutions and over 70 local learning centres. Lews Castle College (LCC) is one of the UHI colleges and is based in the Outer Hebrides with its main campus in Stornoway on the Island of Lewis. The music courses discussed in this case study are delivered from a small LCC learning centre on the Island of Benbecula.

Between 2001 and 2019, the University music courses delivered from these islands have undergone multiple transitions. This has necessitated adaptation from the communities of practice—academic, student, wider

S. Bradley • A. W. Stevenson (✉)
Lews Castle College UHI, Lionacleit, Isle of Benbecula, Outer Hebrides, UK
e-mail: Simon.Bradley@uhi.ac.uk; anna.stevenson@uhi.ac.uk

A. Campbell et al. (eds.), *Located Research*,
https://doi.org/10.1007/978-981-32-9694-7_11

located and virtual. In addition, the local non-academic community has had to adapt as the music courses have been linked, since their inception, to important community initiatives closely related to cultural identity. Since 2012, the UHI Bachelor of Arts Applied Music Degree has been delivered from this Island base, although it is networked throughout the Highlands and Islands of Scotland. This is achieved through video conference and online technologies, and in partnership with other colleges, creating 'virtual' communities of practice in a Blended Learning delivery model.

Reflections on resilience have been enabled by detailing the transition of this delivery over nearly 20 years from the exclusively 'real' (face to face) to the development and addition of virtual and distributed communities of practice. Of particular importance has been the difficulty of conveying this transition to the surrounding host community, revealing the significant complexity of multiple and differing speeds and levels of change.

Context: Place, People and University

The Outer Hebrides are also known as the Western Isles and Na h-Eileanan Siar in Gaelic. These islands off the north-west coast of Scotland form a 130-mile arch from north to south. 'Uist' refers to the central section of the Outer Hebrides stretching from the islands of Berneray in the north to Eriskay in the south via causeways connecting the separate islands of North Uist, Grimsay, Benbecula and South Uist over a distance of 60 miles. The landscape of Uist is dramatic. Treeless, barren peatlands predominate on the east side, patched with thousands of waterways and sea lochs, while the West side is home to swathes of white sandy beaches and a biodiverse coastal machair grassland, which is of international conservation importance. Winters host long nights and awe-inspiring Atlantic storms, and summers can be—though short—a tourist's paradise.

The landscape is also a rich source of inspiration for artists and archaeologists and a place with few distractions for a practicing musician. The island location offers a sense of community where Gaelic is spoken by 60% of the islanders, the highest percentage in Scotland, and where the traditions of bagpiping, song and dance are strong (Scotland's Census,

2011). Since the 1700s, the islands have been of interest to historians and folklore collectors, highlighted by the work of the University of Edinburgh's School of Scottish Studies and archives such as Tobar an Dualchais, 'Kist O' Riches' (online, http://www.tobarandualchais.co.uk/)

Lewisian Gneiss forms the Pre-Cambrian Uist bedrock and is among the oldest in Europe and named after the island of Lewis itself. World Heritage Site St. Kilda can be seen with the naked eye on a clear day from Uist and the significance of the island's archaeology is seen as offering potential growth for the economy by bringing tourism to the islands. Religious identity is important and significant in the Outer Hebrides with the lowest percentage of people reporting 'no religion' in the Scottish census of 2011. This feature of identity ranges from Presbyterian Sabbath observance and Gaelic psalm singing (Precenting) in the north to native Scottish Catholicism dating to pre-Reformation times in the islands to the south (Reddy, 2018). Crofting is a traditional farming system unique to the Highlands and Islands and is prevalent and intimately woven into the Uist identity and way of life. This low input cultivation also supports globally significant populations of breeding waders and other birds that attract conservationists and bird watchers alike.

Uist is both beautiful and challenging as a place to live, work and study. Uist experiences the highest levels of household fuel poverty in Scotland with full-time average earnings 20% below the Scottish average. (Outer Hebrides Community Planning Partnership Local Outcome Improvement Plan, 2017–27: 6). It is estimated to have an enormous wind power potential with the capacity to produce 20 times the domestic demand, although the infrastructure to capture and export this resource is lacking (Outer Hebrides Community Planning Partnership Local Outcome Improvement Plan, 2017–27: 7). The population has been in long-term decline, having lost 40% over the last century, and the current demographics report a 'population gap' in the 15–30 age range, (Outer Hebrides Community Planning Partnership Single Outcome Agreement, 2013–2023) which is also seen across the Highlands and Islands region, averaging at 25% lower than Scotland as a whole (Simco & Campbell, 2011).

One of the 'core purposes' of UHI, which was supported by Regional and European funds, was to help address this problem by providing learning opportunities for people in these areas so they would not have to leave. This demographic issue is a concern at the Outer Hebridean community planning level with a priority to stabilise the population by attracting 'new Hebrideans' to live and work in the islands (Outer Hebrides Community Planning Partnership Local Outcome Improvement Plan, 2017-27: 6). Seasonal tourism is significant to the Uist economy during the summer months, and the academic calendar neatly covers the rest of the year when there is less demand. This makes the presence of music students a natural fit with the needs of the Uist economy and social life, especially as they learn about the indigenous music and culture and take part in local performances and ceilidhs.

Face to Face 'Real' Communities of Practice

The first music course in 2001 grew from community development initiatives and combined the study of music and Gaelic equally, and often concurrently, with the music placed in a Gaelic context. The market research for this programme delivery was based around local Gaelic arts organisation Ceòlas, established in 1996 to celebrate, promote and nurture the indigenous traditional music, language and heritage. This course, a Further Education National Certificate (FE NC), was accessible to all with the entry requirement being simply the desire to learn. The cohort was largely drawn from local adult interest groups forming a strong though finite base. In 2004, the addition of the Higher National Certificate (HNC) to the academic portfolio demanded a stronger focus on musical and academic skill. This qualification also functioned as a foundation year for access to music college and university degrees on the mainland and proved to be the more popular course attracting applications from largely school leavers because of its outward-looking stance. Over the next few years, demand for the lower-level FE NC course Traditional Music with Gaelic waned. Several high schools, due to their own sustainability issues, were now offering the FE NC course delivered as core 6th year units, but this meant the college could no longer accept such applicants as this would be a duplication of level and certification, thus rendering them ineligible for funding. By 2011, the decision was made to discontinue the FE NC, which saw the loss of Gaelic assessment in the framework. Arguably, with the withdrawal of the linguistic link, the music courses had become music *in* rather than *of* the community.

The emphasis on performance practice in the HNC was linked to the expertise and professional networks of the staff, performers and composers from the field of traditional music in their own right. Opportunities such as Celtic Connections Festival in Glasgow proved attractive to students and attracted a higher calibre of entrants, raising the level of musicianship on the course. The reputation of Uist as a destination for the study of traditional music flourished, and this positive feedback loop saw increasing student numbers coming to study on the islands.

The Student Experience and Its Impact on Student Numbers

The experience for a young student moving to Uist contrasts radically with a move to Glasgow, for example, and is not to everyone's taste. There is no student accommodation as such on Uist, so the solution is that local houses are rented ad hoc. The affordable rented housing stock is poor with high levels of fuel poverty and cold and damp conditions being a challenge. In summer the natural beauty of the island landscape is spectacular, but students are forced to vacate their accommodation to make way for lucrative holiday lets, and sometimes before the end of term, which is problematic. Internet availability, while improving, has often lagged behind the connection rates common on the mainland. Uist is both rural and remote with the double cost of distance and time applied to any physical interaction. There are a few small conurbations but most housing and facilities are widely spaced, and this, along with very low population density, is a major hurdle to the provision of effective public transport. The distances between destinations means that walking and even cycling is unlikely to meet students' travel needs, resulting in a car-based mobility dependency. Travel off-island is also an issue, with ferries sometimes unreliable due to the extreme weather conditions typical in winter months. There is also poor connectivity with buses and trains on the other end and many days are lost due to travel plan failure. Flights are available to Glasgow and Inverness although tickets are very expensive and beyond the budget of most students, even with the Air Discount Scheme for residents. Reliance on ferries and flights also introduces unpredictability and risk to any plans, including accessing performance and teaching opportunities out with the islands.

These challenges do, however, create a sense of adventure, as students are constantly problem solving and cooperating to find solutions to these everyday issues (Jobson, 2019). Local age group peers have often moved to the mainland to study or find employment. There are few public houses/hotels and the cinema is mobile and visits every 10-12 weeks. In general, there are events and social clubs but these tend to be geared to and attended by an older demographic. Some students join the local pipe band or choir, although in general the bonds are much stronger between students than with the community. Bonds with the community are greatly aided by the presence of 'local' children electing to take the course who can help integrate the visiting students into community settings and experiences. Several bands have formed from these cohorts and most are named after local mountains or other island features. Strong relationships are formed, and the lack of distractions has furnished the more serious students with a great opportunity to focus and develop musical as well as social bonds. The momentum and reputation built gained national media attention through numerous student concerts at festivals; brokered by the staff and residential student numbers grew to a peak of 35 in 2013.

Creagorry Music Hub: A Temporary Solution

By 2013, there were too many students for the LCC learning centre, which led to the use of a local and mostly unused hotel—repurposed in the academic term as 'Creagorry Music Hub'—in order to provide additional teaching and rehearsal space along with accommodation close to the college campus. Weekly music sessions, '*Ceòl agus Curry*' (Gaelic for Music and Curry), held in the hotel bar and led by staff, attracted members of the community and provided regular performance and ensemble opportunities for the students to engage with. Several formal concerts at Creagorry Music Hub attracted large audiences and received media coverage.

While there were clear benefits to offering students a single location close to the learning centre in which to live and socialise, financial investment was required to bring the hotel up to the full requirements as student accommodation and agreement between the college and hotel owner was not reached. After one year, the decision was made to discontinue the hiring of the hotel. A stopgap was found for student accommodation in council housing, although this was over an hour each way on the bus to the college learning centre, even though free travel was secured. This accommodation proved unpopular with students and only lasted two years.

BA Applied Music (BAAM) and the Addition of 'Virtual' Communities of Practice

Since 2004, when LCC in Uist first offered the HNC Music, more colleges began to offer this qualification. This trend has continued and currently there are over ten and rising comparable courses in Scotland in 2019 whereas there were only three in 2004. From 2012 onwards, degree and master's level music courses were developed and delivered from Uist. These have proved to be successful and attractive to students, although this placed onerous demands on available staff resources.

The BA Applied Music degree was validated in 2012 and represented an innovative solution to the UHI remit of offering equal access to tertiary education across the Highland and Islands region, without the requirement to leave one's own community and location (Campbell & Simco, 2011). The delivery was 'networked' across various colleges within the UHI banner with staff from these other centres working with Uist LCC staff to provide this Blended Learning opportunity to students across Scotland. This includes video conference technology that is used to deliver the bulk of classes and is complimented by four residentials per year, bringing students together in various and varying locations throughout the Highlands and Islands. Individual performance practice specialist tutors are externally sourced and contracted to provide students with bespoke, individual, face to face tuition at a location negotiated between student and elected teacher. The residentials form a crucial part of the student experience, and surveys conducted after each one evidence the important role they play in forming communities of practice. All students meet in September at the start of the UK academic year in Inverness where they undertake educational projects, collaborate and perform. This quickly forms the bonds required to maintain the virtual communities between residencies.

The second and third residencies are held in multiple locations so as to respect the cohort's geographical distribution. The final residency takes place in Stornoway, the island base of LCC where performance assessments and final recitals take place. Internal surveys illustrate the generally inclusive and functioning communities of practice formed, with subsequent online collaboration on projects assessed and demonstrated at residential performances.

The video conference online classes are conducted with some students attending 'live' and contributing to the seminar-style 'flipped classroom', where students read material distributed in advance and then discuss this in the sessions led by the tutor. Others prefer to watch the recordings and forfeit the ability to steer discussion and content during the VCs. This was a new style of teaching and was a learning curve for staff and to a lesser extent for students too. Littlejohn and Pegler (2007: 224) note the major impact and adaption in working practice was required from teaching staff in this transition. Panciroli, Engstrand, Graham, and Clarke (2015), in their review of the literature on Blended Learning, note the importance of 'warmth' in the lecturer style that facilitates in bridging the virtual gap, which is significant in that students primarily look for social interaction whereas the tutor most often focuses on knowledge transfer. This tutor role of 'social presence', along with cognitive and teaching presence proposed in the community of inquiry model (Garrison, Anderson, & Archer, 2000), considers the importance of establishing trust through open communication and emotional expression for the purpose of group cohesion, and a sense of 'real' people existing within the virtual community of practice. This meant that new techniques were required on the part of the staff in this transition of delivery.

Meanwhile, internal BAAM student surveys from 2017 reveal that the flexibility offered in this programme was the most popular aspect of this delivery, whereby students were not 'time or place bound' and indeed only a small minority (10%) accessed these classes from a campus setting. The appeal to students of the Blended Learning delivery method is mirrored in other student surveys conducted around the world (Moskal, Dziuban, & Hartman, 2013; Taylor & Newton, 2013). The flexibility facilitated by the delivery allowed BAAM to access students around Scotland and also appealed to working and experienced musicians, thus increasing the potential catchment without placing demands on physical infrastructure.

UHI attained University status in 2011 and BAAM was one of the first 'networked' degrees to be delivered. The BAAM is multigenre, again widening the catchment of students, and the syllabus reflected this with lessons on Classical and Popular music context and history. The study also necessitated academic research and writing and use of technology to record and collaborate remotely, with file sharing and asynchronous compositional and arranging tasks. For such a small learning centre to be awarded leadership of this course was a controversial coup for Uist, as there were other UHI colleges with much bigger music departments and developed infrastructure. This demonstrated commitment on the part of the university to their 'Curriculum for the 21st Century' (C21C) vision of distributed teaching and learning (Simco & Campbell, 2011) and also recognised the momentum and success developed and demonstrated by the Uist music staff.

MA Music and the Environment (MATE) was validated in 2013 and is a further progression route offered by LCC and delivered and led by staff from the Uist learning centre. This addition to the portfolio of courses is based on the 'off-campus' model from C21C (Panciroli et al., 2015), where the delivery is entirely online. This has facilitated students from oversees joining the 'virtual' communities of practice. This UHI 'four model' conceptual framework of delivery style ranges from a) local to b) site-specific to c) networked to d) off-campus. All four can have elements of blended delivery but decrease in geographical fixedness and increase in flexibility for the student. From a) to d) there is an increase in the virtual and a decrease in the physical communities of practice.

Impact of Community Connectivity and Engagement Across the Highlands and Islands

Each student is required to engage with their own local communities where they live and develop ensemble or teaching experiences and opportunities. This roots the student cohort within their individual geographical base, and so while BAAM does not exclusively meet the needs of the Uist community, the cumulative and entire effect is one of significant coverage and impact within the Highlands and Islands region and indeed beyond, with many students electing to base themselves in Glasgow.

An important network within which the real and virtual student communities of practice operate is the range of agencies, organisations and community groups that engage and collaborate with individual students and the course in general through projects, employment, shadowing and work placements. For those choosing to base themselves locally within Uist, students perform regularly in local community concerts and ceilidhs and contribute to informal teaching provided in schools and take part in workshops and fèisean classes. This student-community engagement is replicated throughout Scotland and also in non-traditional music settings such as community choirs and orchestras. Volunteering at festivals and arts organisations is common and helps students develop appropriate networks and experience. Together then, the music courses provide and facilitate multiple overlapping and both real and virtual communities of practice that impact all over Scotland and root the provision within multiple geographical and demographic communities.

Overall student numbers enrolling onto the new music courses with Lews Castle College have significantly increased and been matched with Uist-based staff increasing from 1 to 3.6 full-time between 2012 and 2019. However, students based in Uist peaked in 2013 and have decreased since then (Fig. 11.1).

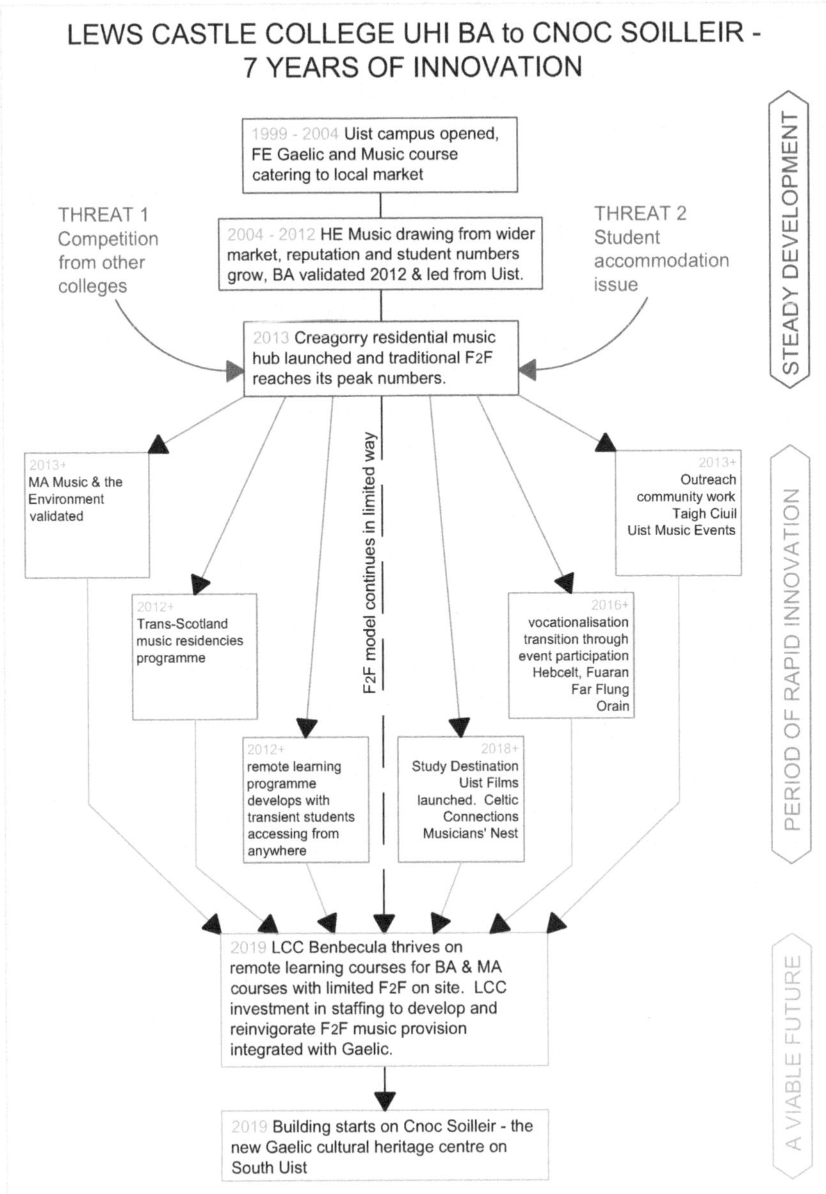

Fig. 11.1 Seven years of innovation

Tensions and the Lack of a Narrative

While the skills developed through BAAM are highly applicable and apt to the professional demands within the sector, the needs and expectations of the local community in Uist were not as aligned as had previously been the case. Teaching was focused 'elsewhere' and somewhat diluted and re-directed away from the local towards the regional accessible offering.

Logistically, student concerts needed to be based around Scotland as did the residentials, for which Uist proved an expensive and time consuming destination for the majority of students. Community understanding of the shift from Face to Face to Blended Learning with resulting student mobility was vague, and although effort was made to continue local engagement, tensions were arising that were not easily resolved. Moskal, Dziuban and Hartman (2013) highlight the success and popularity of Blended Learning from the perspective of students, faculty and institution, captured in their major survey of close to a million student's responses. Moskal, Dziuban and Hartman also point out the need to develop narratives and vocabulary to communicate the disruption to the status quo moving to this model from that of conventional Face to Face delivery. Interestingly, this study of the blended learning experience in the United States does not include the need to communicate this transformation with the surrounding local community, although it does suggest a bespoke approach.

As a case study location, Uist has many unique features, and the duality of both welcome and unwelcome past interaction with outside forces illustrates why this narrative could be challenging. This is succinctly alluded to by poet Aonghas MacNeacail from the neighbouring island of Skye in his short poem 'The gift of anger', where he asks whether the 'foreigner' has crushed or saved 'us' (MacNeacail, 2007). Nonetheless, the staff in Uist continue to raise the profile of activity associated with living on the islands, seeking and delivering a range of highly successful national and international partnerships to deliver performances. These include:

- Fèisean nan Gàidheal at Celtic Colours Festival in Canada,
- Festival performances in Spain in partnership with Albastur Cultural Exchange,
- Working with the Soundstorm and Wave's award-winning music education agency to deliver workshops to 4000 young people in England,

- Working with the Far Flung Collective and
- The development of a Gaelic song app in partnership with the Multi Media Unit of the Comhairle nan Eilean Siar (Western Isles Council).

The work of this department has been award-winning, demonstrated by the granting of honours such as the 'MG Alba Scots Trad Music Tutor of the Year', University of the Highlands and Islands Student Association 'Best Video Conference Lecturer' and as finalists in the Herald Higher Education Awards for Community Engagement (Denholm, 2017). More specifically and pointedly, in 2018 the production by music staff of an alumni CD (*Nead nan Ceòladair—The Musicians' Nest*), the brokering of a Uist Music Scholarship offering financial incentive to study in Uist and the production of eight short films—'Study destination Uist'—supplemented ongoing media presence, advertising the courses and Uist as a base.[1]

Yet, even in these actions there are tensions. The films celebrate and promote Uist as a destination for study, while the CD celebrates the 'success' measured by national and international performance success which in itself requires 'leaving' the island.

Why Has This Narrative Not Been Successfully Conveyed at the Community Level?

From the community's perspective, the period from 2001 to 2011 was a comfortable validation of musical and linguistic heritage, formally captured in an academic setting and embedded in community life. From 2004 to 2013, there was steady growth in student numbers and musical

[1] Credits: Production: Trixpix Media and co-produced by Anna-Wendy Stevenson.
LCC Music Courses: 'Do Something Different'—Study Destination Uist.

Film 1 https://vimeo.com/251624866
Film 2 https://vimeo.com/250104259
Film 3 https://vimeo.com/251551234
Film 4 https://vimeo.com/user9802221
Film 5 https://vimeo.com/250109210
Film 6 https://vimeo.com/251552790
Film 7 https://vimeo.com/250111698
Film 8 https://vimeo.com/251554076

proficiency with increased media attention and high-profile, well-attended performances. The failure of the Creagorry Music Hub as a solution was never openly explained to the community and the subsequent period from 2013 to 2019 will have appeared, on the surface, to be a slow decline of musical and linguistic life. However, it was in fact a period of intense rapid innovation that produced success at a regional and national scale, while developing an impressive portfolio of curricular and academic capacity building from the expanding music staff. This development also generated significant student numbers and associated finance for the college and university.

This has occurred in a context in which there has been a marked shift in applying market forces on University Governance in the UK since 2012, when the funding of study was transformed along with much higher fees. In the English context, the National Audit Office explicitly look to 'market mechanisms' to deliver Government objectives for the HE sector "in particular relying more on student choice and provider competition to improve quality, and value for money." (National Audit Office, 2017). McGettigan (2015), points out that these changes have also meant associated reductions in funding received by Universities from Government. In *'Dismantling the Curriculum'*, Hall and Smyth pose the question as to whose benefit higher education is ultimately for, and questions how this sits with the requirement for the generation of 'surplus value' through 'marketisation and financialisation' (Hall & Smyth, 2016: 2). The *'Competitiveness of Higher Education in Scotland'* review (Scottish Government, 2004) asked whether this sector was ready for global competition in the marketplace, and by 2013 Universities Scotland in *'Our values and our value added'* pledged global competitiveness as well as maximising accessibility (Universities Scotland, 2013).

There is an increasing onus on universities in receipt of public funding to be seen to be in public ownership by wider communities (Murray, McAlpine, Pollock, & Ramsay, 2013). These 'generational institutions' are of regional significance and influence, and while it was noted in a Government Report (Scottish Government, 2012) that there was no template for such community involvement in governance (Section 6.3), Universities Scotland's submission to Government in 2017 *'Going for growth'* pledged to support the transformation of these same local and wider host communities (Universities Scotland, 2017).

Blended Learning with its distributed communities of practice can be financially 'lean' but seem 'mean' to locations wishing the student cohort to themselves. UHI's C21C demands that networked degrees such as BAAM share this human resource, and attempts by any one college, even a lead partner, to corner and commandeer these numbers through marketing can be resisted and discouraged at the institutional level. Equal access has to be seen to be offered across the network. In the case of Uist as a marketed student location, even this was shrouded in complication as the college (LCC) is named after a location in another island and the headquarters also reside elsewhere.

Ideas about locality are important to learning practices more broadly, but to the workings of communities of practice more specifically. This in turn has impacts on the choices made by students in more remote regions. The accessibility and mobility offered to those choosing to study with the University of the Highlands and Islands allows young islanders to study in their home location; however, not all young people wish to elect this route and, of course, 'going to college' is part of a wider learning experience. A study by Highlands and Islands Enterprise (HIE) in 2015 canvassing the attitudes and aspirations of young islanders (aged 15–30) found divided viewpoints on their desired location in which to live and study. When asked, a combined percentage of 45% of young people in Uist and the neighbouring island Barra described themselves as 'committed stayers' while only 31% were 'committed leavers'. Reluctant leavers, those potentially encouraged to stay given favourable conditions, represented only 19% of those surveyed. The same survey reports nearly 75% feel included within their community, but only 40% feel it is 'ok to be different', lower than the HIE average at 51%. Clearly, intervention is necessary if regional areas are to retain young people who wish to continue their studies (Highlands and Islands Enterprise, 2015).

An Opportunity for Resolution?

Uist has pressing concerns with demographic sustainability, economic viability and fraught efforts to preserve the Gaelic language, not to mention culture (Fig. 11.2). The narrative of academic development has been

Fig. 11.2 Musicians' Nest

complicated, continually evolving and full of tensions and contradictions. As far back as 1998, the local newspapers in the area reported that the Uist Building Preservation Trust were hoping to secure funding to convert a disused school building in South Uist into a 'Gaelic language, music and cultural centre with recording studio' (Buildings at Risk Register for Scotland, 2015). This proposal was intimately tied to the success of the newly formed Ceòlas festival. The subsequent 'Uist 2000' development project linked the economic regeneration of the area and Gaelic language preservation to the need for such a centre, but these plans did not bear fruit. Nevertheless, the Ceòlas festival has continued to grow and consolidate, and with the success of the college music courses, a formal partnership was established between Ceòlas Uibhist and LCC in 2015.

The possibilities offered by teaching and learning centres and institutes in regional and remote locations can be found elsewhere. For example, the University of Tromsø in northern remote rural Norway is an example of positioning an educational institution in such areas in order to help address population decrease (Gaski, Halvorsen, Aaraas, & Aasland, 2017). The Ole Bull Akadamiet, also in Norway, where university students are provided with access to an immersion course in traditional folk music and dance, is an example of investment at the university level in the national cultural heritage. Ceòlas has found inspiration from the latter and other models as innovative solutions to fill a new centre and provide a portfolio of activity required to ensure financial sustainability. This direct working link between community and college is pioneering in the UK, although precedents can be found elsewhere.

A recent report into Leadership in the Arts (Stevenson, Plimmer, Kay, Amstell, & Pitceathly, 2018) found that places of periphery are often places of opportunity and creativity, a safe space for experimentation. The partnership with Ceòlas has enabled the development of a shared aspiration for purpose-built premises—a centre for teaching, performance, community activity, archive and recording facilities. Such a partnership supports government strategies such as the Gaelic Language (Scotland) Act, 2005, which had legally secured official language status (Scottish Parliament, 2005). As well as facilitating the roll out of infrastructure to promote the survival of this language, all public bodies in Scotland now had to prepare Gaelic language plans. The location for the proposed building in South Uist is in the strongest speaking Gaelic area per head of population. The language itself, however, is fragile. Numbers of speakers and the transmission to future generations have been stabilised after many years of decline, but much remain to be done if the language is to survive as a genuine community asset. These concerns are echoed in the Lews Castle College UHI 2018-23 Gaelic Language Plan, which pledges commitment to 'the objectives set out in the National Plan for Gaelic and has put in place the necessary structures and initiatives to ensure that Gaelic has a sustainable place at the centre of college life' (Lews Castle College UHI, 2018: 3). UHI also has a Gaelic Language Plan 2014–2018 and is 'intensely aware' of its obligation as the only university covering the Gaelic-speaking areas (University of the Highlands and Islands Gaelic Language Plan, 2019:2). So while the rapid curriculum innovation and development had taken up LCC music staff time, this had facilitated the forging of a partnership that was then able to draw up a business case and building designs and secure funding. This centre was to be named Cnoc Soilleir ('the bright hill' in Gaelic), and at the time of writing has secured funding for the first phase of this exciting project (Comhairle nan Eilean

Siar, 2019). Such a development offers the possibility of attracting students to the islands by providing up-to-date facilities, new additional curriculum and in turn a sense of student community that comes from increased numbers on the ground.

There are several initiatives and national policies that support the development of the Cnoc Soilleir project. Initiatives such as the Community Empowerment Bill (Scottish Parliament, 2015) and The Islands Act (Scottish Parliament, 2018) recognise the distinct and sometimes additional challenges these environments face, and are seen as a possible solutions to the increasingly fraught economic and social sustainability of remote rural communities.

Timing and Resilience: Final Thoughts

Resilience implies an ability to reorganise, learn and adapt to perturbations while maintaining structure and function (Resilience Alliance, 2015). Multiple speeds of change are one of the key features of this case study. Staff willingness and commitment to respond to rapid change and innovation in multiple and concurrent scenarios has offered resilience to the music courses delivered from Uist. Up until now, the technological changes that have impacted on this case study have occurred at a faster rate than the capacity to develop the local physical infrastructure.

Adaptability is a key indicator of success in employment for today's musician, no matter the genre or locality, as acknowledged in various industry and government reports (Musicians' Union, 2012, 'The Working Musician'; Australia Council for the Arts, 2017, 'Making Art Work: A summary and response by the Australia Council for the Arts'). The BAAM, being a practical degree delivered using a blend of technology and face to face delivery, while presenting considerable logistical challenges, has demonstrated through graduate success that it is a model fit for purpose and an attractive offering to students. The flexibility it affords has enabled students to practice professionally while studying, and the mobility has also enabled students to engage with and draw upon the inspiration and opportunity of being located in the Uists for periods of time during their studies. The face to face teaching option afforded by the HNC has provided a means to develop a deeper located understanding of a wider cultural and physical environment in the Gaelic-speaking heartland of Uist.

BAAM has provided unintended and unanticipated consequences to the local community from where the programme is led. However, through the financial income it has generated, it has also afforded the most exciting opportunity to take all that has been learned and achieved and develop anew—with the development of the Cnoc Soilleir Project being the opportunity to consolidate a Uist identity and home for the learning of music and facilitate a deeper integration into the community. This chapter has shown that adaptability is key for academic staff and that change is a constant in this fast-paced and technologically mediated world, and that working simultaneously on different platforms, engaging with different audiences and communities, is arguably necessary to ensure sustainability.

'Education either functions as an instrument which is used to facilitate integration of the younger generation into the logic of the present system and bring about conformity or it becomes the practice of freedom, the means by which men and women deal critically and creatively with reality and discover how to participate in the transformation of their world' (Freire, 2000: 34).

References

Australia Council for the Arts. (2017). Making art work: A summary and response by the Australia Council for the Arts. Retrieved March 23, 2019, from https://www.australiacouncil.gov.au/workspace/uploads/files/making-art-work-companion-repo-5a05105696225.pdf

Buildings at Risk Register for Scotland. (2015). Lochboisdale School and Schoolhouse, Lochboisdale. Historic Environment Scotland, Edinburgh. Retrieved March 23, 2019, from https://www.buildingsatrisk.org.uk/details/893343

Comhairle nan Eilean Siar. (2019). Comhairle Secures Government Funding for Major Investments. Retrieved September 20, 2019 from https://www.cne-siar.gov.uk/news/2019/march/comhairle-secures-governmentfunding-for-major-investments

Denholm, A. (2017). Finalists announced for the third annual Herald Higher Education Awards. Retrieved March 23, 2019, from https://www.heraldscotland.com/news/15323613.finalists-announced-for-the-third-annual-herald-higher-education-awards/

Freire, P. (2000). *Pedagogy of the oppressed.* New York: Continuum.

Garrison, D. R., Anderson, T., & Archer, W. (2000). Critical inquiry in a text-based environment: Computer conferencing in higher education model. *The Internet and Higher Education, 2*(2–3), 87–105.

Gaski, M., Halvorsen, P. A., Aaraas, I. J., & Aasland, O. G. (2017). Does the University of Tromsø – The Arctic University of Norway educate doctors to work in rural communities? *Tidsskr Nor Legeforen.* Retrieved March 23, 2019, from https://tidsskriftet.no/en/2017/08/originalartikkel/does-university-tromso-arctic-university-norway-educate-doctors-work-rural

Hall, R., & Smyth, K. (2016). Dismantling the curriculum in higher education. *Open Library of Humanities, 2*(1), e11.

Highlands and Islands Enterprise. (2015). Our next generation young people in outer Hebrides attitudes and aspirations. Retrieved March 23, 2019, from www.hie.co.uk/regional-information/economic-reports-and-research/archive/young-people-and-the-highlands-and-islands%2D%2Dattitudes-and-aspirations-research.html

Jobson, J. (2019). Eabhal's Nicky Kirk grateful for move to Scottish Island Uist. *The National Scot.* Retrieved March 23, 2019, from https://www.thenational.scot/news/17548486.eabhals-nicky-kirk-grateful-for-move-to-scottish-island-uist/

Lews Castle College UHI. (2018). 2018-23 Gaelic Language Plan. Retrieved March 23, 2019, from https://www.lews.uhi.ac.uk/t4-media/one-web/lews/about-us/publications/LCC-GLP-2018-23.pdf

Littlejohn, A., & Pegler, C. (2007). *Preparing for blended e-learning.* London: Routledge.

MacNeacail, A. (2007). *Laoidh an Donais Oig (Hymn to a Young Demon) poems.* Edinburgh: Polygon.

McGettigan, A. (2015). The treasury view of HE: Variable human capital investment. *Political Economy Research Centre*, Papers Series 6. Retrieved March 23, 2019, from http://www.perc.org.uk/project_posts/perc-paper-the-treasury-view-of-higher-education-by-andrew-mcgettigan/

Moskal, P., Dziuban, C., & Hartman, J. (2013). Blended learning: A dangerous idea? *Internet and Higher Education, 18*, 15–23.

Murray, C., McAlpine, R., Pollock, A., & Ramsay, A. (2013). The Democratic University A proposal for University Governance for the Common Weal. *The Jimmy Reid Foundation*. Retrieved March 23, 2019, from http://reidfoundation.org/wp-content/uploads/2013/11/Democratic-Universities.pdf

Musicians' Union. (2012). The working musician. Retrieved March 23, 2019, from https://www.musiciansunion.org.uk/Files/Reports/Industry/The-Working-Musician-report

National Audit Office. (2017). Department for Education, The Higher Education Market. Retrieved March 23, 2019, from https://www.nao.org.uk/report/the-higher-education-market/

Outer Hebrides Community Planning Partnership. (2013). Single outcome agreement 2013–2023. Retrieved March 23, 2019, from http://www.ohcpp.org.uk/index.php?option=com_content&view=article&id=121&Itemid=175

Outer Hebrides Community Planning Partnership Local Outcome Improvement Plan 2017–27. Retrieved September 20, 2019 from http://www.ohcpp.org.uk/images/OH_CPP_Local_Outcomes_Improvement_Plan_2017-27.pdf

Panciroli, E., Engstrand, S., Graham, E., & Clarke, S. (2015). 'Blended' learning at the University of the Highlands and Islands: A case study in self-awareness and policy making. *Journal of Perspectives in Applied Academic Practice, 3*(1), 39–48.

Reddy, K. (2018). Ùrnaighean nan Gàidheal: Text and context of Gaelic Prayer in South Uist, 1880–1960. MRes thesis Glasgow University. Retrieved March 23, 2019, from http://theses.gla.ac.uk/9144/1/2017reddymres.pdf

Resilience Alliance. (2015). Retrieved March 23, 2019, from https://www.resalliance.org/resilience

Scotland's Census. (2011). Gaelic report national records of Scotland. Retrieved March 23, 2019, from https://www.scotlandscensus.gov.uk/documents/analytical_reports/Report_part_1.pdf

Scottish Government. (2004). The competitiveness of Higher Education in Scotland *Phase 3, Higher Education Review*. Retrieved March 23, 2019, from https://www.webarchive.org.uk/wayback/archive/20180517222615/http://www.gov.scot/Publications/2004/03/19029/34063

Scottish Government. (2012). Report of the review of higher education governance in Scotland. Retrieved March 23, 2019, from https://www.webarchive.org.uk/wayback/archive/20170701172847/http://www.gov.scot/Publications/2012/02/3646/downloads

Scottish Parliament. (2005). Gaelic Language (Scotland) Act 2005. *The National Archives*. Retrieved March 23, 2019, from http://www.legislation.gov.uk/asp/2005/7/contents

Scottish Parliament. (2015). Community empowerment (Scotland) Act 2015. *The National Archives*. Retrieved March 23, 2019, from http://www.legislation.gov.uk/asp/2015/6

Scottish Parliament. (2018). Islands (Scotland) Act 2018. *The National Archives*. Retrieved March 23, 2019, from http://www.legislation.gov.uk/asp/2018/12/enacted

Simco, N., & Campbell, G. (2011). Developing blended learning in higher education – A case study of the University of the Highlands and Islands. *Practitioner Research in Higher Education, 5*(1), 3–8.

Stevenson, A., Plimmer, J., Kay, N., Amstell, S., & Pitceathly, J. (2018). *Leading from the edges*. Enquiry Report Extend Leadership Programme 2017–18. Retrieved March 23, 2019, from https://www.lews.uhi.ac.uk/t4-media/one-web/lews/about-us/find-us/benbecula-campus/FromTheEdges_Report_2018.pdf

Taylor, J. A., & Newton, D. (2013). Beyond blended learning: A case study of institutional change at an Australian Regional University. *Internet and Higher Education, 18*, 54–60.

Universities Scotland. (2013). Higher education in Scotland: Our values & our value added. Retrieved March 23, 2019, from https://www.universities-scotland.ac.uk/wp-content/uploads/2015/09/HE-Our-Values-and-our-Value-Added-FINAL.pdf

Universities Scotland. (2017). *Going for growth*. Retrieved March 23, 2019, from https://www.universitiesscotland.ac.uk/campaigns/going-for-growth/

University of the Highlands and Islands Gaelic Language Plan. (2019–2022). Retrieved September 20, 2019, from https://www.uhi.ac.uk/en/t4-media/one-web/university/about-uhi/facts-and-figures/publications/pdfs/gaelic-language-plan.pdf

12

The Dynamics of Place-Based Virtual Communities: Social Media in a Region in Transition

Naomi Smith and Susan Yell

Introduction

This chapter takes the Hazelwood mine fire as a case study to examine how a community used social media (specifically Facebook) during and after a disaster with health impacts, as the community underwent the complex process of recovery. It examines the strengths and pitfalls of social media use in relation to community empowerment and engagement. Further, it demonstrates how groups within this community used Facebook to work through their response to crisis, change and transition.

In early 2014, a fire in the Morwell open-cut coalmine adjacent to the Hazelwood power station in the Latrobe Valley, Victoria, burned for approximately 45 days, shrouding surrounding communities in smoke. As authorities struggled to put out the fire, the nearby communities became increasingly concerned about the perceived health risks of exposure to the smoke, particulate matter and gas emissions from the burning

N. Smith • S. Yell (✉)
Federation University Australia, Churchill, VIC, Australia
e-mail: n.smith@federation.edu.au; sue.yell@federation.edu.au

A. Campbell et al. (eds.), *Located Research*,
https://doi.org/10.1007/978-981-32-9694-7_12

coal. Members of the community noted increased ill health in family members, neighbours and friends (Teague, Catford, & Petering, 2014). This fire was unlike the more typical experience of an Australian bushfire. The Hazelwood mine fire, initially treated as a fire emergency, 'evolved into a chronic technological disaster … and a significant and lengthy environmental and health crisis' (Teague et al., 2014: 28). There was a strong community call for an investigation into the health impacts, culminating in a petition with over 21,000 signatures. In response, the Victorian Department of Health (now Department of Health and Human Services) made a decision to fund a ten-year longitudinal Health Study (the Hazelwood Health Study).

In the aftermath of the fire, there were no obvious or usual recovery activities to be undertaken by groups such as the Community Recovery Committee, a group more used to working with the aftermath of bush fires. As noted in the submission made by the Latrobe City Council to the second mine fire inquiry:

> the mine fire event was quite unlike other fire events in that no community assets were lost, no homes were lost and there was little damage to social and community infrastructure. … there has been no 'traditional' resilience work which has presented itself, such as the rebuilding of community halls, re-establishment of community walking tracks and paths. (Latrobe City Council, 2015)

As one local journalist interviewed for this study put it: '[it's] hard to pinpoint what you're rebuilding when you can't actually see what's been lost. It's like the difference between a broken leg and a mental illness'.

The Hazelwood mine fire of 2014 was a disaster that impacted upon an already-disadvantaged community, both economically and in terms of health. For example, a 1996 study into the burden of disease in Victoria showed that 'the Gippsland region, and in particular the Latrobe Valley, had a higher than state average of healthy years lost due to disease' (Teague et al., 2014: 250). In addition, the Latrobe Valley's history is one of economic decline after privatisation of the largest employer in the region, the State Electricity Commission. Subsequent restructuring of the coal mining and electricity generation industries has led to major job losses and population decline that 'transformed the Valley into the most disadvantaged location in regional Victoria by most social and economic indicators' (Tomaney & Somerville, in Duffy & Whyte, 2017: 423–424).

Community, Place and Social Media

The effects of the Internet on interpersonal relationships, and particularly our sense of community, have long been a topic of scholarly interest. Early scholarship by Miller and Slater (2000: 5) argues that it was most appropriate to 'treat the internet as continuous and embedded in other social spaces'. This stance is also reflected in the empirical work examining social network sites (SNS) and finds that SNS are primarily aimed at supporting the maintenance of pre-existing social networks (boyd & Ellison, 2007). boyd and Ellison define SNS as:

> web based services that allow individuals to (1) construct a public or semi-public profile within a bounded system, (2) articulate a list of other users with whom they share a connection, and (3) view and traverse their list of connections and those made by others within the system. (2007: 210)

What differentiates SNS from previous technologies of connection is not that they enable users to meet strangers, but that they articulate and make visible their offline relationships and networks. The structure of SNS primarily facilitates connections between people who also share an offline connection (boyd & Ellison, 2007; Ellison, Lampe, Steinfield, & Vitak, 2011). Unlike other online spaces, on SNS user's identities are frequently anchored in shared relationships, institutional affiliations and physical proximities in a way that mirror the offline aspects of people's lives (Hargittai & Hsieh, 2011).

Aspects of this are apparent in the Buy, Sell and Swap pages that proliferate on Facebook. Previous research has identified that these avenues for buying and selling outside of commercialisation consumption are important avenues to facilitate local face-to-face contact (Rufas & Hine, 2018). Rediscovering the ways in which social media can be used to foster place-based connection and engagement is an important contribution to discussions that often over-emphasise the isolating effects of technology (Dotson, 2017; Turkle, 2011). Such discussions often nostalgically invoke the supportive-place-based communities of previous generations as fundamentally broken by the advent of smart phones and social media (Hampton & Wellman, 2018). There are a number of different critiques of this argument, including that of 'networked individualism' where Rainie and Wellman (2012) argue that new technological developments like social media enabled multiple, partial and far flung social networks, in turn, from the dense, connected place-based networks of the past. However, more often social formations are a mix of online and offline connections, including those based around place.

Often place-based networks are realised through social media in times of disaster or crisis. The ongoing contaminated water crisis in Flint, Michigan, is one such example of the importance of social media for place-based activism (Moors, 2019). In this instance, and in the Hazelwood Mine Fire, social media activism preceded journalistic coverage (Moors, 2019). Social media is then not only a way of raising awareness and communicating during crisis but also a way for local communities to exercise some narrative control over their community and its identity as a place (Moors, 2019). As Willems (2019) argues, many of the debates regarding activism online have neglected to address the ways in which physical and online spaces intersect and are co-constitutive. Examining how communities utilise digital spaces alongside physical ones to engage in activism in times of crisis and transition is one way in which we can highlight how information and action circulates across online and face-to-face contexts (Willems, 2019).

In addition to allowing communities a level of control over their representation, empirical research demonstrates that the Internet increased levels of social capital (Wellman, Haase, Witte, & Hampton, 2001). Social capital is a necessary precursor to collective efficacy, that is, the ability for community to band together to solve local problems (Coleman, 1988). Collective efficacy then is particularly important for communities undergoing change and transition, particularly as a result of crisis.

A Case Study in Social Media, Crisis and Change

In Australia, social media such as Facebook and Twitter emerged as significant methods of disseminating information during the 2010–2011 floods in Queensland (Bird, Ling, & Haynes, 2012; EMV, 2014), and have been used by impacted communities in many disasters and crises since.

Social media's advantages include timely information exchange and promotion of connectedness (Taylor, Wells, Howell, & Raphael, 2012), qualities which are particularly important to their users during a crisis. Social media can fulfil an important role in enabling users to share their own stories when they do not feel heard by authorities or see their experiences of a disaster reflected in reporting by traditional media (Duffy & Yell, 2018; Murthy & Gross, 2017; Neubaum, Rösner, Rosenthal-von der Pütten, & Krämer, 2014; Tandoc & Takahashi, 2017). Social media offers 'a means of communal expression and ... a mode of self-therapy' (Murthy & Gross, 2017: 368). Such emotional support is vital when, through disaster, people are in crisis and their lives disrupted (Perez-Lugo, 2004).

Social media play a role in assisting communities to cope during a crisis and to recover after a crisis, in other words, in developing resilience. A recovery handbook developed by the AEMI (2011: 28) states: '[a] resilient community: predicts and anticipates disasters; absorbs, responds and recovers from the shock; improvises and innovates in response to disasters'. Thornley, Ball, Signal, Lawson-Te Aho & Rawson (2015: 23) confirm this trend and state that 'within the disaster recovery literature, resilience had been defined as the capacity of people, communities and societies to prevent, respond to and recover from the consequences of disaster'. Social media, including Facebook, can play an important role in this process. The pre-existing communication networks that operate in the community—for example, Facebook, texting, newsletters—are important contributors to a sense of social connectedness both pre and post disaster (Thornley et al., 2015).

Disasters can negatively impact existing social networks, social connectedness and sense of community. Poor social connectedness and a lack of social trust, personal agency and social resources exacerbate a community's capacity to recover; 'In short social capital is drained from impacted communities, and individuals become overwhelmed by the process' (Picou, 2010: 5).

However, not all disaster impacts are negative, with many communities experiencing a surge in community connectedness in the aftermath of such events (Thornley et al., 2015). Cline, Orom, Berry-Bobovski, Hernandez, Black, Schwartz, & Ruckdeschel (2010: 3) found that greater sociality is characterised and facilitated through 'a shared identity among victims, a sympathetic stance toward those most affected, and a shared understanding of the disaster's effects and victims' needs'. Nonetheless, the strength of pre-existing and emerging social networks and the range of social supports available to community members will affect the success of recovery efforts and underpin community resilience (Cline et al., 2010; Takazawa & Williams, 2011; Thornley et al., 2015).

Affected communities may gain resilience by 'replacing their helplessness with dignity, control as well as personal and collective responsibility' (Keim & Noji, 2010: 47). Greater access to information through both formal and informal networks on social media increases the flow of information, reducing vulnerability to disasters by strengthening community connections (Belblidia, 2010). Social media can provide a means for empowering communities to help themselves 'through provision of accurate, timely and relevant information and a mechanism to connect with others' (Taylor et al., 2012: 26). In this context, Twitter is most useful for the dissemination of frequent and timely updates, whereas Facebook enables users to connect and share their experiences throughout different phases of a disaster (Anikeeva, Steenkamp, & Arbon, 2015: 23–24).

There is a considerable body of research on the use of social media in disasters and emergencies, which it is beyond the scope of this chapter to review comprehensively. With the brevity of messages and use of searchable hashtags on Twitter, it presents a more accessible research option, meaning there has been greater analysis of Twitter use by citizens experiencing disasters than other forms of social media (Reuter & Spielhofer, 2017; Simon, Goldberg, & Adini, 2015). A quick search on database EbscoHost shows that there are almost double the number of articles on disasters or emergencies and Twitter, than on Facebook. However, there is a clear need to examine Facebook when researching how communities deal with crisis and change, especially given the uses and affordances of Facebook differ from those of Twitter. This is particularly apparent when we consider Facebook's user base, which it reported in the fourth quarter of 2018, was 2.32 billion monthly active users (Statista, 2019). Comparatively, Twitter has 326 million monthly active users (Omnicore Agency, 2019). Facebook has also taken a much more active role in disaster response, implementing a disaster response button where users can mark themselves as safe if they are in the area of a disaster. This has been deployed in response to natural disasters, such as bushfires, floods, landslides and earthquake, as well as crisis situations such as terrorist incidents and mass shootings. The disaster response button also functions as a public forum in which people can request and offer assistant. These affordances speak to Facebook's importance as a community hub during crisis.

Research Methods

The case study research for this chapter was conducted in the context of a larger study of the health impacts of the Hazelwood mine fire (the Hazelwood Health Study).[1] Within this larger study, a programme of work was carried out with a focus on the impact of the smoke event on community wellbeing. Along with more traditional social research methods (qualitative analysis of interviews and focus groups), this research incorporated social media research to enrich the data and to provide a snapshot of community members' experiences and responses to the crisis and its aftermath, on an immediate day-by-day basis.

The decision to focus on Facebook rather than other social media platforms (e.g. Twitter) was made in part because Facebook is the most popular social media platform in Australia with 15 million active monthly users (Civic Web Media, 2019). Facebook enables the formation of groups, and can be used by community members to interact with each

[1] The Hazelwood Health Study is funded by the Victorian Department of Health and Human Services and was set up in response to community concerns, in order to investigate potential health impacts resulting from the smoke from the fire (http://hazelwoodhealthstudy.org.au/).

other within a shared space, while Twitter is not strongly linked to specific communities, and can be used by those from other locations outside the community to talk about the event. In essence, while Twitter is used to talk about others, Facebook is used by people to talk among themselves. Also, as Twitter is designed for short messaging and has limited characters, it lends itself to broadcasting factual info and commentary, rather than building networks/relationships and strengthening communities. Twitter is not designed to connect with people you know, and is less focused on mapping offline social connections in to digital space. In addition, in the context of crisis communication, emergency managers have observed that 'Twitter is best suited for broadcasting frequent updates', whereas 'Facebook is well-suited to establishing online communities where users can share their own experiences, ask for help and provide assistance where needed' (Anikeeva et al., 2015: 23).

Mid-Level Analysis

The landscape of public Facebook pages and groups in the Latrobe Valley was gathered in two ways. Firstly, a keyword search was conducted for 'Latrobe Valley'. Facebook returned a mix of the most popular groups and pages in the areas, as well as results that were closest to the researchers' location (Churchill). These results include recent news articles and geotagged photos. From these results, we narrowed down the most popular groups and pages by number of members or likes. As this research was also interested in community transition after the mine fire, we also searched for relevant community groups that may have had some role during and after the mine fire. These results included Voices of the Valley and Reactivate Latrobe Valley. The results of this are discussed in greater detail below.

Micro-Level Analysis

For a finer-grained picture of how a community uses social media in times of crisis and change, it was necessary to focus in more depth on the dynamics and texture of social media activity. For the case study, three Latrobe Valley Facebook groups were chosen. Criteria for selecting these Facebook groups were that they were active during the mine fire event, were initiated by the community, were focused on the mine fire, and were used by members to discuss and react to the mine fire. These groups don't necessarily represent the views of the entire population of the Latrobe Valley, but they do provide a useful case study to show how place-based social networks are used during a period of crisis and transition.

Facebook posts from these groups were collected during the initial six-week period that the fire was active, as well as for an extended period after the crisis (from 2014–2017). In addition, interviews were conducted with four of the administrators of these Facebook groups. Qualitative analysis was used to identify prominent themes in the social media posts, and in the responses of the interviewees. Approval for the interviews was gained through Federation University's Human Ethics Committee (B15-067). Informed consent was gained from all interviewees.

A limitation of conducting this research was the time-consuming nature of data collection. In order to carry out thematic analysis of the content of posts, posts in a selected time period (weeks one to six of the crisis) were captured as screen shots in PDF format. These posts were then saved by date and post number, as full text records for qualitative analysis. In the period after the fire was declared safe, the research assistant used the Facebook search function to identify relevant posts. Posts were recorded and entered into a database, noting the date and the topic.

Semi-structured interviews were also conducted with administrators of the Facebook pages. Four social media administrators were interviewed, from three different Facebook groups, for their views on the role of Facebook during and after the mine fire. We used thematic analysis to interpret the data, a form of textual analysis which identifies prominent themes in texts (such as media texts, interview transcripts).

The Ecology of Facebook in the Latrobe Valley

Social media is often perceived as existing beyond place, connecting people from geographically disparate places. However, previous research demonstrates that the Internet and social media can be powerful tools of community formation in place (Wellman et al., 2001). This is also evident in the social media ecology of the Latrobe Valley. There are a variety of formal and informal place-based pages that exist to either foster communication between formal governance structures like the local council and the community, promote community activism or facilitate informal economic activity. Community groups that facilitate the exchange of goods in a 'buy, swap, sell' format are particularly popular in the Latrobe Valley. Three groups, the Gippsland Trading Post; Buy, Swap, Sell (Latrobe Valley, Gippsland, Churchill) and Buy/Sell/Swap Morwell, Moe, Traralgon, Warragul and Surrounding Areas each have thousands of members and tens of new posts a day.

Crime-focused Facebook groups and pages are also popular in the valley, and with Eye Watch, the official Victoria Police Facebook page for the Latrobe Valley having 23 k likes and the Facebook group Latrobe Valley Crime Watch having 2.5 k members.

Pages that provide information on the governance of Latrobe Valley are also well represented on Facebook. The Latrobe City Council has 9.5 k likes. There are also new governance structures emerging in response to the mine fire of 2014, such as the Latrobe Valley Health Assembly, funded by the Department of Health and Human Service. The Health Assembly is part of a multi-tiered governance structure, which includes the Health Assembly, the Health Innovation Zone and an evaluative arm, and these are all spread across multiple Facebook pages. There are also Facebook pages for the various community research projects being conducted in the valley including Reactivate Latrobe Valley (3.5 k likes).

There are also related grass-roots community groups, initially prompted by the mine fire, that engage with ongoing issues associated with the transition of Latrobe Valley from a coal-based economy. Groups like Voices of the Valley (1997 k likes), and the associated Voices of the Valley News (332 likes) are a way for the Latrobe Valley community to share and discuss hyper-local concerns, outside of one-to-many distribution forms such as the Latrobe Valley Express (a bi-weekly newspaper, whose Facebook page has 3.1 k likes).

Social Media Use During and After the Hazelwood Mine Fire

For a fine-grained analysis, three Facebook groups were chosen: The Air That We Breathe, Occupy Latrobe Valley and Voices of the Valley. The Air That We Breathe Facebook page was started by a Morwell resident and former *Big Brother* contestant with the specific purpose of generating discussion and media interest regarding the impact of the mine fire on the Latrobe Valley community. It was active from Day 10 of the fire. The 'About' page reads: 'A page to create awareness & action for the people of Latrobe Valley suffering from very poor air quality due to the Hazelwood Mine Fire'. It grew from 100 followers to 3500 within a week, and currently has 3.1 k likes. Activity on this site decreased significantly after the 45-day fire period.

The Occupy Latrobe Valley Facebook page was started by an anonymous member of the Latrobe Valley community as part of the broader Occupy Movement, a grass-roots pro-democracy movement which protests against social and economic inequality. The page was not originally mine fire specific, but became a site of activity for residents wanting to communicate during the mine fire period. It was active from Day 12 of the fire, and the number of followers grew to about 5000 during the fire. The page is now no longer active.

The Voices of the Valley Facebook page was formed on 8 March 2014 out of the Disaster in the Latrobe Valley Facebook group to raise awareness about the impact of the mine fire. It was active from Day 28 of the fire and is ongoing. It is a public group with current membership of over 2000. It played an important role in collecting and collating information, surveys and data on death rates in the Latrobe Valley during and after the fire, to contribute to the Hazelwood Mine Fire Inquiries, and continues to advocate for initiatives to improve the health, and social and economic wellbeing, of the community after the smoke event.

The social media analysis carried out for our case study, along with supplementary interviews with administrators of these Facebook pages and other community members, showed that there were complex dynamics at work in the ways this community used social media during and after the crisis.

Initially, in response to the crisis, members of the local community turned to social media as an alternative space in which to share information, tell their stories and organise for the purpose of activism. Community members organised public rallies, created a social media presence and network and began to demand answers to their questions and concerns.

A key concern was the poor communication from authorities, and the lack of advice and conflicting advice received regarding the health impacts of the smoke permeating Morwell and surrounding communities for such a lengthy period. As one interviewee noted:

> No one knew … where to go, … what help was available. … What we were getting from the media and other services seemed to contradict each other. (social media administrator)

Analysis of social media posts and interviews with social media practitioners showed that an issue causing significant distress within the community was the lack of accurate, accessible and timely information from authorities about the crisis.

Conflicting advice, as well as what was interpreted as silence from authorities on important matters, fuelled suspicion and lack of trust due to questioning of the accuracy of information provided through official channels. Rather than relying on the mainstream media, or on government authorities involved directly in emergency management for relevant information, social media users turned to other online sources as well as eyewitness accounts from members of the local community.

However, the social media administrators of the Facebook groups in this study highlighted the risks in providing information from such sources because of the difficulty in verifying its accuracy. As explained by two active members of one Facebook group who were receiving posts regarding conditions inside the mine

Social media administrator 1: It becomes hard to know what to share and how do you verify it's true?

Social media administrator 2: And protecting the people that were giving this information as well; the last thing you want to do is put them at risk.

While social media enabled community voices to raise important concerns shared by many, some questioned the authority of a few community members to speak on behalf of the range and diversity of people in Morwell and the wider Latrobe Valley. Disagreements occurred over who could speak for the community, and whose experiences were 'real', 'true' and representative. Despite their success in drawing public attention to the crisis, the organisers of the three Facebook groups in this case study were not necessarily seen as representing the community's views. During and after the mine fire, some community members questioned whether those voices emerging strongly via social media could speak for the community. As one interviewee explained

> I think that groups like Voices of the Valley have gained a real credibility with government and have almost become some sort of de facto spokesperson for the Latrobe Valley community. I think the media certainly has a responsibility to take there. I think and it comes back to that point I made before about in the absence of being able to have other people to speak to you're constantly going to the same people—their profile inevitably gets lifted. (media professional)

It is not simply that there was an apparent few who seemed to have greater exposure on social media, but that this highlighted the divisions within the local community. In this instance, the crisis of the Hazelwood Mine Fire served to highlight the fractures and tensions in Morwell community. Place-based identities are like other forms of identity, multiple and contingent. While we are sensitive to this across other forms of identity, social media illustrates the contingent nature of place more clearly. Our interviewees felt that social media made these divisions more obvious but didn't create them.

> And on social media I've noticed a big divide of just local community, between people, there's a huge divide … it also caused in a lot of ways—oh not the page didn't cause it but … it became apparent in the community there were people who thought we should have just sat back, shut up and dealt with it. We got … blamed for the downfall of Morwell … that people don't want to come here anymore, it was all our fault. (social media administrator)

A consequence of using social media as a platform to highlight inadequacies in the emergency response and recovery is that those speaking out may be seen as exacerbating the difficulties the community is experiencing, despite the fact that their efforts may indeed lead to necessary actions to address shortcomings. Conflict and disagreements arose over who was genuinely affected by the event, and whether or not it was legitimate to complain and to criticise the emergency response by authorities. Some in the community and on these sites saw this as 'whingeing':

> And the criteria to get funding, so then anybody that got funding to leave town, oh yep meh, meh, meh they got bagged out and anyone who couldn't get it was whingeing and complaining and bagging. It was just … it was like them and us and none of us could be in it together, they [the authorities] created these divisions. It was social and geographical. (social media administrator)

Thus, social media inadvertently contributed to divisions already present within the community.

Nevertheless, social media took on an important function in empowering the community to self-organise in response to the crisis. Social media was an important avenue for members of the community to question and challenge the poor response from government and other authorities. Some of those involved in these social media groups drew attention to their use to fulfil a 'watchdog' function, holding government, private companies, and other organisations to account. As one interviewee observed:

> Unfortunately, the people, the watchdogs that are supposed to do it have failed, so the communities had to … take it back and do it themselves. (social media administrator)

The relative intimacy of social media meant that community members could feel more comfortable speaking in that forum, when some wouldn't go to the mainstream media with a problem or issue. As a result, new voices have emerged within the community, as some community members affected by the Hazelwood mine fire have become better at speaking out, and have discovered they have a voice that is listened to. This illustrates the importance of social media in exercising narrative control and narrative agency, which can then be harnessed to call for further, formal action. This was particularly apparent with the role played by the Voices of the Valley, where, as one journalist explained, this became a strong avenue through which calls for government and industry responsibility and culpability were made:

> Is this community happy with the recommendations [of the Hazelwood Mine Fire Inquiry]; are they satisfied that they address the terms of reference; that they go far enough and I think we probably saw that not necessarily from I think the media trying to do that but from a very well organised—I don't want to call them activist group but the Voices of the Valley in their pushing for some sort of avenue to address what they didn't see—what they say as I guess failings of the first inquiry and that it didn't address the alleged deaths and the increase in deaths. (media professional)

Of the three social media groups in this study, VOTV was the only one which continued to speak on the mine fire's impacts and the recovery process after the crisis was officially over. As well as organising public protests, facilitating public meetings, and distributing information, VOTV conducted a health survey completed by 650 community members, which was then submitted to the first Hazelwood Mine Fire Inquiry (2014).[2]

The formation of VOTV demonstrates the importance of social media in the formation of collective efficacy, self-help and agency, as well as capacities for resilience inherent in communities under stress (Bach,

[2] In its report, the Hazelwood Board of Inquiry commended the organisers of VOTV, stating, 'the Board commends those responsible for the establishment of Voices of the Valley and the actions of this group in disseminating important information to the local community and advocating on their behalf during the emergency' (*Hazelwood Mine Fire Inquiry Report*, 2014: 403).

Kaufman, & Dahns, 2015). As Wattchow (2016: 22) pointed out, the work of groups like VOTV

> wasn't just about the fire anymore. This was about what the local economy needed to thrive after an industrial disaster, after privatisation, after coal. This was, as President of Voices of the Valley, Wendy Farmer says, about jobs and hope. Two things the Valley had been in want of for a long time.

VOTV's effectiveness in grass-roots activism relied to a large degree on their use of Facebook. This intimate and rapid form of communication helped reactivate social relationships and constitute a base to support a range of community projects. The emergence of informal community groups such as VOTV provided a space for considering 'the way forward'. In doing so, they can promote a community's disaster resilience, defined as the ability to 'bounce forward' after a disaster (Dufty, 2012). The activities undertaken by VOTV also demonstrate the important role a community group like this can play not only in advocating on behalf of others but also in the potential to partner with government authorities to support effective crisis communication with the community. The work of VOTV has shifted to a focus on broader community issues, most notably with transition plans to diversify the Latrobe Valley's economy beyond brown coal mining and power generation (Wattchow, 2016).

Conclusion

Understanding how social media, community and place intersect is valuable from both practical and scholarly perspective. The implication for community in the age of social media is still an open question. There are two schools of thought—either community is experienced in the vein of 'networked individualism' which sees community expressed through non-local relationships, facilitated and mediated by the internet, mobile phones and social media (Rainie & Wellman, 2012). In this argument, place is largely irrelevant to community and a focus on place-based community runs the risk of reinforcing the nostalgia around imagined communities past (Gruzd, Jacobson, Wellman, & Mai, 2016).

Alternatively, focusing wholly on data and web-based methods for understanding communities can be a matter of form over substance; mapping connections and edges via community detection algorithms can run the risk of mistaking connectivity for community (Gruzd et al., 2016). Furthermore, online-focused approaches neglected the ways in which physical places still shape daily life and are reflected in our online behaviours. While the Internet and social media can create communities of interest across spaces and time, the rhythms of daily life are intimately connected to the places we live in.

On a practical level, social media can give both researchers and policy makers insight into community sentiment and concerns. Social media can also help highlight the stories that communities wish to tell about themselves, outside of mainstream or traditional media outputs. While examining social media does not take the place of long-established methods such as ethnography, it does make community dynamics visible to researchers in new ways. Community fractures and tensions become hyper-visible and observable to researchers, and the community itself in new ways. Continuing to engage in a qualitative examining of Facebook pages remains important in the age of big data.

Big data is most often gathered by querying a social media application programme interface (API). However, access to APIs by researchers is contingent on the goodwill of social media companies. Most recently, Facebook has severely restricted access to its API and it is now no longer a viable way for researchers to collect data on Facebook use. Qualitative research, particularly research regarding place and crisis, is more effective if researchers are invited into the online spaces communities are using, guided by a rich understanding of how they are connected to and implicated in offline activities. It is only through sustained and multiple avenues of engagement that we can begin to assess how social media platforms affect community practice.

References

Anikeeva, O., Steenkamp, M., & Arbon, P. (2015). The future of social media use during emergencies in Australia: Insights from the 2014 Australian and New Zealand disaster and emergency management conference social media workshop. *Australian Journal of Emergency Management, 30*(1), 22–26.

Australian Emergency Management Institute. (2011). *Community recovery* (handbook 2) Australian Emergency Management Institute, Part of the Commonwealth Attorney-General's Department, Australia. Retrieved March 23, 2019, from www.aidr.org.au/media/1488/handbook-2-community-recovery.pdf

Bach, R., Kaufman, D., & Dahns, F. (2015). What works to support community resilience? In R. Bach (Ed.), *Strategies for supporting community resilience: Multinational experiences* (pp. 309–340). Crismart: The Swedish Defence University, Multinational Resilience Policy Group.

Belblidia, M. (2010). Building community resilience through social networking sites: Using online social networks for emergency management. *International Journal of Information Systems for Crisis Response Management, 2*(1), 24–36.

Bird, D., Ling, M., & Haynes, K. (2012). Flooding Facebook – The use of social media during the Queensland and Victorian floods. *Australian Journal of Emergency Management, 27*(1), 27–33.

boyd, d. m., & Ellison, N. B. (2007). Social network sites: Definition, history, and scholarship. *Journal of Computer-Mediated Communication, 13*(1), 210–230.

Civic Web Media. (2019). Retrieved March 23, 2019, from https://www.civicwebmedia.com.au/australias-most-popular-social-media-sites-2019/

Cline, R., Orom, H., Berry-Bobovski, L., Hernandez, T., Black, C., Schwartz, A., et al. (2010). Community-level social support responses in a slow-motion technological disaster: The case of Libby, Montana. *American Journal Community Psychology, 46*, 1–18.

Coleman, J. S. (1988). Social Capital in the creation of human capital. *American Journal of Sociology, 94*(Suppl), S95–S120.

Dotson, T. (2017). *Technically together: Reconsidering community in a networked world*. Cambridge, MA: MIT Press.

Duffy, M., & Whyte, S. (2017). The Latrobe Valley: The politics of loss and Hope in a region of transition. *Australian Journal of Regional Studies, 23*(3), 421–446.

Duffy, M., & Yell, S. (2018). Community empowerment and trust: Social media use during the Hazelwood mine fire. *Australian Journal of Emergency Management, 33*(2), 66–70.

Dufty, N. (2012). Using social media to build community disaster resilience. *The Australian Journal of Emergency Management, 27*(1), 40–45.

Ellison, N. B., Lampe, C., Steinfield, C., & Vitak, J. (2011). With a little help from my friends: How social network sites affect social capital processes. In

Z. Papacharissi (Ed.), *A networked self: Identity, community, and culture on social network sites*. New York: Routledge.
Emergency Management Victoria. (2014). National review of warnings and information: Literature review. EMV/IPSOS. Retrieved May 23, 2016, from http://fire-com-live-wp.s3.amazonaws.com/wp-content/uploads/National-Review-of-Warnings-Information-Literature-Review-FINAL-Oct2014.pdf
Gruzd, A., Jacobson, J., Wellman, B., & Mai, P. (2016). Understanding communities in an age of social media: The good, the bad, and the complicated. *Information, Communication & Society, 19*(9), 1187–1193. Retrieved March 23, 2019, from. https://doi.org/10.1080/1369118X.2016.1187195
Hampton, K. N., & Wellman, B. (2018). Lost and saved … again: The Moral panic about the loss of community takes hold of social media. *Contemporary Sociology, 47*(6), 643–651. https://doi.org/10.1177/0094306118805415
Hargittai, E., & Hsieh, Y. P. (2011). From dabblers to omnivores: A typology of social network site usage. In *A networked self: Identity, community, and culture on social network sites*. New York: Routledge.
Keim, M. E., & Noji, E. (2010). Emergent use of social media: A new age of opportunity for disaster resilience. *American Journal of Disaster Medicine, 6*(1), 47–54.
Latrobe City Council. (2015). Submission to the Hazelwood mine fire inquiry. 10 August 2015. Retrieved September 18, 2019, from http://hazelwoodinquiry.vic.gov.au/wpcontent/uploads/2015/08/Latrobe-City-Council_Submission.pdf
Miller, D., & Slater, D. (2000). *The internet: An ethnographic approach*. London: Bloomsbury Academic.
Moors, M. R. (2019). What is Flint? Place, storytelling, and social media narrative reclamation during the Flint water crisis. *Information, Communication & Society, 0*(0), 1–15.
Murthy, D., & Gross, A. (2017). Social media processes in disasters: Implications of emergent technology use. *Social Science Research, 63*, 356–370.
Neubaum, G., Rösner, L., Rosenthal-von der Pütten, A. M., & Krämer, N. C. (2014). Psychosocial functions of social media usage in a disaster situation: A multi-methodological approach. *Computers in Human Behavior, 34*, 28–38.
Omnicore Agency. (2019). Retrieved March 23, 2019, from https://www.omnicoreagency.com/twitter-statistics/
Perez-Lugo, M. (2004). Media uses in disaster situations; a new focus on the impact phase. *Sociological Inquiry, 74*(2), 210–225.
Picou, J. (2010). *The BP catastrophe and sociological practice: Mitigating community impacts through peer-listening training*. St Louis: AACS Presidential Address.

Rainie, L., & Wellman, B. (2012). *Networked: The new social operating system*. Cambridge, MA: MIT Press.

Reuter, C., & Spielhofer, T. (2017). Towards social resilience: A quantitative and qualitative survey on citizens' perceptions of social media in emergencies in Europe. *Technology Forecasting & Social Change, 121*, 168–180.

Rufas, A., & Hine, C. (2018). Everyday connections between online and offline: Imagining others and constructing community through local online initiatives. *New Media & Society, 20*(10), 3879–3897.

Simon, T., Goldberg, A., & Adini, B. (2015). Socializing in emergencies – A review of the use of social media in emergency situations. *International Journal of Information Management, 35*, 609–619.

Statista. (2019). Retrieved March 23, 2019, from https://www.statista.com/statistics/264810/number-of-monthly-active-facebook-users-worldwide/

Takazawa, A., & Williams, K. (2011). Communities in disasters: Helpless or helping? *Perspectives on Global Development and Technology, 10*, 429–440.

Tandoc, E., & Takahashi, B. (2017). Log in if you survived: Collective coping on social media in the aftermath of Typhoon Haiyan in the Philippines. *New Media & Society, 19*(11), 1778–1793.

Taylor, M., Wells, G., Howell, G., & Raphael, B. (2012). The role of social media as psychological first aid as a support to community resilience building: A Facebook study from 'Cyclone Yasi update'. *Australian Journal of Emergency Management, 27*(1), 20–26.

Teague, B., Catford, J., & Petering, S. (2014). *Hazelwood mine fire inquiry report*. Vol. 1. Melbourne: Vic Government Printers. Retrieved September 2017, from http://report.hazelwoodinquiry.vic.gov.au/

Thornley, L., Ball, J., Signal, L., Lawson-Te Aho, K., & Rawson, E. (2015). Building community resilience: Learning from the Canterbury earthquakes. *Kōtuitui: New Zealand Journal of Social Sciences, 10*(1), 23–35.

Turkle, S. (2011). *Alone together: Why we expect more from technology and less from each other*. New York: Basic Books.

Wattchow, K. (2016). The future of the Latrobe Valley: Community leadership on just transition. *Chain Reaction, 128*, 22–23.

Wellman, B., Haase, A. Q., Witte, J., & Hampton, K. (2001). Does the internet increase, decrease, or supplement social capital?: Social networks, participation, and community commitment. *American Behavioral Scientist, 45*(3), 436–455.

Willems, W. (2019). The politics of things': Digital media, urban space, and the materiality of publics. *Media, Culture & Society*. Retrieved March 23, 2019, from https://doi.org/10.1177/0163443719831594

13

Collective Hope and Action in a Time of Transition: Kitchen Table Conversations with Gippsland Sustainability Change Agents

Monica Green and Sherie McClam

Introduction: Collaborative Sustainability Research in the Making

I don't do anything else like this; adults who are on the same page it's fantastic and achieving something too is hugely important. I just quietly sit here and think, 'oh yay' when I hear what each of you are doing, it's just fantastic. There's a multiplying effect. Sometimes I think, 'oh we're such a small group we should try and grow' but then listening to everyone, I realize who knows where it all ends. The diagram of a tree. The roots, the branches, where does it end? (Joan, RCE Gippsland committee member)

The research in this chapter emerges from an international sustainability research collective we (Monica and Sherie) belong to. Sherie's contact

M. Green (✉)
Federation University Australia, Churchill, VIC, Australia
e-mail: monica.green@federation.edu.au

S. McClam
Manhattanville College, Harrison, NY, USA
e-mail: sherie.mcclam@mville.edu

A. Campbell et al. (eds.), *Located Research*,
https://doi.org/10.1007/978-981-32-9694-7_13

with Monica materialised as part of her US-based Education for Sustainability (EfS) sabbatical plans that united a group of Australian and US EfS educators and academics to examine the particular conditions affecting EfS educators' capacity for achieving sustainability impact and outcomes. Initial email exchanges between Sherie and Monica revealed significant links between our respective university EfS teacher educator work and shared research interests in the impact of EfS practice in educational and community settings. Sherie discovered Monica on the RCE Gippsland (Regional Centre of Expertise in Education for Sustainable Development) website, an international site hosting over 130 RCE's affiliated with the United Nations charter that mobilises Education for Sustainable Development (ESD) to the local communities where they exist. In keeping with its agenda to unite global sustainability stakeholders, Sherie's contact with Monica is living testament to the UN's intention to create and advance ESD and connect RCE networks globally.

RCE Gippsland is a local sustainability network represented by government organisations, schools, local shires, industry and other sustainability-based groups. RCE Gippsland translates the global objectives of the United Nations Decade of Sustainable Development (DESD, 2005–2014), and its General Assembly new global development agenda, *Transforming our world: The 2030 Agenda for Sustainable Development*, across Gippsland's six municipalities by focusing on climate change and sustainable living. At the global level, UNESCO recognises the importance of local communities in urban, rural and indigenous settings as key drivers of sustainability that are central to the success of sustainability initiatives at the local level (Tichaona & Lupele, 2018). It is in this vein that RCE Gippsland mobilises EfS through its annual *Sustainable School Expo* held across the municipalities to promote and support school-based sustainability. Broadly speaking, the Expos provoke and encourage children's and teacher's sustainability thinking and action.

At a governance level, RCE Gippsland is auspiced under the Gippsland Climate Change Network (GCCN), an incorporated not-for-profit network of 50-plus member organisations across government departments and agencies, private businesses, community groups and other organisations in regional Gippsland. Its mandate is to provide a voice for Gippsland on climate change issues by disseminating information, consultation and facilitation that enables climate change action. RCE and GCCN committees have a diverse representation of community members—EfS educators and practitioners, volunteers and other professionals who are connected to extensive networks across the region and who work towards collective goals for a sustainable future (Braithwaite, 2004a).

In returning to the wider sustainability collective from which our collaboration emerged, Sherie travelled to Australia to spend time with each academic member over a nine-week period. As part of her three-day trip to Gippsland, she and Monica conducted 'kitchen table conversations' with RCE and GCCN committee members that explored sustainability and climate change practice in the region. These conversations form a critical backdrop to the study, emphasising how regional change agents instigate change and transformation.

Gippsland: Regional History, Change and Transition

The region of Gippsland makes up approximately 18% of the Victorian land form. Geographically and demographically complex, it extends easterly from the fringes of the capital Melbourne towards to the New South Wales border and north to the high plains that extend across the two south-eastern states. Gippsland has a diverse geo-demographic make-up boundaried by six local government areas (LGAs) each facing their own set of unique environmental, social, cultural and economic challenges, including rising sea levels, urban development, high unemployment, declining forests and other natural resources, and other climate change considerations (ClimateWorks Australia, 2011). Geographically central is Latrobe Valley, a bioregion characterised by three major brown coal-fired power stations and neighbouring open-cut mines that generate electricity to the region, Melbourne and interstate. Like many industrial regions worldwide, Latrobe Valley experiences intergenerational unemployment, poverty, and hardship, making it a community of disadvantage facing a diverse set of socio-cultural challenges, high levels of social alienation and disengagement, elevated incidences of mental and physical health problems, and developmentally vulnerable children (Needs Assessment Snapshot: Latrobe Local Government Area, 2016).

At the time of commencing our collaboration, Hazelwood, the state's biggest and highest polluting brown coal-fired station in Latrobe Valley, had closed, shedding 450 jobs. Comparable to other regions which have faced post-coal to low-carbon economy transitions (see Tomaney & Somerville, 2010), the region is slowly coming to terms with its climate change vulnerability, grappling with the challenge of what it wants to look like in a climate-adapted world (see Stanley et al., 2013). As Plowright, Green, and Johnson (2019) articulate, these seismic events and shifts are writing new layers of trauma onto the people of Gippsland transitioning into an unknown future. It is within this discourse of transition that RCE and GCCN committees address a range of ecological and economic threats, to create a different future (Braithwaite, 2004b) by taking an active role in a more sustainable, just and carbon constrained future (ClimateWorks Australia, 2011).

To understand the nature and impact of the committees' change agency, we applied the theoretical framework of 'collective hope' (Braithwaite, 2004a, 2004b; Courville & Piper, 2004; Drahos, 2004; McGeer, 2004) as discussed.

Collective Hope and Communities of Practice as Theoretical Tools

In preparation for our first kitchen table conversation, Monica shared her perceptions of the respective committees' impact in driving regional social change. As she did, Sherie found herself reflecting on a previous research project in which she and her colleagues sought theoretical frameworks to better understand the ways a community of learners produced and nurtured change agency and collective action (McClam et al., 2015). The conversation led Sherie to consider if the agency experienced through and sustained by the two Gippsland committees might be the result of or facilitated by a theoretical frame Valarie Braithwaite calls *collective hope.*

Scholars from across social scientific disciplines maintain hope is inextricably linked to agency. Whether conceived as a unifying and grounding force, a crucial component of, or a cognitive pathway to human agency, most contend you cannot have one without the other (McGeer, 2004; Lueck, 2007; Snyder, Lopez, Shorey, Rand, & Feldman, 2003). Using a metaphor that is particularly apt for our EfS work, Courville and Piper argue that 'hope can be seen as a renewable resource for social change' (2004: 57). Collective hope as defined by Braithwaite is a 'shared desire for a better society articulated through a broad set of agreed-upon goals and principles, developed and elaborated through inclusive dialogue' (2004b: 146). Collective hope she argues is a productive hope that enables us to create different futures for ourselves. Citing Courville and Piper, she suggests that collective hope

> acquires its object through shared feelings of injustice, through ideas for change and through a sense of empowerment as the feelings and ideas are shared within a community. But most important is action. Once action is seen as being the companion of hope, social movements gain momentum. (2004: 9)

When thinking about the importance of community in this process, we returned to our questions about how our co-participants experience themselves as agents of change engaged in collective action. With this in mind, we noted Reese and Bamberg's (2014: 467) reference to group identification as a 'powerful driver of social change' that 'encourages people to display commitment to the cause of changing society, to form coalitions with similar groups, and crucially, to act together in order to achieve their political goals'.

Methodological and Epistemological Considerations for Co-creating Sustainability Knowledge

In our new collaboration, we devoted considerable time to explore how best to conduct our local research, taking into account the foundational and ethical principles of our respective research stances. Through shared post-structural sensibilities we found common ground in and aspiration for locating ourselves in the thick of the research picture we were attempting to create. We figured that if we wanted to understand the sustainability issues faced by the region and how Gippsland change agents were attempting to effect change at a time of regional transition, we needed to engage with the participants in a reciprocal relationship. We agreed that such a process ought to involve talking about our own experiences in educating citizens for a globally sustainable and just future; articulating with integrity, our own values, shared meanings, visions and hopes. In our decision to co-create knowledge (Hollinshead, 2010) alongside our participants, we would speak *with* them dialogically.

These research considerations were particularly pertinent to undertaking local research, prompting us to consider not only how local research might be conducted, but also how regional research might be disseminated, accessible and comprehensible to regional communities (Howley & Howley, 2014). Cognizant of our researcher roles, we sought a relaxed and open research context where language, voice, communication and power would not privilege any participant, including our researcher selves. Our approach was positively affirmed by one participant after the first kitchen table conversation who expressed relief at the 'personal' nature of the discussion and 'its lack of academic jargon'. These research sensitivities are noteworthy for researchers more generally, but especially for regional researchers who often live and work in the regions and have ongoing professional relationships with the wider community where research is often conducted (Plowright et al., 2019). Placing ourselves and our co-participants on equal footing is a key facet of the qualitative research paradigm, as highlighted by Yvonna Lincoln and Norman Denzin:

> The qualitative researcher is not an objective, authoritative, politically neutral observer standing outside and above the text. Qualitative inquiry is properly conceptualized as a civic, participatory, collective project, [which] joins the researcher and the researched in an ongoing moral dialogue. (2000: 1049)

Consistent with their position, we committed ourselves to inclusive and participatory research approaches that rejected methods claiming to operate from a dispassionate and purportedly objective stance (Fine,

Weis, Weseen, & Wong, 2000; Heshusuis, 1994; Lincoln & Denzin, 2000). In keeping with qualitative research visionaries, we subscribe to the belief that researchers have an ethical responsibility to include participant participation in recursive processes of data collection and analysis (Fine et al., 2000). For this reason, we used a narrative ethnographic approach that moved us towards 'a deeper level of kinship between the knower and the known' (Heshusuis, 1994: 16).

Narrative Ethnography: Co-constructing Meaning and Knowledge Around a Kitchen Table

We employed narrative ethnography as a key approach because of its emphasis on collaborative participatory inquiry where meaning is co-constructed. Characterising narrative ethnography as the middle ground between realist ethnography and autoethnography, Lisa Tillmann-Healy explains that 'cultural analysis emerges from the character and process of ethnographic dialogue' (2001: 10). This was consistent with our aim to work collaboratively *with* our participants to explore how they effect change for a just and sustainable future in Gippsland. This approach requires researcher and researched to have a trusting, reciprocal relationship that can support and sustain these efforts (Tillmann-Healy, 2001). Between us, we were all too familiar with research projects where researchers took a distanced, dispassionate or objective stance. Our discomfort with such projects led to asking very real questions about representation, which, in turn, led us to reflexively and relationally position ourselves within the study's field of view. Ethically, we knew we needed to share the vulnerability that comes with being researched, which meant opening up ourselves to the same research questions and processes.

Furthermore, we had read more than our share of dispassionate, abstracted explanations for social phenomena that left us cold and unaffected; unable to find a place in which to connect and be stirred to action. Again, Lisa Tillmann-Healy's (2001) description of narrative inquiry offered something we believed would lead to different effects. By moving from conceptions of factual truth to narrative truth, she argues narrative ethnographies should be assessed by what people do with what they've read. Narrative ethnographers, she argues, write for people who want to be engaged with the research. Readers are invited to offer responses because narrative ethnographies are open-ended and encourage multiple interpretations (Tillmann-Healy, 2001). This is what we hope for our readers.

Earlier thoughts of conducting two focus group interviews with committee members at Monica's regional university returned us to our underlying research intention, which was to position ourselves reflexively and relationally in the research through collaborative and participatory approaches. It was in this spirit we opted for shared conversations around two distinctive farmhouse tables as a meeting point for more relaxed exchanges, which better aligned with the research thematic and small participant numbers we were working with. Mindful that kitchen tables have historically played and continue to play an important role in advancing grass roots activism and community engagement (Sarkissian, Hofer, Shore, Vajda, & Wilkinson, 2008), we recognised the kitchen tables' capacity to unite and connect us.

Monica and Sherie's fortnightly Skype meetings had occurred at Monica's redgum kitchen table, a centerpiece fixture in her open-planned loungeroom, which hosted the first kitchen table conversation with seven RCE committee members who brought food, wine and cheese to share. The second kitchen table conversation transpired over lunch around a blackwood table in a farmhouse kitchen at a GCCN member's house. We identified the two tables as synonymous with shared conversation and food, conviviality, warmth and abundant emergent stories. The two recorded and later transcribed conversations commenced with an introduction to Sherie as international EfS colleague and educator, a brief overview of our wider US-Australia research collective, and the study in focus. We also informed participants of our intention to publish, which we explained would be shared with them.

Applying a Post-Humanist Lens to Our Kitchen Table Conversations

In relation to our earlier work in the broader US-Australia collective, we had engaged with several theoretical and methodological frameworks advanced by group members as a way of inspiring and assisting our theorization of sustainability research and practice. Drawing on these experiences, coupled with our shared authorial post-structural dispositions, we sought to apply a post-humanist 'new materialism' lens to our study. We had been inspired by Karin Hultman's and Hillevi Lenz-Taguchi's (2010) methodological approaches to educational research that challenged researchers to shift their gaze beyond the human-centric, while simultaneously moving beyond the 'ruthlessly linear nature of the narrative of knowledge production in research' (St. Pierre, 1997: 179). Further, we engaged with the work of other renowned post-humanist researchers (e.g. Braidotti, 2007; Rautio, 2013; Somerville, 2015) who advance the agential force and capacity of the material world in research settings. From this understanding, we set about exploring the deeper meaning, representation and contribution of the non-human entities and materialities within our research setting, as having agency or agential power.

New materialism stems from a post-structural reading of the world to re-consider the ways humans and more-than-human continually create the conditions for each other's existence (Braidotti, 2007; Panelli, 2010; Taylor, Pacini-Ketchabaw, & Blaise, 2012). No better has this theory been exemplified than in early childhood education discourse where the productive nature of forces and forms are portrayed as vibrant matter in the lives of young children (Duhn, 2014; Lenz-Taguchi, 2010; Rautio, 2013). Underpinning new materialist thinking is the recognition that humans are constituted through complex entanglements with other animate and non-animate co-existing entities and materialities, the vitality of which Jane Bennett refers to as an 'ecology of things' (2010) or 'thing power' with agential powers that 'flow[s] around and through humans' (2004: 349). Feminist philosopher Karen Barad (2007) describes this process as 'intra-active', enabling new understandings of the co-emergence of difference through encounters. In this regard, entities become other than they were before the encounter because they act on each other simultaneously. According to Bennett, instead of inter-action as turn taking by individuals, intra-action considers the forces that are generated by all kinds of encounters between 'vibrant matter' of all sorts (2010).

In observing the intersection of human and more-than-human in the respective kitchen settings, we moved beyond thinking of ourselves and our participants as 'exclusively active and agentic' (Hultman & Lenz-Taguchi, 2010: 527). Instead, we turned our attention to the materialities of the research setting that flowed around and through us (Bennett, 2004), stemming from the tables themselves, their wooden surfaces hosting shared plates of cheese, biscuits, salads, bread and spreads, bottles of wine poured and spilt, a jug of water, notebooks, phones and computers, a recording device transported around the table to capture our collective voice, a vase of freshly picked flowers, dogs underfoot, a fresh outside breeze moving into the house; all brokering intra-active encounters between humans and other non-human bodies (Rautio, 2013), each responsible for transforming the other, and each changed as a consequence of the intra-action.

Becoming Kitchen Table: Making Meaning Through the More than Human

Taking up the new materialism theories we embarked on the process of writing as a method of inquiry (Richardson, 1994) to examine the meaning of the kitchen table intra/inter-actions through a post-humanist lens. This meaning-making process began even before the last words in our two kitchen table conversations were spoken. Instinctively, as we participated in those conversations, we listened for emerging themes and storylines. We shared what we heard along the way and asked questions to investigate them further. With transcripts of these conversations in hand, we then launched fully into the process of writing to explore meaning (McClam & Sevier, 2010).

Our 'writing to understand' process began with individual and shared writing about kitchen table/researcher intra-actions. Monica initially wrote reflections from the tables' perspective and Sherie responded with her reflections of 'meeting' and engaging with each. The writing exchange brought each of us into a new material space where we experienced the two tables—redgum and blackwood—not as pieces of inert furniture, but as active agents in the meanings and significances produced in the two kitchen table conversations (Figs. 13.1 and 13.2).

Fig. 13.1 Redgum table

Fig. 13.2 Blackwood table

From Monica: Redgum Table

Eucalyptus camaldulensis. River red gum. Rich-red tabletop streaked with dark fingers running the length of my oblong crown. Surrounded by six chairs but more when the numbers swell around me, I afford countless conversations. Some are happy and celebratory and involve end of the day home cooked family dinners with freshly picked garden food, and reflective family rituals: naming the best part of one's day, cheering and merriment, birthdays, parties, popping champagne corks and clinking glasses, applause for school reports, awards, graduations and new jobs, travel plans and working meetings. And then there's less happy and difficult conversations, sad and tearful exchanges about life's disappointments, loss and pain, break ups; mindful conversations with candles commemorating anniversaries of the deceased, old and young alike, with delicate discussions that grapple with the puzzling complexities of death and dying. And in the thick of it all is me, sustaining and enabling each and every exchange.

From Sherie in Response

As we sat around you, we experienced one another and our collective strength. We marched with Peter in his purple school uniform as he protested the Vietnam war. We felt Ash's deep compassion for cooking and sharing food. We saw the importance of Raquel's recollections of her nana's veggie filled backyard. We walked with Joan in Arnhem Land as she navigated a seemingly impossible path to forgiveness. We experienced the depth of Nola's agricultural roots. We gained strength from Jessie's youthful commitment to social change, and we shared Monica's deeply held affection for one another. With and through you this would-be interloper sensed the extraordinary effects of rural Australian life and its life changing transitions. I became part of something bigger/other than me, part of a generative collective; on an immanent plane of becoming. And there was you, sustaining and enabling each and every conversation that has taken place through and around you.

From Monica: Blackwood Table

Acacia Melanoxylon. Blackwood. Pride of place is where I sit. At the centre of the kitchen. Squarely in front of a welcoming sky-blue slow combustion stove. I am privy to decades of shared conversations and shared meals—dinner parties, hangover breakfasts, lunches, pots of tea, poppy seed cakes and scones. Around me sit family, old friends, new friends, friends of friends, siblings, cousins, local farmers and loggers, politicians, film-makers, neighbours, religious ministers, activists, conservationists. All in deep and convivial conversations about the state of the world. There are other visitors too. Of the non-human kind. Resident dogs lie beneath me on the cool of the slate tiles, free-ranging chickens make their way through the sliding doors that have been left open. So too an inquisitive summer-time tiger snake, curious birds as well as thousands of meddlesome blowflies. And there at the heart of it all is me. Sustaining and enabling each and every interaction.

From Sherie in Response

What is it about a kitchen table like you that draws us in, bringing us into communion with others? How is it that you generate spaces where hierarchies are disrupted, differences respected, and shared interests found? Is it because every seat has equal weight or importance? Is it because you are a place for nourishing? Is it like Ash said when he suggested that we make ourselves vulnerable when we share our food with others? Is it in that vulnerability that we can put down defenses and make ourselves open to other points of view? Is there something about a peaceful dog lying underneath you that generates peaceful ways of being? I'm not sure, but I felt all of those things as I sat at your side with the others, and it was not hard for me to see how anyone who had sat around you before me felt them too.

Exploring Meaning Through Vignettes

As we moved to explore the kitchen table conversations, we wanted to avoid the fragmenting and disembodying effects that often come with traditional approaches to coding transcripts as qualitative 'data'. In contrast, we applied a 'vignette analysis' that enabled patterns and themes to emerge within the fully embodied conversations we were very much a part of (Van Maanen, 1988). Specifically, we studied conversational storylines and in response wrote dialogic vignettes to make sense of the collectives' work to affect change for healthy, just and sustainable living. At the end of each vignette, we looked towards theories of place, collective hope and new materialism to strengthen our understanding of each vignette, a process that brought us face to face with the emergent effects of intra-actions between a place (the region), concerned citizens and two kitchen tables, which produced a story (a becoming) of collective hope, agency and action.

The following vignettes: Stories of place; Group identity; Shared desire for change; Collective action; and Sense of community are designed to bring readers into our meaning-making process: to literally and metaphorically invite them to take their own place around the respective tables. We ask the reader, what will emerge for you as you bring your own experiences and interpretations to our kitchen table conversations?

Stories of Place and Connection

Redgum Table

After passing Peter the cheese platter, Ash responds to Monica's invitation to talk about how he became engaged in sustainability education. 'I'm involved in sustainability through deep passion really. I grew up worrying about the world since grade 5'.

'What was it that twigged you in grade 5, Ash?' Monica asks.

'I grew up on a farm', he offers. 'Where we grew our own food, so there was a strong connection to the landscape. And I had this image that the cities were just concreting over everything, and I didn't want Australia to become like Japan, like a concrete jungle'.

'I grew up in the country too', Jesse shares. 'Not on a farm, but we grew our own food, and I could see the land getting chewed up by houses. And I could see that we needed the valuable farming land and to protect our environment'.

As platters continue to find their way around the table, Nola joins in. 'I was raised on a farm. We used the fat from the animals as butter, we milked our own house cows and we had our own vegetable garden'.

'I lived in town my whole life as a child and did not understand why my parents didn't have a farm!' Raquel exclaims. 'My Mum's side of the family lived in town too, but they had backyards filled with veggie patches and a chook pen'.

Raquel leans on the table and thinks for a minute. 'I don't know where it came from but I just I didn't understand why we had to live in town, why we didn't have a farm and raise our own animals and have big veggie gardens and things'.

With an understanding nod, Ash suggests, 'It's something that's entrenched in people'.

'I've come to the conclusion', Joan quietly interjects, 'that before Australia can move on, we need to repair the damage we've done to Aboriginal people. We've put so many layers onto what was here originally. If we can get down to understanding Aboriginal culture, we can start to really see what a mess we've made'.

Peter agrees, 'I think conservationists often get back to that point of the Aboriginal heritage and the connection with the land. My father was really involved in the field naturalist organisation and we went out on excursions into the bush all the time. So that was an influence for me and for my activism'.

Blackwood Table

With her beautiful cake sliced and placed on the table in front of us, Liz begins to talk about her journey to sustainability education. 'I grew up following my father Freddie around the veggie paddocks in Melbourne. He would scoop up handfuls of soil and tell me, "You could eat this stuff". Although it made no sense to me at the time, when I became an organic farmer here in Noojee, his words rang true'.

'And I've been here for 33 years', she adds with a wave of her arm. 'My interest has always been around how nature's been impacted right from how we farm right through to the health of ecosystems. I started asking questions locally around the management of forests in this upper catchment of the Latrobe River. As an activist, I didn't want my backyard chopped down'.

All of us according to Margaret Somerville (2010) are embedded in local places wherever we exist. In this vignette, we recognise the influence of place throughout the participants' lives constituted by earlier place experiences in childhood or young adulthood. Place for many was a specific locale with shared relations and interactions/intra-actions (place-making) in that place—a farm, a garden, the land, a protest march, country. As David Gruenewald maintains,

> Place … foregrounds a narrative of local and regional politics that is attuned to the particularities of where people actually live, and that is connected to global development trends that impact local places. (2003a: 3)

Place, Gruenewald (2003b) suggests, is profoundly pedagogical, teaching us about how the world works, how our lives fit into the spaces we occupy, and how particular places with their particular attributes shape identity and possibility. Beyond place as a physical entity, we consider Doreen Massey's (2005) interpretations of place as constituted by relational, dynamic, and spatio-temporal events that afford new possibilities to negotiate ways of knowing, acting, and being in the world. As highlighted, such place-making events are intrinsic and influential factors that shape participants' identity, sense of place, injustice, and fears, as well as their aspirations for protecting aboriginal heritage and the land.

Group Identity and Belonging

Redgum Table

Joan reaches for the grapes and considers Monica's question about what it means to be part of the RCE. 'Well', she says, 'being part of this group is a brilliant way for me to feel that I'm part of something that's doing something'.

Casting a warm smile in Peter's direction, she proudly adds, 'My tribe too, Peter'.

'Yeah', Jess jumps in. 'It's feeling like I've actually done something. Maybe not for now but for the future, and sometimes for now'.

'I'm a bit the same', adds Ash. 'I'd do this for free probably. If they [my boss] said you're not going to RCE, I'd go yeah, okay. Then I wouldn't mention the RCE and I'd just say I'm going to a meeting'.

Everyone laughs.

Feeling connected—even if temporarily—with the warmth and comradery of the group Sherie wonders aloud, 'What is it about this collective?'

Without hesitation, Ash says, 'It's somewhere deep in me, how important this work is. It's something real and tangible that I can get involved in instead of sitting around having meetings upon meetings. It's where I can actually get my hands dirty and have a bigger deeper conversation with likeminded people'.

'Exactly', Raquel affirms. 'With my work in the council, for example, I'm always competing with things that are perceived to be more important. So, it's always nice to come here and know that everyone's on the same page and that we're all trying to work for something better'.

Peter looks up from his cheese and crackers and says, 'The RCE gives me opportunity to work with a team doing something that I do want to do'.

'It's the two things, isn't it?' Joan adds, 'It's sustainability and education that we're all interested in and passionate about'.

Blackwood Table

Liz smiles proudly as she contemplates her involvement with the Catchment Management Authority (CMA). 'We did some really good innovative work', she says. 'The stuff we did on climate change back in 2005, we were the first CMA to actually have a crack at that'. 'As a group', Liz explains, 'we said, "let's just see how climate change is going to impact our region, and let's look at bringing all the players together and teasing this out". From this work, the committee members grew to know how to think for the future, how to think systemically'.

'So, you helped create an organisation', Sherie interjects, 'that could actually take risks and understand systems thinking. And not just systems thinking from a biological/ecological standpoint?'

Liz nods. 'When people understand connectivity & connection and that we're all a part of a system, I think it blows their minds. I've seen it with farmers. When something drops and they get that this whole thing is a system, they get so excited about it'.

In this vignette, participants associate their change agency roles and efforts to a welcomed sense of belonging within the group where 'important' work happens. In this space, they feel confident about fitting in, or as Riley (2017) describes, feeling safe in their identity to generate social and environmental change within the region. A central tenet is the assemblage of the group; a 'tribe' brought together to make collective meaning. According to participants, it is this collective space and its familial sensibility that provides the security to pursue their individual hopes (Braithwaite, 2004b) through an all-encompassing sense of collective hope (Braithwaite, 2004a). In this constructed space, there is latitude to generate ideas and subsequent initiatives that seek to build community relations, moulding a different pathway for the region. This 'hope process', as described by Braithwaite (2004b), generates a shared commitment and connection: a shared sense of purpose for seeking a just and sustainable world. Drawing on similar notions profiled in Rees and Bamberg's (2014) study investigating the determinants of individuals' collective climate action intention, group identity, and belonging loom large in this vignette as a 'powerful driver of social change' (p. 467), willing the participants to think for the future and achieve their goals by acting together.

Shared Desire for Change

Redgum Table

Thinking for a minute about what her role as a sustainability educator should be, Raquel says, 'My sustainable living program focuses on schools and kindergartens because I think that we need to be educating our future leaders rather than trying to educate the adults who are not caring'.

'That's why we [the RCE] are involved in education at the school level', Joan agrees. 'They're the future generations and they have an impact on the current parent generation because they take it back to home'.

Ash sips his wine and ponders what's been said. 'Maybe it's about what we love about Gippsland. We need to start with our values and the reasons why we live here, getting them to think about how climate change might impact that thing you do or that thing that you love. And there's something you can do about it'.

'I think that's an interesting take', Monica says, 'because that whole idea of celebrating place or people in place runs counter to the narrative you hear about the Latrobe Valley'.

'I've done a whole lot of thinking about the Latrobe Valley', Ash responds. 'Its biggest challenge is to unshackle the bad name that it's got. From the stuff I've read and understand, you cannot pull a population up to a better place. It has to come from within'.

Monica pauses to consider Ash's assertion. 'We've heard it a bit today about the influences of people in our lives', she notes. 'So, I want to talk more about the Latrobe Valley Ash because there's such a dilemma around how we move to whatever the next steps are for those communities. I'm not convinced it's got to come from the people. I get your sentiment, but I'm wondering if there's something else at play that might support movement towards something else. In a lot of my research working with kids, there is often a champion or an advocate who is educating them around new ways of thinking and exposure to ideas'.

'Yeah, I guess I see this' Ash confirms, 'because in my work in Traralgon Secondary College I've realised that being exposed to one thing is not going to change someone's outlook or life. Maybe we open their minds a bit to this stuff and then they may be intrigued to take it upon themselves'.

'If this was an easy fix', Jessie offers, 'it would have happened easily, but it's a long-term thing and we can only do so much. It's cultural changes, it's everything. So, we can only keep chipping away'.

Blackwood Table

Monica pours herself a cup of tea, looks at Liz and asks, 'So, what does your work as an activist in a number of community-based action groups mean for you?'

Liz smiles an infectious smile. 'It means you can put your two bobs worth in and people either think they can't imagine it, or they go, "Yeah, let's do it"'.

Liz pauses for a moment to pass Sherie the tea pot.

'It's like that sustainability leadership program we developed', she continues. 'How do we broaden people's skillset so that they can be more powerful in their communities, to lead discussions, to have dialogue in communities with those who are not getting it, or who haven't recognised that there are limits to growth, that the resource space is not infinite. The [sustainability leadership] course was designed to ask, what is a good enough life?'

The shared desire to bring about long-term change as represented in this vignette emerges as a critical ingredient of the respective committees' community-led practice, which relies on individual and collective strengths (Burnell, 2012). For the RCE, the main platform for change is to advance children's and young people's sustainability knowledge and action: to open their minds and to recognise them as crucial and strategic stakeholders who have their own shared desires for change (Green, 2019). The complex themes identified throughout the vignette are explicitly concerned with a shared desire to re-imagine a future that will and must be different from current circumstance. How might these shared desires for change be evoked in Gippsland? Participants proffer multiple possibilities, namely place-making and place-specific activities, changing the way we think, valuing the things and places we love, creating culture change, initiating collaborative community-based and community-led events, programmes and projects, and living 'a good enough life' that genuinely acknowledges the limitations of the planet. Such hopes and desires are testament to how the change agents grapple with their collective desire for change.

Collective Action

Redgum Table

Sitting back in her chair and looking around the table, Monica asks the group about what they think the RCE has achieved. 'What have you seen that's inspired you?'

'Oh, the Sustainable Schools Expos', Jessie exclaims. 'As a region if we can come together like that it really demonstrates to not only the people within the region but the State and Australia that we're serious about sustainability and that we're moving forward.'

Raquel thinks for a moment about Monica's question and what Jessie has just said. 'In 2014, we had an RCE forum where people walked away saying they were definitely going to do stuff about climate change'.

'Yeah', Peter jumps in excitedly, 'and they talked about actual projects'.

'Right', Monica agrees, 'it traversed a few projects around changing the political climate in the Latrobe Valley. There were conversations around transition to low carbon economies and all sort of projects'.

Peter thinks about other effects of his work, 'You know, one of the most delightful things about my job is walking down the street and having kids come up to me and say hello Mr Collins! I don't know who they are, but they're from a school that I've worked with, where I've done something like worm farming. It's just a lovely sense of community that comes from it'.

Ashley smiles at the thought of school children approaching Peter on the street. 'It's because they remember what you spoke about'.

Peter says, 'I think kids are just really good at remembering names'.

Monica laughs and asks, 'Do you reckon it's just about being good at remembering names, or is there something about the nature of your work and its significance and meaning to the children you interact with?'

Looking a little sheepish, Peter responds, 'It's not so much that I'm thinking, "oh, I'm doing important work", but because they've remembered my name, I'm thinking they remember what I talked about'.

Blackwood Table

Liz leans under the table to pet the dog sleeping on her foot and contemplates her accomplishments over the years. 'In the 1990s, I started work working with the dialogue methodology. I worked with the timber industry [loggers] and conservationists, looking at how we find a way together on an issue that's polarising the community, and where there's wholly divergent views'.

Intrigued, Sherie asks, 'When you say the dialogue method, that was method for negotiation and mediation?'

'Yes', Liz confirms. 'There's steps in how to undertake a dialogue. It's to suspend your own judgement, to listen to conversations, don't judge, just hear them [the loggers]. In that [dialogue] space you just listen and ask questions of yourself'.

Monica's face shows her deep understanding of Liz's prowess with the dialogue method.

'So, the loggers and I became incredibly good friends', Liz continues, 'We found places where there were things we could never agree on but there were things we could agree on. These were the places that we kind of focused on and built from there'.

Liz sips on her tea and takes a bite of cake. With a broad smile and a twinkle in her eye, she says, 'I can remember telling them why the forests are so meaningful to me, and the old guys, the old loggers said 'Yeah, I get that'. Those blokes loved the bush. They knew where they'd find endangered species and they knew where the orchids were; they knew the forest better than we did. But it was also their livelihood. They knew what they wanted from their livelihood, and there were some things that they weren't happy with either like overcutting'.

'So', Liz continues, 'meeting with the loggers gave me that interest in people and how to acknowledge sustainability as a people problem. It's not a technical problem or a resource problem. In order to come to find some solutions we have to find better ways of being with people and understanding the psychology of how people work'.

In this particular vignette, we are transported into the realm of the collective action deployed by participants in their various change agency roles. A central element of this theme is concerned with the resounding ways in which the change agents come together to seek ways forward through organised action. The collective action is bolstered by shared goals and trust that scaffolds viable pathways for strengthening community and achieving community outcomes and resilience (Barr & Devine-Wright, 2012). The impact of the collective action for social, environmental, and economic change is brought to light by the participants' stories of action. Courville and Piper (2004) refer this action as being the companion to hope; hope being the renewable resource for social change. In one example, we are exposed to one change agents' ambitious attempt to bring a group of loggers and conservationists together to dialogically explore the complexities of preserving the health and longevity of regional forests and maintaining the forester's livelihood. Other participants illustrate their efforts to establish regional forums designed to unite community via collective discussion, debate and action to envisage a different future. Through the means of their group's collective action, the change agents generate new forms of collective action within other groups of like-minded and less like-minded people to explore potential opportunities, challenges and solutions.

Sense of Community

Redgum Table

Taking a sip of wine, Raquel thinks for a moment about the significance of the RCE. 'A lot of people in the Council see my work as an airy-fairy fluffy job, as unimportant. When I come here you guys all think my work is important and that reinvigorates me. I get a collective group of people who are all on the same page'.

'I don't do anything else like this', Joan asserts 'working with adults who are on the same page and achieving something too is hugely important. I just quietly sit here and think, "oh yay" when I hear what each of you are doing'.

'There's a multiplying effect', Joan explains. 'Who knows where it all ends'.

Monica looks around the table and says, 'I shared with Sherie earlier about that day we went up to the local school. I remember driving out of the carpark after that meeting and thinking something like "oh my god, what a good bunch of people to be part of". It was very uplifting and significant for me'.

'Right', Nola jumps in. 'The four of us went to Warragul Secondary College on a Saturday afternoon. And there weren't just school teachers there, members of the community were there helping in the program too'.

'And especially on a Saturday', Nola explains. 'We all volunteered our time to go out of our way on the weekend to talk about something that was important to us'.

'I just find it inspiring and enjoyable', Jessie offers, 'to come along and plan for sustainability and seeing the impact and the results that it has in the community'.

Monica thinks for a moment. 'It's collective work I could never do on my own', she adds. 'A lot of my work happens in isolation at Uni so, I think it is important for me to be connected to a group of people where I feel like I can make a contribution to something tangible, something really significant. For me it's rich work that has a real bearing on our region'.

Blackwood Table

Thinking about the significance of her work with collectives like the Gippsland Climate Change Network (GCCN), Liz explains, 'We needed a regional approach to climate change that wove in all of the agencies, from councils right through to all of the various departments, environment, agriculture, health. So, we did a lot of pulling agencies work together to understand climate change and its impact. And since then agencies and other bodies have actually developed climate change strategies'.

Thinking about the impact of Liz's work across multiple regional collectives, Monica asks, 'How about the Catchment Management Authority (CMA)?'

Liz smiles as she reflects on that work. 'Farmers and locals worked together to create a regional plan for how natural resources were to be managed. We created bio-links along the whole length and breadth of the catchment, which was important for climate resilience, and the provision of habitat. This was an important climate change strategy that the group developed. It's about reimagining the future [together]'.

'I really adored the people who worked [in the CMA]', Liz continues. 'They all brought something to the group. I think we did some really good and innovative work. In 2005, we were the first CMA to consider how climate change was going to impact our region, and to look at bringing all the players together to tease it out'.

Sherie says, 'So it was a collective of people bringing effort and ideas'.

'Yeah', Liz confirms.

In her paper *Collective Hope*, Valerie Braithwaite reminds us, 'If hope is at the core of our being, the question becomes how do we hope productively, not only as individuals but also as collectives?' (2004a: 7). In light of what it means for a group of change agents to unite in tackling major sustainability issues across the Gippsland region, Braithwaite's question becomes a valuable reference point for exploring the emergent sense of community availed to individuals who hope and act collectively. In this vignette, RCE participants identify the collective as giving them legitimacy, a sense of belonging; affirming them in the collective's work that would ordinarily be unachievable on their own. From their insights, we note the broader sense of solidarity and cohesion they experience through a collective conscience, or a common body of beliefs and sentiments that give them an avenue to act through a 'collective group of people who are all on the same page'. A similar sense of community is illuminated in the blackwood table conversation where multiple collectives of people come together to generate ideas, action, and resilience through establishing a regional climate change plan and other projects where farmers and locals work together to create a natural resource management plan.

Conclusion: Researcher Reflections

In this chapter, we focused on the nature and impact of regional change agents' work. As experienced EfS educators and activists (of multiple decades between us) we appreciate the fullness of the vast challenges faced by climate change and sustainability change agents. In this line of work, there is no greater challenge than the associated sense of isolation. The kitchen table conversations highlighted in this chapter are testament to the enabling capacity of the collective or 'tribe' that anchors and supports regional change agents in their work, binding them together in their quest for a different and just world. From this study, we come to appreciate the sense of identity and belonging afforded to the participants through hoping collectively in the pursuit of shared ideas and actions. Further, we acknowledge the capacity and benefits of the co-participatory approaches that framed the study, particularly for the researchers who actively participated in the shared storytelling. On a final note, we recognise the agential capacity of the non-human forces at play in this study, particularly the respective kitchen tables that were so instrumental in the exchanges that occurred around and through them, all the while challenging and supporting us as researchers to produce different knowledge and to produce knowledge differently.

References

Barad, K. (2007). *Meeting the universe halfway: Quantum physics and the entanglement of matter and meaning*. Durham, NC: Duke University Press.

Barr, S., & Devine-Wright, P. (2012). Resilient communities: Sustainabilities in transition. *Local Environment: The International Journal of Justice and Sustainability, 17*(5), 525–532.

Bennett, J. (2004). The force of things: Steps toward an ecology of matter. *Political Theory, 32*(3), 347–372.

Bennett, J. (2010). *Vibrant matter: A political ecology of things*. Durham, NC: Duke.

Braidotti, R. (2007). Feminist epistemology after postmodernism: Critiquing science, technology and globalisation. *Interdisciplinary Science Reviews, 32*(1), 65–74. https://doi.org/10.1179/030801807x183623

Braithwaite, V. (2004a). Preface: Collective hope. Hope, power and governance. *Annals of the American Academy of Political Science, 592*(March), 6–15.

Braithwaite, V. (2004b). The hope process and social inclusion. Hope, power and governance. *Annals of the American Academy of Political Science, 592*(March), 128–151.

Burnell, J. (2012). Small change: Understanding cultural action as a resource for unlocking assets and building resilience in communities. *Community Development Journal, 48*, 134–150.

ClimateWorks Australia. (2011). Low carbon growth plan for Gippsland. Retrieved April 26, 2018, from http://climateworks.com.au/publication/document/low-carbon-growth-plan-gippsland

Courville, S., & Piper, N. (2004). Harnessing hope through NGO activism. Hope, power and governance. *Annals of the American Academy of Political Science, 592*(March), 39–61.

Drahos, P. (2004). Trading in public hope. Hope, power and governance. *Annals of the American Academy of Political Science, 592*(March), 18–38.

Duhn, I. (2014). Making agency matter: Rethinking infant and toddler agency in educational discourse. *Discourse: Studies in the Cultural Politics of Education*, 1–12.

Fine, M., Weis, L., Weseen, S., & Wong, L. (2000). For whom: Qualitative research, representations & social responsibility. In N. K. Denzin & Y. S. Lincoln (Eds.), *Handbook of qualitative research* (2nd ed., pp. 107–131). Thousand Oaks: Sage Publications.

Green, M. (2019). Frog bogs, turbines and biodiversity: Bringing children's sustainability knowledge to life through hand-made artefacts. In M. Green, S. Plowright, & N. Johnson (Eds.), *Educational researchers and the regional university: Agents of regional-global transformation* (pp. 153–172). Dordrecht: Springer.

Gruenewald, D. (2003a). The best of both worlds: A critical pedagogy of place. *Educational Researcher, 32*(4), 3–12.

Gruenewald, D. (2003b). Foundations of place: A multidisciplinary framework for place-conscious education. *American Educational Research Journal, 40*(3), 619–654.

Heshusuis, L. (1994). Freeing ourselves from objectivity: Managing subjectivity or turning toward a participatory mode of consciousness. *Educational Researcher, 23*(3), 15–22.

Hollinshead, K. (2010). Tourism studies and confined understanding: The call for a 'new sense' of postdisciplinary imaginary. *Tourism Analysis: An Interdisciplinary Journal, 15*(4), 499–512.

Howley, C., & Howley, A. (2014). Making senses of rural education research; art, transgressions and other acts of terror. In S. White & M. Corbett (Eds.), *Doing educational research in rural settings; methodological issues, international perspectives and practical solutions* (pp. 7–25). London: Routledge.
Hultman, K., & Lenz-Taguchi, H. (2010). Challenging anthropocentric analysis of visual data: A relational materialist methodological approach to educational research. *International Journal of Qualitative Studies in Education, 23*(5), 525–542.
Lenz-Taguchi, H. (2010). Investigating learning, participation and becoming in early childhood practices with a relational materialist approach. *Global Studies of Childhood, 1*(1), 36–50.
Lincoln, Y., & Denzin, N. K. (2000). The seventh moment: Out of the past. In N. K. Denzin & Y. S. Lincoln (Eds.), *Handbook of qualitative research* (2nd ed., pp. 1047–1065). Thousand Oaks: Sage Publications.
Lueck, M. (2007). Hope for a cause as cause for hope: The need for hope in environmental sociology. *American Sociology, 38*, 250–261.
Massey, D. (2005). *For space*. London; Thousand Oaks, California: Sage.
McClam, S., Cass, A., Connors, C., Frawley, D., Heald, S., McPartlin, A., et al. (2015). Collaboratively creating and sustaining hope and agency. *Journal of Sustainability Education*, (10), 1–18.
McClam, S., & Sevier, B. (2010). Troubles with grades, grading, and change: Learning from adventures in alternative assessment practices in teacher education. *Teaching and Teacher Education, 26*, 1460–1470.
McGeer, V. (2004). The art of good hope. Hope, power and governance. *Annals of the American Academy of Political and Social Science, 592*, 100–127.
Needs Assessment Snapshot: Latrobe Local Government Area. (2016). Retrieved October 18, 2018, from https://www.gphn.org.au/wp-content/uploads/2014/02/Latrobe-FINAL.pdf
Panelli, R. (2010). More-than-human social geographies: Posthuman and other possibilities. *Progress in Human Geography, 34*(1), 79–87.
Plowright, S., Green, M., & Johnson, N. (2019). Agents of regional–global transformation: Federation University Gippsland Education (FUGuE) researchers. In M. Green, S. Plowright, & N. Johnson (Eds.), *Educational researchers and the regional university: Agents of regional-global transformation* (pp. 1–22). Dordrecht: Springer.
Rautio, P. (2013). Children who carry stones in their pockets: On autotelic material practices in everyday life. *Children's Geographies, 11*(4), 394–408.
Reese, J., & Bamberg, S. (2014). Climate protection needs societal change: Determinants of intention to participate in collective climate action. *European Journal of Social Psychology, 44*, 466–473.

Richardson, L. (1994). Writing as a method of inquiry. In N. Denzin & Y. Lincoln (Eds.), *Handbook of qualitative research* (pp. 516–529). Thousand Oaks: Sage.

Riley, K. (2017). Community collaboration and partnership in volatile times. In P. Earley & T. Greany (Eds.), *School leadership and education system reform* (pp. 139–147). London: Bloomsbury.

Sarkissian, W., Hofer, N., Shore, Y., Vajda, S., & Wilkinson, C. (2008). *Kitchen table sustainability: Practical recipes for community engagement with sustainability*. London: Routledge.

Snyder, C. R., Lopez, S., Shorey, H., Rand, K., & Feldman, D. (2003). Hope theory, measurements and applications to school psychology. *School Psychology Quarterly, 18*(2), 122–139.

Somerville, M. (2010). A place pedagogy for 'global contemporaneity'. *Educational Philosophy and Theory, 43*(3), 326–344.

Somerville, M. (2015). Emergent literacies in 'the land of do anything you want'. In M. Somerville & M. Green (Eds.), *Children, place and sustainability* (pp. 78–91). Basingstoke, Hampshire: Palgrave Macmillan.

St. Pierre, E. (1997). Methodology in the fold and the irruption of transgressive data. *Qualitative Studies in Education, 10*(2), 175–189.

Stanley, J., Birrell, R., Brain, P., Carey, M., Duffy, M., Ferraro, S., et al. (2013). *What would a climate-adapted settlement look like in 2030? A case study of Inverloch and Sandy Point*. Retrieved October 18, 2018, from https://www.nccarf.edu.au/synthesis/what-would-climate-adapted-settlement-look-2030-case-study-inverloch-and-sandy-point

Taylor, A., Pacini-Ketchabaw, V., & Blaise, M. (2012). Editorial. Children's relations with the more-than-human world. *Contemporary Issues in Early Childhood, 13*(2), 81–85.

Tichaona, V., & Lupele, C. (2018). *Accelerating sustainable solutions at the local level in Issues and Trends in Education for Sustainable Development: Education on the Move*. Retrieved January 18, 2018, from https://bangkok.unesco.org/content/unesco-publication-issues-and-trends-education-sustainable-development

Tillmann-Healy, L. M. (2001). *Between gay and straight understanding friendship across sexual orientation*. Walnut Creek: Alta Mira Press.

Tomaney, J., & Somerville, M. (2010). Climate change and regional identity in Latrobe Valley, Victoria. *Australian Humanities Review, 49*, 29–47.

Van Maanen, J. (1988). *Tales of the field: On writing ethnography*. Chicago: University of Chicago Press.

14

Inventing the Green Metropolis Ruhr, Germany: Postindustrial Greening Narratives and Critical History Culture

Pia Eiringhaus

Introduction

Touring the present-day Ruhr Valley, once Germany's heavy industrial heartland and motor of economic growth, visitors may get the impression they have arrived in industrial heritage fairyland. Formerly desolated industrial spaces have been redesigned as industrially shaped cultural landscapes, and various local plants, mining towers, and worker settlements have become representative museum complexes. They are popular hotspots for German and international tourists as well as forming new identification points for local people. Regional theatre, music festivals and art installations are springing up everywhere, representing a clear demarcation from the past imageries of this region that characterised these communities as comprising uneducated and uncultured working-class milieu (Metropole Ruhr Tourismus, 2019). The Ruhr universities, founded in the 1960s, have come to embody the shift from physical to mental work, and thus have become representative trademarks of the

P. Eiringhaus (✉)
Martin-Luther-Universität Halle-Wittenberg, Halle, Germany
e-mail: pia.eiringhaus@paedagogik.uni-halle.de

A. Campbell et al. (eds.), *Located Research*,
https://doi.org/10.1007/978-981-32-9694-7_14

Ruhr region. Formerly desolated and damaged landscapes have been reinvented as the Green Metropolis, a marketing label created to advertise the Ruhr Valley's ecological transformation (Ruhr Regional Association, 2019). The success of this transformation can be seen in the city of Essen, the central and second largest city of the Ruhr, Germany, which, in 2017, was voted the European Green Capital, a prelude to a Ruhr Regional Association programme called the Green Decade 2017–2027 (Grüne Hauptstadt Europas, 2018). Previous perceptions of this landscape as industrial have fundamentally changed and the Ruhr Valley is now widely known as a recreational and natural space, characterised by ecological rehabilitation with new biodiversity, new aesthetics of nature and great opportunities for local recreation.

This phenomenon of what I have characterised as industrial heritage fairyland reached its culmination in December 2018, when the last operating coal pit, Prosper Haniel in Bottrop, a Ruhr city adjoining Essen, closed down. No lesser figure than the Federal President Frank-Walter Steinmeier gave a speech to symbolically bid farewell to the region's industrial past (Meinke, 2018). During the year 2018, countless regional remembrance ceremonies, academic conferences, exhibitions and events took place, as the 80-page event calendar published by the Ruhrkohle AG demonstrates (Ruhrkohle AG Foundation, 2018). In this mad heat of remembering the end of German coal mining industry, the nostalgic celebration of the region's industrial past and its successful restructuring into the present-day "Metropolis of Culture, Knowledge and Nature" was universally celebrated as a dominant narrative of transformation. Rising out of this transition, an homogeneous memory machinery has emerged, in which commercialised culture and tourist industries, as well as institutionalised history, culture and regional industrial heritage research tell the same story: how the uneducated, dirty-grey region with a grim industrial past has successfully transformed into a future-oriented and creatively working region with ecological quality label (Eiringhaus & Kellershohn, 2018).

Drawing on a cultural historical approach, I aim to analyse regional memory discourses as they apply to the Green Metropolis Ruhr, and specifically to investigate the narrative of the Ruhr Valley focused on this idea of "the black past to the green future". With focus on (re)greening the environment and the role this plays in reinventing this region, I argue that postindustrial greening narratives focus on the harmonisation of region and environment, thereby creating a narrative that presents harmonious development that facilitates a sustainable mode of coping with an ecological transformation (Eiringhaus, 2018). Such a narrative, one that is designed around a notion of successful, unambivalent structural change, is highly problematic, since it has reduced, or rather ironed out the complex realities and developments of the past. These sorts of narratives mean that there is a simplification about—or even lack of awareness of—the processes that have led to what is now present in this regional environment. The ways in which this narrative of past and present is harmonised tells a success story that moves from ecological crisis to conciliatory environmental heritage and thus works against both critical reflection and emergence of controversial engagement. Accordingly, the aim here is to highlight these contradictions, ruptures and ambivalences of this transition processes to diversify, problematise and historicise interpretative patterns that dominate the regional research field.

Few studies have examined the significance of greening a postindustrial environment and the importance of how nature is imagined and constructed for those who live and work in the Ruhr Valley (Eiringhaus, 2018). Research on industrial heritage in a regional context is primarily concerned with structural changes in social and economic perspective, as well as the historical localisation of industrial heritage (Bartel, 2008; Blotevogel, 2002; Borsdorf, 2014; Goch, 2002; Wehling, 2016). Current international and comparative works have broadened this research focus, developing critical perspectives on regional history culture (Berger, 2019; Berger & Wicke, 2017; Berger, Wicke, & Golombeck, 2015, 2017; Bluma, 2018; Kellershohn, forthcoming; Stiftung Bibliothek des Ruhrgebiets & Deutschen Bergbau-Museum Bochum, 2014). One avenue of such research is the work of environmental historians who have examined the Ruhr Valley's ecological history and transformation (Brüggemeier, 1992, 2014; Siemann & Freytag, 2003; Uekötter, 2014); however, this body of work has not investigated greening the postindustrial region in the context of cultural history or memory studies. Particularly, the field of landscape theory provides theoretically valuable approaches to address this gap, identifying perceptions of environment and nature as culturally constructed, and the redesign of (industrial) landscapes as sociocultural process (Arndt, 2015; Kühne & Bruns, 2015; Schönwald, 2015; Siemann & Freytag, 2003). For instance, the dissertation of Hillary Angelo is a brilliant study on greening in the Ruhr Valley as cultural practice to cope with postmodern urbanity (Angelo, 2015). Considering this state of the art, my work is located at the interdisciplinary intersection of memory and industrial heritage research, environmental and cultural history, as well as landscape theory. The goal is to bring existing approaches into dialogue to analyse postindustrial greening discourses and their sociocultural dispositions together so as to determine how they have shaped the Ruhr Valley.

For this purpose and as a *first* step, I briefly introduce the concept of industrial nature, the heart of the regional greening narrative invented and popularised during the International Architecture Exhibition Emscher Park (*Internationale Bauausstellung Emscher Park*, IBA),[1] to reveal the constructedness and variable functionality of this neologism. I argue that representations of regional greening are constituted around the idea of harmonising region and environment, which has become the backbone of regional environmental heritage discourses. In a *second* step and to shed light on how such processes of harmonisation are expressed, mediated and made effective, I analyse two rather different, but exemplary application fields, the Duisburg North Landscape Park and the Ruhr Museum, using various sources, material and topographical concepts.[2] The Duisburg North Landscape Park demonstrates the early formation of greening discourses, while the Ruhr Museum, in contrast, is an impactful public and educative institution, which was founded ten years after the International Architecture Exhibition. At both sites, the history of the region and its environment is negotiated through memory discourses. As conflict areas between history and memory, both spaces carry

[1] This was the most influential programme for structural changes in the German Ruhr region from 1989 to 1999 with the goal to ecologically and economically renew the most difficult parts of the Ruhr region.

[2] Relevant sources are IBA publications, information and marketing material from both heritage sides, as well as documents and online material published by the Ruhr Association, a regional network that plans and coordinates development concepts in the Ruhr Valley (Ganser, 1999; Ganser & Dettmar, 1999; Latz, 2016; Regionalverband Ruhr, 2010).

past, present and future layers of time, thus have a genuine temporality (Berger & Seiffert, 2014; Große Kracht, 2014). Cultural memory discourses interact with sociocultural developments, therefore integrating both sheds light on the mechanisms of how collective identity formations are shaped and filled with meaning. In a *third* and integrative step, I intend to show that the greening narrative shows a consistency of semantics, plot and imagery, and this indicates the importance of personal ongoing relations with place and the interpretative frameworks used to make sense of these relations, alongside the powerful effects of knowledge construction and communication through regional agencies. This can be seen most clearly in the public commemorations in 2018 (Ruhrkohle AG Foundation, 2018), and raises concern regarding the necessity to critically reflect upon the nature and manner of how the history of the Ruhr Valley can and perhaps should be remembered.

The Power of New Vocabulary: Industrial Nature Discourse

©Ruhr Musem/Photo: Rainer Rothenberg

The narrative of success of the regional shift "from black to green" owes much to the International Architecture Exhibition. The programme aimed to ecologically and economically renew the Ruhr Valley's most difficult areas. The programme was funded by the Federal State of North Rhine Westphalia with a total of 35 million DM. The programme was conceptualised in a top-down manner by renowned regional directors and politicians (Goch, 2011; Reichert, Niemann, & Uttke, 2010). The goal to enhance the attractiveness of the region required building a regional consciousness, or as explained by the regional historian, Stefan Goch, this process needed the "production or evocation of regional identity" (Goch, 1997: 594). This new commitment to Ruhr regionalism involved integrating regional culture and history as well as the region-specific industrial landscapes. These broader goals characterise the International Architecture Exhibition's political initiative, which invented a new philosophy for transforming the Ruhr Valley.

At the core of this philosophy for ecological image change lies the concept of industrial nature. "Industrial nature" is an ecological term, originally used to describe the new vegetation that appears on industrial production areas (Oligmüller, 2002; Storm, 2014). The term itself and what it came to symbolise, however, were coined and shaped by the International Architecture Exhibition (Internationale Bauausstellung, 1998). The Ruhr photographer, Peter Liedtke, and his work of the 1980s and 1990s highlights the semantic and meaning-related formation of this concept. It demonstrates how both the visual perceptions of the Ruhr Valley and the terminologies used to describe these perceptions have changed. In 1986, the photographer's exhibition on vegetation on industrial spaces was called quite pragmatically: *Ruhr Nature: spoil heaps, landslides, industrial wasteland* (Liedtke, 2015). Only ten years later in 1995, he created another photo exhibition, whose title had a completely different design; *Industrial Nature: Vocabulary of new nature aesthetics of industrially shaped landscapes* (Liedtke, 1996). During the 1990s, the Ruhr Nature was reframed as industrial nature and the image-related impact of the International Architecture Exhibition becomes evident.

By the mid-1990s, the term "industrial nature" was established as label to express and advertise these new perceptions of nature. The International

Architecture Exhibition completed a project called *Route Industrial Nature*, which was made up of a network of various parks, as well as greening and recreation areas in the Ruhr Valley (Internationale Bauausstellung, 1998; Kommunalverband Ruhr, 1999, 2001). Exemplary points of this project include landmarks such as the Duisburg Nord Landscape Park or the Zollverein Park, Essen, Germany. The project was conceptualised as part of the Route Industrial Heritage, covering 13 guided tours with names including, *The desert alive*, *After us, the jungle*, and *From the beach to spoil heaps*. The route is supposed to show the Valley's most beautiful spaces and mediate the special experience of Ruhr nature to visitors. After the end of the International Architecture Exhibition, the route was made permanent with 19 guided tours with the Ruhr Regional Association its leading actor. In line with the realisation of the Route Industrial Nature, there was an explosion in the use of this term in the fields of regional planning, tourist and cultural industries (Dettmar, 1991; Ganser, 1999; Ganser & Dettmar, 1999; Regionalverband Ruhr, 2010). This period of time correlates with the effective period of the International Architecture Exhibition, and demonstrates the influence of this group on how our understanding of landscape arises out of such a cultural construct. Given this, "industrial nature", a compound noun and oxymoron, can be theoretically defined as in terms of an empty signifier, the heart of a discursive formation that gathers manifold elements around without having a genuine meaning on its own.

A quotation from the exhibition, *Other places are beautiful as well*, conceptualised and realised by the *Landschaftsverband Rheinland* in 1999, sheds light on the associations and diverse symbolic meanings ascribed to the concept of industrial nature

> Every erosion, including industrial ones, has its wonders beyond that of birch trees and old brickwork, bushes and dried grasses. The IBA would not be the inventive IBA, if they did not also locate this phenomenon and find a way to derive benefit from it: they call it "industrial nature". (Landschaftsverband Rheinland, 1999: 10)

The reference to "industrial wonders" beyond old brickwork suggests that the location and characterisation of the "phenomenon" located here is enough to harmonise the damaged history of the Ruhr Valley and its environment. The suggestion that "industrial erosion" can be wonderful hints at the intention to write a positive meaning into environmental destruction caused by industrialisation. Therefore, the concept of industrial nature plays a vital role in what is understood, communicated and advertised as the Green Metropolis Ruhr. It arises out of the relationship between the semantics of reconquering nature, industrial heritage and smokestack nostalgia, new nature aesthetics and romanticised imagery, and finally greening as an ecological and economic promise of successful structural change (Eiringhaus, 2018). These symbols function as a basis for what I identify as an industrial nature discourse. This discourse acts with industrial heritage, because of the interplay between both landscape as industrial nature and history as industrial heritage, facilitates the invention of the Ruhr Valley as both ecologically and culturally sustainable. Even so, while as Ulrich Borsdorf rightly argues, "industrial heritage still has to prove itself" (Borsdorf, 2014: 90), the same applies to these ideas about industrial nature.

From Ugly to Innovative? Industrial Nature and the Duisburg North Landscape Park, Germany

The Duisburg North Landscape Park is a flagship project for the ecological transformation of the Ruhr Valley. It is one of the Ruhr's most prominent industrial heritage sites as well as representative event for many culture, arts and music festivals. The space covers a total of 180 hectares with the former steel works *Duisburg Meiderich* at its heart and attracts more than one million visitors a year (Dettmar, 1999; Winkels & Zieling, 2010). The beginnings of today's park complex can be traced back to a bottom-up movement of the IG Nordpark citizens' initiative, which advocated preservation of the area after it was closed in 1985. As early as 1989, the city of Duisburg registered the former steel works and its surrounding area as project for the International Architecture Exhibition. An international planning competition had to decide how to redesign and reinterpret the industrial wasteland and how to give new meaning to the spaces. The successful design, *Networks of industrial infrastructure and remains of a production plant becomes landscape* (Latz, 2016), was proposed by landscape architect Peter Latz who was made responsible for the architectural and landscape-related reconception. The work began in 1991 and three years later, the park was opened (Dettmar, 1999; Hasselberg, 2011; Hemmings & Kagel, 2010). As public area, the Duisburg North park complex addresses a heterogeneous audience. While the open area is public and without an entrance fee, it is used for leisure and sport activities. In addition, attending cultural events like art installations, readings, cinema and concerts, requires an entrance fee.

Harmonising Difference in Topography and Narrative

Industrial nature's topographical manifestations are modelled around the postindustrial greening narrative of harmonisation. Visiting the former steel works at the heart of the landscape park evokes the impression that industry and nature not only exist *next* to each other, but *with* each other. Plants and bushes are growing unregulated in the middle of industrial relicts, both are not separated, but visibly arranged together. In the entrance area, several fruit trees were planted directly in front of the blast furnace, each of them fixed with sectional steel. The architect explained, "the trees share their naturalness, while the steel works shares its technical existence" (Latz, 2016: 117). In a similar manner, former storage bunkers were redesigned as classic gardens, having roses, herbs, ferns and ivy vines. This contrasting design also evokes, as the architect explains, "specific associations, since cultivated gardens are located next to industrial relicts" (Latz, 2016: 116). Both park sections demonstrate that nature and industry are rearranged, intentionally put into a harmonic, almost productive relationship.

Design and material manifestations express what I have theoretically captured as harmonisation narrative, a harmonising topography that aims to maintain the place's infrastructure, making it visible, and highlighting its processual character. The historical industrial landscape is preserved and, at the same time, reassigned with new symbolism and meanings. Landscape elements and topographical concepts are arranged in a self-contradictory manner. In this way, the dichotomy between industry and nature is visibly neutralised, the damaged relation between the Ruhr Valley and its environment harmonised (Hasselberg, 2011; Hemmings & Kagel, 2010). At the same time, a technically shaped nature is legitimised as one of many forms of nature. Not only was the formerly ugly and dirty industrial landscape *designed* to be identified with, but was also staged as the role model of a new, artificially artistic urban nature.

How is this environmental harmonisation mediated textually? The narrative representation of industrial nature in this region's information and press material is captured in the phrase that nature has reconquered formerly natural areas—a prominent phrase in all advertisements and information material on industrial nature (Ganser & Dettmar, 1999; Kommunalverband Ruhr, 1999, 2001; Regionalverband Ruhr, 2010). Marketing flyers and visitor boards, as well as specialist publications, emphasise the new symbiosis between these formerly "incompatible elements" (Landschaftspark Duisburg Nord, 2011). The idea of reconquering nature, however, disguises the actual artificiality of regional greening by postulating instead its naturalness. The semantics of reconquering or recapturing evoke the imagery of a powerful, clearly romanticised nature (Ganser & Dettmar, 1999), which in fact does not correspond to the hybrid nature in the park complex (Keil, 2002). For instance, in the bilingual publication, *From Ironworks to Theme Park*, the idea of the "amazement of the unyielding strength of nature" (Winkels & Zieling, 2010: 52) is emphasised, staged in such a way that "the age-old cycle of nature is repeated: growth, decay and new growth" (Winkels & Zieling, 2010: 52). Such a narrative demonstrates the consistency of culturally formulated critical perceptions of nature, and shows a new way of considering what nature is. What could be perceived as an inherent paradox reveals that this concept of nature is renegotiated in times of change and pushed in specific directions.

New Aesthetics and Regional Identification

Across Western areas and cities, industrial ruins are ambivalent, but significant markers for regional identification (Edensor, 2005). Reinterpreting industrial wastelands requires dismissing, or at least aestheticising deformed landscapes, as Susanne Hauser has stated (Hauser, 2003, 2005). The International Architecture Exhibition provided the initial impetus to refrain from classically beautiful landscapes, trying to redefine relevant categories of nature and landscape aesthetics in the context of the postindustrial landscape. This meant changing not only the idea of what is beautiful in landscapes but also our cultural viewing habits. Such aesthetic reinterpretation are readily found when looking at the Duisburg guided tour, *Cycling the steel works*, which is advertised with the slogan of the "particular beauty and fascination of the strangely unconventional link between industrial relicts and nature" (Tour de Ruhr, 2015: 2), a description dominated by a semantics of peculiarity. For example, the Tour de Ruhr, a dominant regional tourist agency, depicts the industrial–natural interplay as an "expression of unconventional, individualistic aesthetics" (Tour de Ruhr, 2015: 11). An identical choice of words is found in other regional publications, such as in the various editions published by the Ruhr Regional Association. There, the park is presented as "park of characteristic peculiarity" or as "place of surreal impression" (Regionalverband Ruhr, 2010: 60). These particular semantics work with contrasting sets of symbolism and attributions, as a consciously antagonistic design trope, pairing the unusual link between relict and nature, or the "sharp contrast between 'young vegetation' and 'old iron'" (Ganser, 1999; Middendorf & Hamann, 1999).

This imagery of postindustrial attractiveness derives from ideas of radical plurality and postmodern aesthetics, where differences are not neglected, but staged as enrichment (Kühne, 2006; Zima, 2016). Following landscape theorist Olaf Kühne, postmodernity is characterised by a "primacy of aesthetics", since reality is primarily constructed aesthetically (Kühne, 2006: 6). Industrial nature in the Landscape Park expresses and manifests this tendency, since existing structures, here the industrial wasteland, are recontextualised and reinterpreted through the use of new symbols, creative arrangements and a new aesthetics of the unusual. What was once deformed and unpresentable is now creative and innovative. Former dichotomies between "beautiful" and "ugly", "clean" and "dirty", "natural" and "industrial" became obsolete. They are transformed into a meaningful contradiction relevant to this aesthetic construction of reality: innovative vs. non-innovative. This corresponds to the International Architecture Exhibition's attempt to be "particularly innovative in a non-innovative milieu" (Goch, 2011: 74). Inventing new classifications allows industrial landscapes to become *aesthetically shaped* industrial–cultural landscapes. In this context, the aesthetic staging reveals the variable functionalities of the new Ruhr nature: its man-made hybrid nature enhances the invention of an innovative self-image, thus demonstrates the Ruhr landscapes' new potential for regional identification.

In regional research, the aesthetisation of industrial spaces was and is being evaluated by academics. Already during the International Architecture Exhibition, critical voices emphasised that "industrial nature forces an extremely positive evaluation of industrially contaminated sites", arguing that a new vocabulary of "diverse and beautiful landscapes is rather subject to political symbolism" (Hard, 1991: 3).[3] Accordingly, during the planning phase, the unease with relabelling environmental damage as industrial nature to improve the image of the region was discussed critically. In recent publications, this unease with the concept has disappeared. Researchers are increasingly concerned with the problematics of aestheticising former working spaces, and along with it, working-class culture and labour. They argue that the focus on aesthetic pleasure reduces former work environments to "commercialised amusement parks", and thus leads to the "disappearance of a sense of history" (Hemmings & Kagel, 2010: 253). This criticism, sometimes referred to as "smokestack nostalgia" or "ruin porn", mainly carries the idea that the actual history to tell is being undermined, namely the history of labour and working class (Strangleman, 2013). Even so, while the aestheticisation

[3] In 1991, IBA director Karl Ganser organised an interdisciplinary academic symposium to discuss possibilities and limits of industrial nature (Ganser, 1991). The geographer Gerard Hard was one of the most critical participants.

of work is debated, a corresponding criticism of the mystification of nature and regional greening remains unconsidered. While environmental historians suggest there needs to be some redefining of the alleged past dichotomy between economy and ecology, this is a significant challenge to the powerful postindustrial success story of Western environmentalism (Uekötter, 2003). These works seek differentiated perspectives on ecological developments; however, regional knowledge holders remain resistant to criticism, while they reproduce and advertise one-dimensional greening narratives based on the radical distinction between the "black past" and the "green present"—and it is this green shift between past and present that allows the reinvention the Ruhr Valley to be a successfully aestheticised role model for postindustrial landscapes.

Industrial Nature Discourse in the Ruhr Museum, Essen, Germany

The cultural institution of the Ruhr Museum is neither a space to discover nature, nor is it a materially designed landscape. Unlike the Landscape Park, it is part of the Zollverein Coal Mine Industrial Complex, a large former industrial site in the city of Essen. The site has been added to the UNESCO list of World Heritage Sites since December 14, 2001, and is one of the anchor points of the European Route of Industrial Heritage. Accordingly, the museum is located on an industrial heritage complex and has international status. The history of today's Ruhr Museum can be traced back to 1904, when it was founded by the city of Essen. After several relocations and renaming in the following years, it was opened in 1984 as *Ruhrlandmuseum*. Along with ongoing structural changes, the *Ruhrlandmuseum* was again redesigned. The new Ruhr Museum was opened in Zollverein in the year of the European Capital of Culture, 2010. The goal was to create a representative and educative place that narrates the history of the region, culminating in the present-day Metropolis Ruhr (Borsdorf & Grütter, 2010; Dettmar, 2003). It addresses a national and even international audience, which makes it more exclusive than the popularly visited Duisburg Landscape Park.

The Ruhr Museum is a significant mediator of the industrial nature discourse and delivers certain discourses that support both the scope and reproducibility of the International Architecture Exhibition's greening narratives. Industrial nature is an integrative component of the 2010 permanent exhibition titled "Nature, Culture, History" (Borsdorf & Grütter, 2010), and thus is a significant part of what is officially preserved and presented in the self-declared "memory of the region" (Borsdorf, 2003: 106). The way the museum explains and exhibits regional greening initiatives shapes the accompanying interpretative frameworks, through which mechanisms of regional memory politics become effective (Macdonald, 2006b). In this way, the invention of the Green Metropolis as a regional project is not only made manifest in outdoor areas, but also within an institutionalised memory culture.

Harmonisng Difference in the Ruhr Museum's Permanent Exhibition

The Ruhr Museum's permanent exhibition integrates the nature and history of the Ruhr Valley, covering the time span of ecological formation, pre-modern times and industrialisation, up until the present time. The exhibition has three parts and located on three different levels in the building formerly used for coal washing. Starting at the top level, visitors literally descend through the history of the Ruhr Valley, beginning with the section "present time", continuing on to "memory", and ending with the "history of industrialisation in five acts" (Borsdorf & Grütter, 2010; Dettmar, 2003). Already this broader arrangement of museum space demonstrates that the explanation of historical developments is oriented towards the present-day transformation of the Ruhr Valley.

This impression is reinforced when analysing the staging of industrial nature in the museum section "Present: Phenomena". It shows various "mythical imageries typical for the Ruhr" (Borsdorf & Grütter, 2010: 54). Here, industrial nature is exhibited together with display areas showing *typical* cultural phenomena, such as worker settlements, industrial architecture, football, kiosks or religious life. Photographs of such alleged typical Ruhr sceneries are shown next to various plants, blossoms and leaves collected in herbaria (Fuchs & Guderley, 2014). While the industrial landscape used to be a problematic aspect of the Ruhr Valley in the past, the notion of an industrial nature has succeeded both as concept that works with industrial heritage and as a form of regional identification. Furthermore, this section identifies the regional museum as a "powerful myth maker and storyteller", a role that every museum plays, as Sharon Macdonald has argued in her research on museum theory (Macdonald, 1999: 5). Despite the Ruhr Museum has set itself the objective of deconstructing regional myths, the explicit grouping of Ruhr phenomena, based on stereotypes and clichés, rather enhances respective regional commonalities and thus a fixed and normative set of identification points. The self-fashioning of the region as the Green Metropolis is one of them.

The museum links constructions of regional identification to the idea of ecological harmonisation. This becomes particularly evident when looking at the exhibition section "Present: Signs of the Times". Prehistoric exhibits are displayed together with personal, everyday objects, originating from the time of high industrialisation. For example, a preserved worker's black lung is put next to a slice of a tree, whose annual rings function as a symbol of growth and change. Individual memories are set up in contrast to the long-term memory of nature, so enabling engagement in dialogue with "geological witnesses, who function to display the eternity of the geological era" (Borsdorf & Grütter, 2010: 161). This fairly innovative approach places human interference with natural ecosystems in a completely new context of meaning. As the exhibition catalogue states:

> In the Ruhr, two layers of time are emerging, a cultural and a natural, a subjective and an objective, a short and an eternal. This entanglement can be hardly found in any other place. (Borsdorf & Grütter, 2010: 161)

This integrative object arrangement blurs contrasts and dividing lines between individual cultural memories and geological history, which, in a broader sense, leads to reflections on the relations between culture and nature. Objects from everyday life are embedded in the broader history of the earth, which allows presenting culture and nature as two sides of the same coin and thus corresponds to a narrative of harmonisation. The focus on individual memories emphasises the museum's postmodern exhibition approach, pluralising and multiplying possible modes of interpretation. This display strategy is strongly reminiscent of the Landscape Park's contradictory design concepts (Macdonald, 2006a), again demonstrating the powerful effect of creativity and innovation for mediating industrial nature and landscapes.

What does this section tell us about a regional memory landscape? The interpretative pattern of regional greening is extended on a global level. Space and time-transcending elements are used to narrate the end of the industrial age in the example of the local history of a region. This "glocalisation" (Robertson, 1998) of industrial nature shows that a greening discourse has been adapted as an exemplar, and is integral to the UNESCO world heritage site, Zollverein. Stating that this specific entanglement "can be hardly found in any other place" (Borsdorf & Grütter, 2010: 161) implies that especially, eventually *solely* in the Ruhr Valley, the damaged relations between region and environment were successfully restored.

Adapting the Success Story of Regional Greening

The exhibition's companion volume makes the concept of nature an issue of discussion and states that Ruhr's nature cannot be understood as "original nature", but rather as "cultural nature" or "nature-identical form" (Borsdorf & Grütter, 2010: 150). This indicates that the industrially and culturally constructed spaces of nature are a critical reflection of this location and its history, and in presenting this, the museum as an academic and educational institution refrains from the commercialised imageries of wild, untouched nature referred to earlier. However, what follows the considerations are the repetitive phrases of "nature reconquering the formerly naturalness spaces", "new biodiversity" and "red list species" (Borsdorf & Grütter, 2010: 150). Again, the industrial nature discourse is subjected to the overarching paradox of continuing romantic imageries of nature and its hybrid, culturalised reality. What becomes evident is that such catch phrases and symbols are reproduced unchallenged, even in the most reputable museum of the Ruhr Valley. In this way, it reinforces an environmental success story instead of critically reflecting upon the historical ambivalences of ecological development.

During the planning period, the museum concept of "Nature, Culture and History" was discussed at a regional conference with the aim of exchanging different disciplinary perspectives on the museum's reconceptualisation (Aufbaustab Ruhr-Museum, 2003). In the conference report, industrial nature is discussed and referred to as "problematic buzz word to label regional greening" (Blotevogel & Prossek, 2003: 20), but is at the same time attributed with a quality to "express the final result of industrialisation and deindustrialisation" (Blotevogel & Prossek, 2003: 20). Here, the problematic dimensions of strategically aestheticising ecological deficits are not considered, or at least discussed as separate from industrial nature. For example, species protection and biodiversity are widely marketed as a regional feature, yet activists from the "Nature and Biodiversity Conservation Union" are constantly facing conflicts with property owners who claim back those areas once exclusively devoted to nature (Hingmann, Tomec, & van de Sand, 2016). In addition, continued mining operations have resulted in huge underground cavities and elevation differences, and these require costly pump systems to prevent the Ruhr Valley from sinking (Dörries, Balser, & Büschemann, 2013). This severe problematic is a rarely mentioned context of the success story of industrial nature.

Conclusion

By considering what I have defined as industrial nature discourse, I have analysed representations and practices of regional greening in the Ruhr Valley. This discursive field is centred on the International Architecture Exhibition's neologism of industrial nature and fully evolves around the idea of harmonising region and environment. This concept of harmonisation bundles regional transformation and reinvention, the reshaping of regional identities and the invention of new categories of landscape and nature, and it fully engages with global norms of ecological ethics. Yet, the industrial nature discourse manifests differently in both application fields, the Duisburg North Landscape Park and the Ruhr Museum. While aspects of regional identification, new nature aesthetics and ecological sustainability differ in form and function, they are of vital importance for both sites. This site-specific, but consistent modelling of narratives and imageries, allows telling the present-day success story from an ecological crisis to that of environmental heritage. It culminates in the highly marketable label of the Green Metropolis.

Based on the reciprocal relations between regional memory narratives and sociocultural processes, this study has revealed three main functions of harmonisation as dominant plot line for greening in the Ruhr Valley that tells a positive story. *First*, such a narrative is made ecologically future-proof while maintaining relevant elements of the industrial past significant for the formation of Ruhr identities. *Second*, the depiction of regional greening as a solely positive development functions to overcome former image problems and helps to invent the Ruhr Valley as postindustrial role model for ecological transformation. *Third* and with a view to regional identity formation, narratives of postindustrial greening are successfully embedded into a set of future-oriented regional characteristics. Such alleged commonalities are not only marketed and performed, but made obligatory and over determined.

Nevertheless, I would like to stimulate some provoking thoughts on the Ruhr Valley's consensual memory landscapes that are visible among others in the invention of the Green Metropolis. The chapter has shown that both perceptions of the Ruhr Valley and representations of landscape have been changing: the formerly grim, industry-protecting region has become an "excellently suited place for the necessary revision of nature concepts in the 21st Century" (Middendorf & Hamann, 1999: 44). The Green Metropolis is presented as an ecological role model of extreme transformation, eventually serving as pioneer for other regions in need of remaking industrial wastelands to postindustrial idylls. When it is claimed that regional sustainability requires "becoming-self-aware" (Blotevogel, Prossek, & Reulecke, 2009: 13), the Ruhr's self-confident dealing with regional greening demonstrates that ecological self-awareness has already gone beyond the original objective. With view to future developments of regional research, it is unclear to what extent narrative repetitions of self-referential success stories can continue to drive forward controversial discussions. The crucial question is how viable these interpretative frameworks are in the long term, and to what extent they may run into danger of subtle stagnation? Inventing nostalgic hero stories as moment of compensation has certainly enabled regional identification and met societal requirements in critical times of transformation allowing and facilitating structural change over decades. Research into the future requires greater reflexivity, allowing deeper interrogation of the production of the Ruhr Valley's pervasive, homogeneous memory landscapes.

Present-day narratives were mainly constructed and promoted by a specific academic milieu that was and is firmly anchored in the Ruhr Valley and closely entangled with the challenge of politically, economically and culturally coping with processes of structural changes. What happens, however, when the industrial past is de-subjectified, fully converted into cultural heritage and individual (nostalgic) memory that is not the reference point of an historical culture anymore? The Ruhr Valley is in the midst of this process. Facing this new challenge certainly requires more space to tell and question these contrasting histories and help facilitate alternative experiences and interpretations. Future projects could include recalling explicitly unpleasant and anti-heroic aspects of regional development, so that eventually room is created for satire and humour that go beyond regional stereotypes and cliché. Satirical and humorous engagements with the region's homogeneous memory culture could help to deconstruct cultural patterns of interpretation. This approach is, for instance, beautifully realised in the small, local exhibition "*Herzkammer*", "core piece" (Gaß, 2017), a project that has received too little attention in the regional marketing and PR machinery. Also have recent contributions moved into this direction; they call attention to the nostalgic dimensions of memory landscapes in the Ruhr Valley (Berger, 2019), they suggest a more critical engagement with regional historical culture and concepts of identity (Berger et al., 2017; Berger & Wicke, 2017; Eiringhaus, 2019; Eiringhaus & Kellershohn, 2018), and contrast the image of the "Knowledge Metropolis Ruhr" with the history of education and knowledge (Kellershohn, forthcoming). These recent contributions also engage with representations of work and the workers' bodies (Bluma, 2018), offer people from the Valley an opportunity to tell their individual histories (Stiftung Bibliothek des Ruhrgebiets & Deutschen

Bergbau-Museum Bochum, 2014), and make visible histories beyond that associated with coal and steel (Steuwer, 2018). These works are important approaches to bring forward "the other histories" of the Ruhr Valley.

References

Angelo, H. (2015). *How green became good: Urban greening as social improvement in Germany's Ruhr Valley*. New York.

Arndt, M. (2015). Umweltgeschichte, Version: 3.0. *Dokupedia-Zeitgeschichte*.

Aufbaustab Ruhr-Museum. (2003). *Grosser Ratschlag: Stellungnahmen aus der Wissenschaft zu einem Ruhrmuseum auf Zollverein, Dokumentation der Tagung*. Essen.

Bartel, B. (2008). Das Unbehagen in der Industriekultur. Ein Zentralbegriff im Diffusen – ein Kurzinventar. *Forum Industriedenkmalpflege Und Geschichtskultur, 1*, 73–79.

Berger, S. (2019). Industrial heritage and the ambiguities of nostalgia for an industrial past in the Ruhr Valley, Germany. *Labor Studies in Working-Class History of the Americas, 16*(1), 37–64.

Berger, S., & Seiffert, J. (2014). *Erinnerungsorte: Chancen und Grenzen und Perspektiven eines Erfolgskonzeptes in den Kulturwissenschaften*. Essen: Klartext.

Berger, S., & Wicke, C. (2017). Introduction deindustrialization, heritage, and representations of identity Stefan Berger. *The Public Historian, 39*(4), 10–20.

Berger, S., Wicke, C., & Golombeck, J. (2015). Erinnerung, Bewegung, Identität: Industriekultur als Welterbe im 21. Jahrhundert. *Forum Industriedenkmalpflege Und Geschichtskultur, 2*, 15–23.

Berger, S., Wicke, C., & Golombeck, J. (2017). Burdens of eternity?: Heritage, identity, and the "great transition" in the Ruhr. *The Public Historian, 39*(4), 21–43.

Blotevogel, H. (2002). Die Region Ruhrgebiet zwischen Konstruktion und Dekonstruktion. *Westfälische Forschungen, 52*, 453–488.

Blotevogel, H., & Prossek, A. (2003). Das Ruhrgebiet und seine Identität. In A. Ruhrmuseum (Ed.), *Grosser Ratschlag: Stellungnahmen aus der Wissenschaft zu einem Ruhrmuseum auf Zollverein* (pp. 16–24). Essen: Klartext.

Blotevogel, H., Prossek, A., & Reulecke, J. (2009). Das Ruhrgebiet und seine Identität. *Forum Industriedenkmalpflege Und Geschichtskultur, 2*, 11–17.

Bluma, L. (2018). Funktionen von Repräsentationen des Arbeiterkörpers: Objektivierende und idealisierende Darstellungen im industriellen Steinkohlenbergbau. In K. Andresen, M. Kuhnhenne, J. Mittag, & S. Müller (Eds.), *Repräsentationen der Arbeit. Bilder – Erzählungen – Darstellungen* (pp. 23–40). Dietz: Bonn.

Borsdorf, U. (2003). Das Ruhr Museum: Die Grundidee. Eingangsreferat zum Grossen Ratschlag. In A. Ruhrmuseum (Ed.), *Grosser Ratschlag: Stellungnahmen aus der Wissenschaft zu einem Ruhrmuseum auf Zollverein* (pp. 105–111). Essen: Klartext.

Borsdorf, U. (2014). Industriekultur versus Geschichte? In L. Nordrhein-Westfalen (Ed.), *Industriekultur 2020: Positionen und Visionen für Nordrhein-Westfalen* (pp. 90–94). Essen: Klartext.

Borsdorf, U., & Grütter, T. (2010). *Ruhr Museum: Natur. Kultur. Geschichte.* Essen: Klartext.

Brüggemeier, F.-J. (1992). *Blauer Himmel über der Ruhr: Geschichte der Umwelt im Ruhrgebiet: 1840–1990.* Essen: Klartext.

Brüggemeier, F.-J. (2014). *Brüggemeier, Franz-Josef: Schranken der Natur: Umwelt, Gesellschaft, Experimente 1750 bis heute.* Essen: Klartext.

Dettmar, J. (1991). *Exkursionsführer zum Fachsymposium Neue Natur auf Industrieflächen, eine Veranstaltung der Internationalen Bauausstellung Emscher Park.* Gelsenkirchen.

Dettmar, J. (1999). Landschaftspark Duisburg-Nord: Vorbild für den Wandel der Industrielandschaft. In W. Hoppe & S. Kronsbein (Eds.), *Landschaftspark Duisburg-Nord* (pp. 9–19). Duisburg: Mercator.

Dettmar, J. (2003). Das Ruhrgebiet. Die Zwischenstadt. In A. Ruhrmuseum (Ed.), *Grosser Ratschlag: Stellungnahmen aus der Wissenschaft zu einem Ruhrmuseum auf Zollverein* (pp. 26–29). Essen: Klartext.

Dörries, B., Balser, M., & Büschemann, K.-H. (2013, November 26). Uhr Folgen des Kohlebergbaus Löcher im Revier. *Süddeutsche Zeitung.*

Edensor, T. (2005). *Industrial ruins: Space, aesthetics and materiality.* Oxford & New York: Berg.

Eiringhaus, P. (2018). *Industrie wird Natur. Postindustrielle Repräsentationen von Region und Umwelt im Ruhrgebiet.* Essen: Klartext.

Eiringhaus, P. (2019). Zukünfte in der Metropole Ruhr. Überlegungen zur Problematik erinnerungskultureller Homogenität. In S. Goch et al. (Ed.), *Die Stadt der Städte: Das Ruhrgebiet und seine Umbrüche* (pp. 526–532). Essen: Klartext.

Eiringhaus, P., & Kellershohn, J. (2018, August 18). Und wer zahlt die Zeche? *Frankfurter Allgemeine Zeitung.*

Fuchs, R., & Guderley, E. (2014). Checkliste der Gefäßpflanzen des UNESCO-Welterbes Zeche Zollverein und Kokerei Zollverein in Essen. *Elektronische Aufsätze Der Biologischen Station Westliches Ruhrgebiet, 25*, 1–23.

Ganser, K. (1991). *Ergebnisse des Symposiums "Neue Natur auf Industrieflächen", protocol 15.10.1991* (No. AfSB IBA 426 A).

Ganser, K. (1999). *Liebe auf den zweiten Blick: Internationale Bauausstellung Emscher Park*. Dortmund: Harenberg.

Ganser, K., & Dettmar, J. (1999). *IndustrieNatur: Ökologie und Gartenkunst im Emscher Park*. Stuttgart.

Gaß, J. (2017, April 24). *'Herzkammer' zeigt das Leben der Menschen im Ruhrgebiet*. Ruhr Nachrichten.

Goch, S. (1997). "Der Ruhrgebietler": Überlegungen zur Entstehung und Entwicklung regionalen Bewusstseins im Ruhrgebiet. *Westfälische Forschungen, 47*, 585–629.

Goch, S. (2002). *Eine Region im Kampf mit dem Strukturwandel: Bewältigung von Strukturwandel und Strukturpolitik im Ruhrgebiet*. Essen: Klartext.

Goch, S. (2011). Sinnstiftung durch ein Strukturpolitikprogramm: Die Internationale Bauausstellung Emscher Park. In G. Betz, R. Hitzler, & M. Pfadenhauer (Eds.), *Urbane Events* (pp. 67–84). Wiesbaden: Springer.

Große Kracht, K. (2014). Große Kracht, Klaus: Zwischen Gedächtnis und Geschichte: Erinnerungsorte aus kulturwissenschaftlicher Sicht. In S. Berger & J. Seiffert (Eds.), *Erinnerungsorte: Chancen und Grenzen und Perspektiven eines Erfolgskonzeptes in den Kulturwissenschaften* (pp. 57–72). Essen: Klartext.

Grüne Hauptstadt Europas, E. (2018). *Press Release: Grüne Hauptstadt Europas - Essen 2017 ist voller Erfolg*. Essen.

Hard, G. (1991). *Kommentar zu dem Text "Ergebnisse des Symposiums Neue Natur auf Industrieflächen", 4.12.1991*. (No. AfsB IBA 426 A). Duisburg.

Hasselberg, T. (2011). *Parks und Gärten auf Brachen Umnutzung industrieller Flächen im Ruhrgebiet*. Worms: Wernersche Verlagsgesellschaft.

Hauser, S. (2003). Ästhetik der Revitalisierung. In D. Genske & S. Hauser (Eds.), *Die Brache als Chance: Ein transdisziplinärer Dialog über verbrauchte Flächen* (pp. 3–26). Berlin, Heidelberg: Springer.

Hauser, S. (2005). Denkmalpflege und industriekulturelles Erbe: Von den Schwierigkeiten der Begründung von Industriekultur im 20. Jahrhundert. *Forum Industriedenkmalpflege Und Geschichtskultur, 1*, 20–24.

Hemmings, S., & Kagel, M. (2010). Memory gardens: Aesthetic education and political emancipation in the Landschaftspark Duisburg-Nord. *German Studies Review, 33*, 243–261.

Hingmann, S., Tomec, M., & van de Sand, W. (2016, November 8). Interview: Industrial Nature in the Ruhr Valley.

Internationale Bauausstellung. (1998). *Entwurf Route der Industrienatur, Vorlage für den IBA Lenkungsausschuss am 20.4.1998 (No. AfsB IBA 1178 A)* (pp. 1–5). Gelsenkirchen: Archiv für Soziale Bewegungen.

Keil, A. (2002). *Industriebrachen: Innerstädtische Freiräume für die Bevölkerung. Mikrogeographische Studien zur Ermittlung der Nutzung und Wahrnehmung der neuen Industrienatur in der Emscher Region*. Dortmund: Dortmunder Vertrieb für Bau- und Planungsliteratur.

Kellershohn, J. (forthcoming). Mobility and the crisis of intelligence. Mining industry and the negotiation of knowledge under "deindustrialisation". In T. Meyer (Ed.), *Boom – Crisis – Heritage*. King Coal and the energy revolutions after 1945.

Kommunalverband Ruhr. (1999). *Eröffnungsveranstaltung Route Industrienatur im Rahmen der Duisburger Radfahrtage am 13.6.1999 im Landschaftspark Duisburg Nord, press release, 6 June 1999*. Essen.

Kommunalverband Ruhr. (2001). *Themenflyer zur Route Industrienatur*. Essen.

Kühne, O. (2006). *Landschaft in der Postmoderne: Das Beispiel des Saarlandes*. Wiesbaden: Springer.

Kühne, O., & Bruns, D. (2015). Gesellschaftliche Transformation und die Entwicklung von Landschaft: Eine Betrachtung aus der Perspektive der sozialkonstruktivistischen Landschaftstheorie. In K. Gawronski & J. Hernik (Eds.), *Transformation und Landschaft: Die Folgen sozialer Wandlungsprozesse auf Landschaft* (pp. 17–34). Wiesbaden: Springer.

Landschaftspark Duisburg Nord. (2011). *Einer für alles – alles für Sie!* Duisburg.

Landschaftsverband Rheinland. (1999). *Katalog zur Ausstellung "Schön ist es auch anderswo…"*. Oberhausen.

Latz, P. (Ed.). (2016). *Rost Rot: Der Landschaftspark Duisburg-Nord*. München: Hirmer.

Liedtke, P. (1996). *Industrienatur: Vokabeln einer neuen Naturästhetik in einer industriell geprägten Landschaft. Begleitband zur Ausstellung 'Industrienatur'*. Mülheim.

Liedtke, P. (2015). Kulturlandschaft Montan Europe: Fotografie als identitätsprägendes und -stiftendes Medium. *Forum Industriedenkmalpflege Und Geschichtskultur, 2*, 30–35.

Macdonald, S. (1999). Theorizing museums: An introduction. In S. Macdonald & G. Fyfe (Eds.), *Theorizing museums: Representing identity and diversity in a changing world*. Oxford: Wiley-Blackwell.

Macdonald, S. (2006a). Collecting practices. In S. Macdonald (Ed.), *A companion to Museum studies* (pp. 81–98). Oxford: Wiley-Blackwell.

Macdonald, S. (2006b). Expanding museum studies: An introduction. In S. Macdonald (Ed.), *A companion to museum studies* (pp. 1–12). Oxford: Wiley-Blackwell.

Meinke, U. (2018). *Bundespräsident Steinmeier verneigt sich vor dem Ruhrgebiet*. WAZ.

Metropole Ruhr Tourismus. (2019). Extraschicht. Die Nacht der Industriekultur. Retrieved from https://www.extraschicht.de

Middendorf, J., & Hamann, M. (1999). Die Route Industrienatur. *Forum Industriedenkmalpflege Und Geschichtskultur, 1*(1), 43–47.

Oligmüller, R. (2002). Natur und Landschaft im Wandel. Potenziale, Inwertsetzung, Belastung. In H. Heineberg, A. Mayr, W. Seidel, & K. Temlitz (Eds.), *Die Emscher-Lippe-Region: Der Kreis Recklinghausen, die Stadt Bottrop, die Stadt Gelsenkirchen (Städte- und Gemeinden in Westfalen)* (pp. 9–33). Münster: Aschendorff.

Regionalverband Ruhr (ed.). (2010). *Unter freiem Himmel. Emscher Landschaftspark, Under the Open Sky. Emscher Landscape Park*. Basel, Berlin, Boston: Birkhäuser.

Reichert, C., Niemann, L., & Uttke, A. (Eds.). (2010). *Internationale Bauausstellung Emscher Park: Impulse: Lokal, regional, national, international*. Bottrop: Klartext.

Robertson, R. (1998). Glokalisierung: Homogenität und Heterogenität in Raum und Zeit. In U. Beck (Ed.), *Perspektiven der Weltgesellschaft* (pp. 192–220). Frankfurt am Main: Suhrkamp.

Ruhr Regional Association. (2019). Metropole Ruhr. Online. Retrieved April 17, 2019, from https://www.metropole.ruhr/

Ruhrkohle AG Foundation. (2018). *Glückauf Zukunft 2018. Alle Veranstaltungen rund um das Ende des Steinkohlenbergbaus*. Essen: Klartext.

Schönwald, A. (2015). Die Transformation von Altindustrielandschaften. In O. Kühne, K. Gawronski, & J. Hernik (Eds.), *Transformation und Landschaft: Die Folgen sozialer Wandlungsprozesse auf Landschaft* (pp. 63–73). Wiesbaden: Springer.

Siemann, W., & Freytag, N. (Eds.). (2003). *Umweltgeschichte: Themen und Perspektiven*. München: C.H.Beck.

Steuwer, J. (2018, August 26). Jenseits von Kohle und Stahl. Ein altes Buch lädt ein, das Ruhrgebiet neu zu entdecken. Geschichte der Gegenwart website.

Retrieved from https://geschichtedergegenwart.ch/jenseits-von-kohle-und-stahl-ein-altes-buch-laedt-ein-das-ruhrgebiet-neu-zu-entdecken/
Stiftung Bibliothek des Ruhrgebiets, & Deutschen Bergbau-Museum Bochum. (2014, 2017). Digitaler Gedächtnisspeicher: Menschen im Bergbau. Retrieved from. https://menschen-im-bergbau.de/impressum/
Storm, A. (2014). *Post-industrial landscape scars*. New York: Palgrave Macmillan.
Strangleman, T. (2013). 'Smokestack nostalgia', 'ruin porn' or working-class obituary: The role and meaning of deindustrial representation. *International Labor and Working-Class History, 84*, 23–37.
Tour de Ruhr. (2015). *Landschaftspark Duisburg-Nord: Führungen*. Duisburg.
Uekötter, F. (2003). *Von der Rauchplage zur ökologischen Revolution. Eine Geschichte der Luftverschmutzung in Deutschland und den USA 1880–1970*. Essen: Klartext.
Uekötter, F. (Ed.). (2014). *Ökologische Erinnerungsorte*. Göttingen: Vandenhoeck & Ruprecht.
Wehling, H.-W. (2016). Annäherungen an die Industrielle Kulturlandschaft Ruhrgebiet: Prozesse und Strukturen – Zonen, Achsen und Systeme. *ICOMOS: Hefte Des Deutschen Nationalkomitees, 62*, 88–102.
Winkels, R., & Zieling, G. (2010). *Landschaftspark Duisburg-Nord: From iron-works to Theme Park*. Duisburg: Mercator.
Zima, P. (2016). *Moderne/Postmoderne: Gesellschaft, Philosophie, Literatur* (4th ed.). Tübingen: UTB GmbH.

Part III

Challenges for the Future

15

Watery Places: Stories of Environmental and Community Renewal

Jessica Reeves, Fern Hames, and Michelle Graymore

Introduction

Water is important for everyone; we are all connected by our need for water to drink, to cleanse ourselves, to produce food and cook and to renew ourselves. Water reminds us that we are connected to nature—both through our dependence on this vital resource and our ability to impact it. In Australia, the history of manipulation of our waterways extends back over 6600 years through, for example, the volcanic plains of western Victoria where the Indigenous Gundjitmara people used basalt rocks and boulders to control the flow of water for food harvesting (McNiven et al., 2012). Post-European impact on our waterways is also evident, through vegetation clearing; removal of instream woody habitat; construction of locks, dams and weirs to hold water; channel straightening; over-extraction

J. Reeves (✉) • M. Graymore
Federation University Australia, Churchill, VIC, Australia
e-mail: j.reeves@federation.edu.au

F. Hames
Arthur Rylah Institute for Environmental Research, Melbourne, VIC, Australia
e-mail: fern.hames@delwp.vic.gov.au

A. Campbell et al. (eds.), *Located Research*,
https://doi.org/10.1007/978-981-32-9694-7_15

and contamination (Walker, 2019). The 'control' of our waterways has enabled our society to thrive, through navigation—allowing the movement of people and produce of our inland areas to reach distant markets—and through irrigation and water diversion, to support establishment of towns and ensure that food is grown in abundance, withstanding the challenges of a highly variable climate. Our waterways have long formed boundaries—not just of geography, but of culture, land management and civic borders. But they also unite us, giving us a sense of identity and connection through the catchment from the headwaters in the uplands, through the tributaries, to the estuaries of the coast.

Here we present three case studies of regional townships in Victoria, Australia, and their connections to water through stories documenting their different stages of renewal (Table 15.1, Fig. 15.1). The towns vary significantly in terms of population size, geography, rainfall and major industry. Each of the towns has seen shifts in major industry since European settlement and suffered from significant social and economic impacts in the past decade. The researchers have worked with these communities in various projects related to water supply, preservation of biodiversity and ecological restoration, associated with the waterways in these towns.

This study is timely, following the hottest summer (2018–2019) in Victoria on record, with below average rainfall and drought conditions in

Table 15.1 Characteristics of three case studies of regional townships in Victoria

Town	Marysville	Maryborough	Morwell
Population[a]	394	7921	13,771
Indigenous Custodians	Taungurung	Dja Dja Wurrung	(Gunai)kurnai
Major Industry	Ex. Supply Town Tourism	Ex. Gold Mining Wool Highland Festival	Agriculture Coal Mining Power Generation
Waterway	Leary's Creek, Robertson's Gully	Four Mile, Tullaroop and Flat Creeks	Morwell River Latrobe River
Average Rainfall[b] (mm)	1342	527	722
Rainfall 2018 (mm)	1166	418	499
Researcher	Fern Hames	Michelle Graymore	Jessica Reeves

[a]Population at 2016 Census (ABS, 2019)
[b]Mean annual rainfall 1984–2019 (Bureau of Meteorology, 2019)

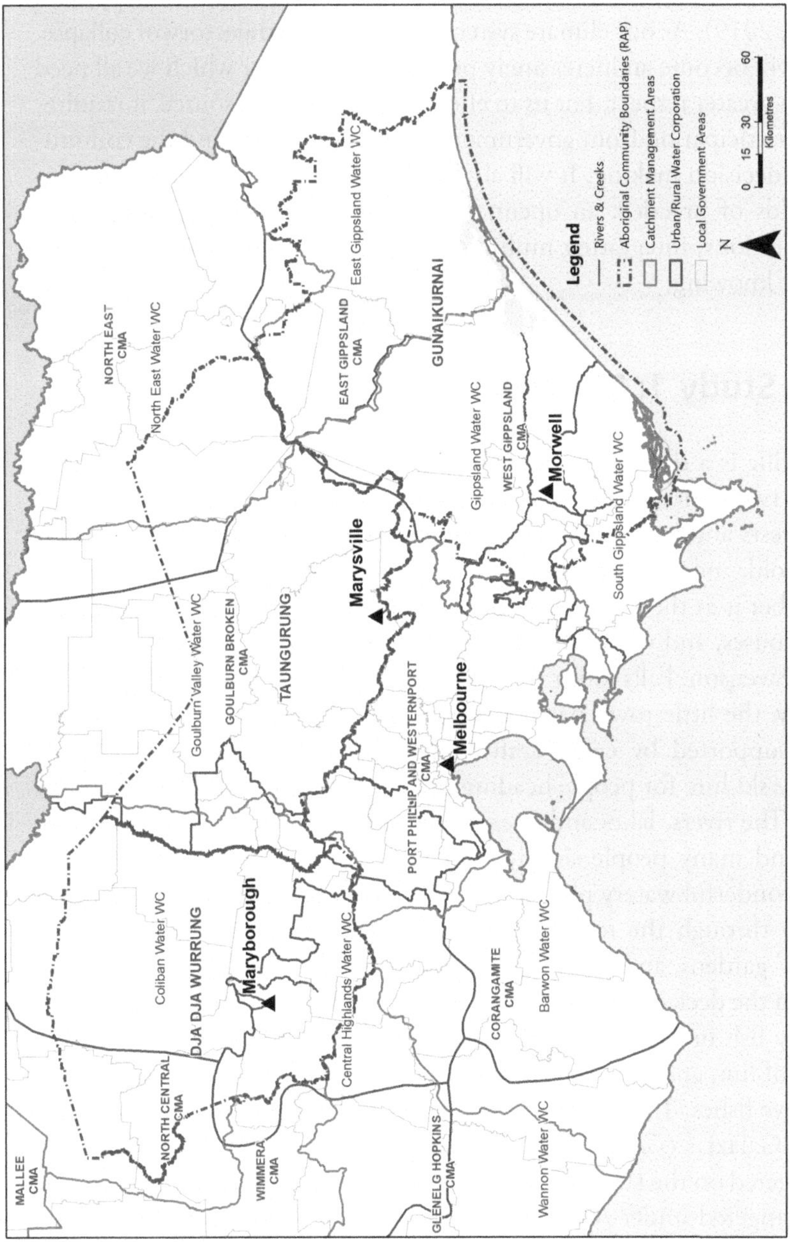

Fig. 15.1 Locality map showing towns and watercourses discussed in this piece. Also depicted are some of the boundaries that define water and land management in the regions, including relevant Aboriginal Community Boundaries, Catchment Management Areas, Water Corporations and Local Government Areas

the south and east of the state, and extensive bushfires across Gippsland (BOM, 2019). As our climate system continues the trajectory of collapse, water will become an increasingly precious resource for which we all need to learn greater respect. For us to effectively share this resource, it requires active participation from government, water authorities and the community in decision making. It will also require a transformation in our various fields of practice; an openness to new ways of doing things, new collaborations, integrating multiple disciplines, and embracing multiple ways of knowing.

Case Study 1: Marysville

Marysville is a tiny town, nestled in the foothills of the Great Dividing Range, two hours north east of Melbourne. It's surrounded by Mountain Ash forests and pockets of cool-temperate rainforest, and ribboned with clear, cold, mountain streams. Many people of my parents' generation remember it as the place they honeymooned, in rambling pseudo-Tudor guest houses, and where they made weekend day trips to see the cascading Steavenson Falls drop 400 feet into the tree ferned gully. More recently, the little town has flourished with a kind of 'weekender economy'; supported by cafes, craft shops, overnight cottages, and cross-country ski hire for people heading up to the nearby Lake Mountain ski resort. The rivers, lakes and streams also attract recreational fishers, kayakers and many people simply enjoying the amenity and aesthetics of these wonderful watery places. One tiny stream; Leary's Creek, meanders directly through the town, through the playground, at the bottom of peoples' gardens, and below a popular patisserie.

From the deck of that patisserie, many people have delighted in watching tiny fish in the stream. The local streams are beautiful, and provide plenty of fun, and they are also critically important for endemic freshwater native fishes. The fish which people could watch in Leary's Creek were Barred Galaxias (*Galaxias fuscus*). Barred Galaxias are listed as Critically Endangered on the IUCN Red List of Threatened Species (IUCN, 2019), as Endangered under Australia's *Environment Protection and Biodiversity Conservation Act 1999*, as Threatened under *Victoria's Flora and Fauna*

Guarantee Act 1988 and as Critically Endangered under the *Advisory List of Threatened Vertebrate Fauna in Victoria* (DSE 2013). You could say they're in trouble. They are very beautiful little fish, only found in these higher elevation, cool streams of the Upper Goulburn river system. They are about as long as your longest finger and the colour of rich gold or ochre. As their name suggests, they are distinguished by markings of thick dark bars. The main reasons they are in trouble are 'drought and trout' (Raadik, 1993). The streams they live in are small and increasingly subject to drying. And the small size of Barred Galaxias means that they have become snacks for introduced, predatory trout (Raadik, 1993; Raadik, Saddlier, & Koehn, 1996).

Both Rainbow Trout (*Oncorhynchus mykiss*) and Brown Trout (*Salmo trutta*) have been introduced to Australia for recreational fishing purposes, and both species are found in the streams around Marysville. They support a very popular fishery, for local people and for fishers from Melbourne and elsewhere. For a few decades, this triggered a divide between trout anglers and native fish conservationists. Trout anglers wanted freer rein to fish for trout throughout the region. Fish scientists wanted streams with Barred Galaxias in them to remain free of the predatory trout. There were plenty of angry words and threats, campaigns with misinformation and zero trust between the two 'camps'. Around 2007 a small group of people, with representatives from both 'sides', began a conversation to attempt to resolve the divide. The group included representatives from the Australian Trout Foundation, the Victorian recreational fishing 'peak body' VRFish and our Arthur Rylah Institute for Environmental Research (part of Victorian Department of Environment, Land, Water and Planning). We shared individual objectives, recognised each other's values, clarified misunderstandings and identified a shared common ground. We all shared a vision for healthy waterways and committed to identify ways to work together to improve the relationship, and to agree on priorities for particular waters, recognising that some waters support valuable recreational fisheries and some are important in supporting native fishes.

At the same time, we were working to raise the visibility of the significance of the Barred Galaxias locally, with the Marysville community. It's pretty special to have such an important, endemic species in your 'backyard' and we wanted to help the local community know more about this golden treasure in their local creek, and hopefully advocate for it and its habitat.

Then 'Black Saturday' happened. On February 7, 2009, following ten years of relentless drought, a huge wildfire engulfed almost half a million hectares (more than a million acres) across Victoria. A total of 173 people died; including 45 in Marysville alone (Parliament of Victoria, 2009). The township of Marysville was effectively obliterated; around 90% of the buildings were destroyed; only 14 houses remained. This was a catastrophe, a tragedy. It is impossible to overstate the impacts of this historic fire on Marysville, indeed across the state of Victoria.

As well as the enormous human impacts, there were massive impacts to natural values too. Around Marysville, several threatened species were affected, including Leadbeater's Possum, Sooty Owls and Barred Galaxias. Although direct heat can be an impact of wildfires on fish (Isaak et al., 2010), the biggest likely impacts follow after the fires. When the first post-fire rains fall, they can wash large volumes of ash and silt into waterways. This massive mobilisation of sediment effectively 'clogs' streams, infills the spaces between rocks and pebbles in stream substrate, and can clog the gills of fish (Lyon & O'Connor, 2008). Fish simply cannot breathe. Their habitat is smothered, aquatic insect food sources are smothered; the capacity for fish to breed collapses (Lyon & O'Connor, 2008). On 'Black Saturday', the fires were so intense and extensive that we anticipated such sedimentation would be a major issue. We decided to undertake an 'insurance measure': a 'fish rescue', by removing a proportion of the fish population from their streams, before the rains came, and housing them in chilled aquaria at our state government research institute in Melbourne. We would return the fish when their habitat recovered sufficiently to support their survival (Raadik, Fairbrother, & Nicol, 2009).

But Marysville was still a crime scene. Police closed the roads to prevent access. The few people still living within the fire affected area (those who hadn't evacuated) were traumatised and suspicious of vehicles and people entering the area. Securing access was possible, but how to work in this area amongst a traumatised community? How to avoid further impacts? Could we find a way to not only avoid impact, but actually support the community?

As part of our training to guide emergency agency and government response to these fires, across a range of contexts, we were provided with advice from trauma psychologists (Gordon, 2009; Hawe, 2009). In summary, it was clear that there were five key things we should consider when working with an affected community, to support recovery. These included a sense of:

1. Safety and calm
2. Hope
3. Connectedness
4. Control
5. Things 'getting back together'.

We thought carefully about how we could incorporate these elements into our activities in Marysville and surrounding areas. We planned a suite of safe, calm activities to give people a sense of connectedness with each other and the environment, we enabled their ownership of decisions in activities such as design of a new interpretive bridge about the fish (control), and we provided regular updates about fish recovery (hope and things 'getting back together') (Hames, 2012). We deliberately linked with events organised by the community themselves, provided a choice of activities across a spectrum of involvement (from simply a fun display at a music festival to active involvement in habitat recovery). We circled back to the Australian Trout Foundation and recognised an opportunity for us all to work constructively together; we all planted trees, shrubs and grasses along the banks of Leary's Creek to support riparian recovery; anglers, scientists and local community members; together.

As habitat began to recover across the district and we began to return fish to their natal streams, residents recognised the synergy in their shared story. People's homes were destroyed on Black Saturday in Marysville; they moved to Melbourne and other places whilst their homes were rebuilt, then many returned. On Black Saturday, the Barred Galaxias' habitat was destroyed; they were moved to Melbourne while their 'homes' recovered, and then returned. This shared story and journey connected people with their little local fish. People integrated the fish story in local community actions. At the celebration opening of a new adventure playground, the community organised a 'parade' of fish art sticks, representing the Barred Galaxias, through the town. As the parade of more than 100 people wound its way towards the stream, participants sang a song (created by the local kindergarten) about the fish. When they reached the stream, people thrust their fish sticks (triumphantly, it seemed to us) into the stream bed and declared; 'the fish are back, and so are we!' This temporary art installation activity was completely created by the community; they had taken on the story of their resilient little residents; the Barred Galaxias. The new playground also included a series of braille plaques; one of these had a drawing of a fish, and the words 'Barred Galaxias' in text and braille. The community played a key role in designing an interpretive bridge in the town, describing the special status of the fish, and its story of rescue and recovery. A new art gallery in Marysville is called 'Little Fishes', after the Barred Galaxias, and the owner shares brochures and stories about the fish with visitors.

We witnessed a process of paired recovery, in which news of the recovery of the Barred Galaxias fish buoyed people; provided hope and a sense of things getting back together and supported their emotional and psychological recovery. In return, this connectedness with the story, and the fish, surfaced advocacy for the fish in various ways. So, natural recovery supported human recovery, which supported further natural recovery, and so on.

The Australian Trout Foundation (ATF) also took on an extended advocacy for the Barred Galaxias. They committed to act as 'Guardians of the Barrier' maintaining the physical barrier in Leary's Creek which prevents upstream movement of trout into the section which supports Barred Galaxias. The ATF also celebrated their advocacy; they produced organisational banners and other materials with photos of their efforts, wrote about it in various publications, and proudly described it, as an indicator of their evolved position. These efforts demonstrate that the ATF is a contemporary recreational fishing organisation with a high level of regard and recognition for native biodiversity values. This position has since extended even further, with their active involvement in support for the Endangered Spotted Tree Frog (*Litoria spenceri*) and alliance with Native Fish Australia.

We see opportunities for other communities and contexts, particularly an opportunity for identifying other stories which resonate with people about local natural values, build synergies, strengthen connections with place and local waterways. In addition, we anticipate there will be increasing opportunities to adopt the practice of integrating psychological principles of supporting disaster-affected communities in recovery programs. In a world with increasing intensity and frequency of extreme events such as fire and flood (BOM, 2019), there will likely be increasing need to support traumatised communities; communities in need of safety, calmness, control, and sense of return to (new) "normal life". Connecting river communities with natural values and their recovery can play an important role in hope and renewal.

Case Study 2: Maryborough

In water scarce environments, such as Maryborough, connection to water is strong. People are acutely aware of the impacts of drought; particularly on water available for use in homes, gardens and farms, in waterways, in green space like gardens and parks, as well as for the animals we care for and with which we share the environment. Consequently, it is in these places where rainwater tanks and rain gauges are common. People meeting in the street discuss rain, how long it has been since it last rained, when it might come, how much fell when it did, the damaging effects when it arrives all at once in intense events, as well as the effects the lack of rain has on the community and its environment. We expect climate breakdown will amplify the importance of water to communities, through longer periods of dry conditions, heatwaves that sap the landscape of its greenness and the community of its energy, and through more intense rainfall events.

For rural communities in areas of low average rainfall, climate collapse is impacting on the security of water supply to towns and farmers alike. From drought conditions reducing availability, to heatwave conditions causing cyanobacteria blooms which render the water unusable for both people and animals, to fire risk where water resources may be allocated to fight fire or drinking water, catchments may be threatened. Thus, the climate crisis is making the job of providing clean water to small towns and farms more challenging for water authorities. Community and regional water managers (all agencies involved in water management including water authorities) have to work closely together to ensure there is water available for those who need it most during the driest times, such as hospitals, fire brigades and community for essential uses. Communities need to trust that the water managers are making decisions that have the best interests of communities in mind as they prioritise how much water is allocated to drinking, farming, environment and recreation when there is not enough water to service all needs. Yet in my water management research in Victorian rural communities, I have seen that many people in the community do not trust the regional water managers. This trust appears to have been eroded by a range of factors: perceived mismanagement of the water supply; perceptions that decisions on water allocations occur without consultation; the lack of knowledge in the community about water management and supply; the centralisation of water management into large water corporations or authorities in the early 1990s which resulted in the loss of the local water manager. This has led to some people in the community feeling as if their needs and values are rarely considered when decisions on water management are made. Further, distrust can manifest in places, such as in Maryborough and its surrounding areas, where the community feels they are provided with substandard water—water that tastes and/or looks unfit to drink or bathe in (Graymore & Schwarz, 2016).

Water has always been scarce in Maryborough. The local Indigenous people, the Dja Dja Wurrung, recognised this and built a series of water wells in the sandstone south of Maryborough to capture rainwater for drinking (CGS, 2019). The discovery of gold in the surrounding area brought people from around the world to Maryborough. The gold brought wealth and the buildings in the town are a monument to this wealth. Fast forward 150 years, the gold is long gone and many of the industries that provided jobs have the left the area. Unemployment is high and the community is now one of the most disadvantaged in Victoria (Vinson, Rawsthorne, Beavis, & Ericson, 2015). Domestic violence, suicide, alcohol and drug abuse and mental health issues are all too common for the people in Maryborough (Department of Health and Human Services, 2015; Vinson et al., 2015). There is also a high level of cancer and other health issues (Department of Health and Human Services, 2015). The water supply, until recently, had high levels of hardness and total dissolved solids from the source water and chlorine residues from the disinfection process (Central Highlands Water, 2015). This caused the town water to have an unpleasant taste and smell. This quality issue is a result of the available water sources for the town; however, the Millennium drought made it worse as the water available in storages became scarce and groundwater became the predominant resource.

Yet 45 minutes away life appears better in towns like Ballarat and Bendigo. They have 'good tasting water' (from a different source to Maryborough), plenty of jobs and seemingly less social-related issues. The water corporation that supplies Maryborough, Central Highlands Water (CHW), is located in Ballarat, although there is a small customer service office in Maryborough's main street. Many residents travel to Ballarat for work or shopping. They know the water is different and better for those who live in Ballarat. This has led to a feeling in the community that the water corporation does not care about Maryborough and its community, they only care about Ballarat, adding to the overall sense of isolation, abandonment (by industry and government services) and victimisation in the community. At the time of this case study, CHW was about to upgrade the water treatment plant in Maryborough (Graymore & Schwarz, 2016) to enable them to provide better quality water to the town.

To understand how this community related to the town water supply, I spent time there drinking and bathing in the water and listening to people's water stories. The water smelt bad. The first time I showered in it, I wondered what chemicals the motel's cleaners had used in the bathroom as the chemical smell was so strong! My skin seemed to suffer, it was drier than normal and itchy. The water tasted bad too, it left a chalky feel in my mouth and made my tea undrinkable. At the CHW office in town, the staff were drinking water from a cooler sourced from another water supply. This spoke volumes to me about their feelings about the local water supply, and I suspect the rest of the community felt it too.

I asked the Maryborough community to complete a survey about how they used water in their houses, gardens and businesses, what sources they used and their understanding of the water supply and water managers (see Graymore & Schwarz, 2016 for full results which I have discussed here). Once people realised I was not from the water corporation they were keen tell me what they thought. Most people did not trust that the water was safe to drink, with around 80% of people drinking water from other sources. They bought bottled water or drink rainwater from tanks attached to their house, some people even got water from other towns to drink. They were worried about the chemicals in the water and the impact they have on their health. Around half of the people surveyed felt that there were health risks associated with drinking the water, with people discussing their fears around the water causing skin irritations and rashes and other health problems, as well as suspicions it was causing the high level of cancer rates in the town. Many people had experienced the same dry, irritated skin that I experienced which they felt was related to bathing in the water. Nearly everyone had experienced scaling on their water using appliances including hotwater systems and kettles caused by the high level of hardness of the town water supply. People needed to replace these appliances much more often in Maryborough than in other towns, causing a greater financial impost. The high level of hardness meant that the hospital filtered their water to enable them to use it for sterilisation and other uses to prevent damage to their equipment, as even marine grade steel was affected by the hardness of the water. Many people in the community talked about how they were embarrassed by the quality of the water, telling visiting friends or family not to drink it, and providing them with either bottled water or water from other sources. Some cafes provided a bottle of water free to customers.

Most of the community did not understand what was causing the water to taste bad; they were unaware of where the water comes from, what was in it that causes the taste and smell, what the purpose of the treatment process is, or in many cases who the water manager is in the community. Historically it was the council (local government), and some people still think it is. This lack of knowledge around the water supply has disconnected people from their water supply and led to many feeling it is causing health issues. Those that do know the water corporation have little trust in them to supply water that is safe to drink and bath in. This is why so many people bought bottled and/or installed rainwater tanks for drinking and bathing. For a community that is in the top 5% in Victoria for social disadvantage; buying water at an average $2.75 a litre, filters, rainwater tanks and the regular replacement of water using appliances is an added financial burden they can ill afford. Maryborough now has access to good quality water thanks to CHW's $5.6 M upgrade of the water treatment plant in early 2017. However, to what extent the community have started using tap water needs further investigation.

When the results of this study were presented to CHW, they were surprised at the level of distrust in the community and the high proportion of people who felt the water was bad for their health. They did not realise the community felt they did not care about Maryborough, nor that their campaign 'tap is best' designed to increase tap water use, was making the community feel CHW had no idea how bad the water was to drink in Maryborough. CHW were keen to try a different engagement process with the Maryborough community going forward to build their trust and understand their needs, which they planned to implement along with the upgrade. Further research is being carried out at the time of writing to see how effective their engagement was at building trust and encouraging people to use the town water supply.

This study demonstrates that water managers are not always well-connected to the needs of the communities they serve. Thus, there is an opportunity for regional water managers to reimagine how they engage and connect with small communities as the challenges of water supply and management intensify with the climate crisis. They must build trust with residents to ensure people comply with water restrictions when water is scarce. They need to build knowledge about water supply, treatment, allocation, management and availability to enable community to understand and participate, where possible, in the decisions that impact their lives. Water managers also have to develop innovative ways to bring communities into water allocation decisions to enable them to better understand the hard decisions about water management, give them opportunities to put forward their needs and values so that they can be included in the decision making process, providing them a sense of control over the situation. Further, water authorities could consider ways to have a local presence in these small communities, either with regular community two-way chat sessions, a pop-up or drop-in office, or a local champion/advocate who has a direct line between the community and water managers. Water managers will need to reconsider how they engage with rural communities in order to build trust and connection with these communities and their water supply to enable management of this precious resource to become more collaborative as the challenges of water management increase in a changing climate.

Case Study 3: Maryvale (Morwell)

Ask anyone over the age of about 60 in the Latrobe Valley how they spent their summers as a child and they will tell you: swimming in the Morwell River, catching eels and freshwater crayfish. With its headwaters in the beautiful Strzelecki Ranges surrounded by towering Mountain Ash and tree ferns, the Morwell River has been the great connector for the Latrobe Valley. A place of peace and recreation, but also essential in facilitating its major industries: the coal-fired power stations, timber plantations and downstream, the Australian Paper mill and factory at Maryvale. This river once teemed with life, as it meandered its way across the valley floor before joining the Latrobe River, which meets the Thomson River, then enters the internationally significant, Ramsar-listed Gippsland Lakes. There are still areas of the Morwell River where people go to relax and actively engage with nature. The Morwell National Park is truly a hidden gem, where you are dwarfed by towering trees. The Morwell River Falls are popular with locals looking to escape the summer heat (but don't tell too many people!). The Latrobe Valley itself has deep scars of the coal industry that supported Melbourne to expand and prosper over the second half of the twentieth century. And it is undergoing another transition as we move away from this resource. But what comes next?

Coal mining has been a feature of the Latrobe Valley since the late 1890s, commencing at Yallourn North. The river has always played its part. In the early days, the Morwell River was used to sluice away the overburden and send it downstream, vastly increasing sedimentation rates and turbidity. However, the river often had its own way, flooding the mine pit on numerous occasions. Coal mining scaled up in the Latrobe Valley in the 1940–1950s with the opening of the coal-fired power stations at Hazelwood and Yallourn. These are vast operations and the Morwell River was straightened and channelised, diverted around the Morwell mine pit and through the centre of the Yallourn open cut, dividing it into two giant lobes. The river fought back against constriction, breaking the levees and flooding the mine for 18 months in 2011. A giant pipeline was constructed across the mine to constrain 3.8 km of the river while a new channel was engineered (Gem Industrial Solutions, 2019). The water from the river is also used in the power station operations, such as in the cooling towers of the Yallourn Power Station.

When Hazelwood Power Station was operational, it discharged into the Hazelwood Pondage. This was a natural waterway (ironically, called Eel Hole Creek) that was impounded to capture the hot water coming from the power station, resulting in a permanently warm water pool. This warm pool in the otherwise cool-temperate valley became a very popular recreation spot, housing a caravan park, yacht club, fishing club and water-skiing club, which proudly held a mid-winter event in the toasty waters. The Pondage also hosted tropical fish, including illegally traded exotic cichlid fish, native to Lake Tanganyika in Africa which presented a threat to native fish, and was stocked with barramundi in 2015 to support recreational fishing.

The Hazelwood Power Station closed somewhat suddenly in March 2017. Many jobs ceased and the community mourned for a loss they knew was coming—but didn't expect to be quite so soon. The turning off of the power station also meant the cooling of the Pondage. First the water skiing stopped, then, not surprisingly, the artificially stocked tropical fish died. A recent engineering audit found the dam's retaining wall to be unsafe, necessitating the water level in the Pondage to be dropped (Engie, 2019). The yacht club was asked to leave, the caravan park shut and now the Pondage itself has been permanently closed to the public. I have been to many public meetings about the closure of the power station, the future of coal mining in the Valley and the plans for rehabilitation, but I have never seen so many angry and upset people as when they were told they could no longer use the Pondage. Losing their jobs was one thing, but losing the amenity of the Pondage was too much.

The power stations are closing—so will the brown coal mines. The vast cavities they leave behind will need to be filled, in order to be left 'fit for purpose'—but what that purpose will be is up for debate. The essential remit for mine closure is that the areas are left safe, stable and sustainable (DEDJTR, 2018). The most direct and cost-effective way to achieve this is through filling them with water. The volume of water required to fill all three large coal pits in the Latrobe Valley is at least 2200GL—or '4 Sydney Harbours', in the common vernacular, in addition to annual top-up to compensate for evaporation (DEDJTR, 2018). Modelling has demonstrated that this will take around 80 years to accomplish, with minimal impact on the surrounding towns and major infrastructure, such as the highway.

Climate is changing: the last two summers have seen drought in central and east Gippsland. The Latrobe Valley community is nervous about seeing that much water being used to drown the coal-face that once supported the towns, when it could equally be used for agriculture in the productive region or to lessen the impact of the salt wedge that is making its way steadily up the Gippsland Lakes system (EPA, 2015). There is frustration that the decision to fill the pits has already been made without the community really comprehending how this will be achieved, why it is necessary and just how long it will take before the sites can be 'useful' again. There is still much work to be done to determine how to achieve the 'safe, stable, sustainable' outcome—not only for the pits themselves, but also for the surrounding towns and all other water users in the Valley and downstream. There is equally as much work required to involve the community in the process—both the risks and opportunities, so truly informed decision making can be made. Social license can only be gained when the community understands the limitations and can be given the chance to engage in the rebuilding process, within these boundaries.

The discussions of rehabilitation of the area surrounding the coal mines have focussed on returning what the region was before mining—to rich pasture for cattle grazing for beef and dairy herds. But that was not the original state. The Latrobe Valley is home to the Brayakaulung people of the (Gunai)kurnai. This well-watered, fertile region supported a strong population of people, who used the river and floodplain in a different way. They feasted on Short-finned Eels (*Anguilla australis*) from the river, used the Water Ribbon (*Triglochin procerum)* tubers for food and Spiny-headed Mat Rush (*Lomandra longifolia*) leaves for basket-weaving, and Black Wattle (*Acacia mearnsii)* for crafting boomerangs and nulla nullas (Park Tracks, 2015). Many of these are depicted in the paintings of Ronald Edwards along the Waterhole Creek Cultural Trail in Morwell.

The landscape was quite different (EVC 55: Plains Grassy Woodland). The Kurnai elders feel the scars in the country that the mines have caused and the land clearing before that. They want to see their country healed—and they also want meaningful lives and employment for their people. There is work underway by the Gunaikurnai Land and Water Aboriginal Corporation in partnership with the West Gippsland Catchment Management Authority on cultural mapping of the Latrobe Valley from the fragments of knowledge that have been retained. The landscape has been so extensively altered, that these links to the past are hard to distinguish—but they do provide an opportunity to better plan for the future.

Once a thriving environment, society and economy, the Morwell River and its townships and people have suffered from decades of (ab)use and neglect. Is there an opportunity here to do better? Is it possible to return the river to health and in doing so, return it to the people? Can we then go one step further and find ways to once again reap the rewards of the healthy river sustainably and share this produce and learnings for responsible economic benefit also? I think there is a way.

If the mine pits are filled to create lakes, there will be a need to manage any water coming from these pits before they enter the main waterways and flow down to the Gippsland Lakes, as anoxic waters or algal blooms are probable, given the depth and potential stagnation of the waterbodies. If we look to the environmental and cultural heritage of the region with regards to indigenous aquatic plants and fisheries, we can reimagine the river, by returning native fish—in particular eels—and plants to the region and developing economic and educational outcomes. How? By Indigenous-driven enterprises of sharing the knowledge of how these resources were utilised and developing them for community-led economic development. Re-wilding the Morwell River provides an opportunity that brings about a truly sustainable business model, addressing environmental health, empowerment of Indigenous and non-Indigenous communities and economic benefit, whilst celebrating the resources of the river. This is a long term project that begins with the rehabilitation of the waterway, but works through community knowledge sharing, through to the development of a viable niche product, targeted at high-end export as well as education and tourism opportunities. Eels have long been fished in the

Morwell system, both by Indigenous and non-Indigenous people. But this is a resource currently under-utilised. Here in the Morwell River, we have the opportunity to put eels and the associated aquatic plants back on the table, as a symbol of bringing the river back to health and returning it to the community. This project can only progress with the support and inclusion of the Registered Aboriginal Party, the Gunaikurnai Land and Waters Aboriginal Corporation and the broader Latrobe Valley community.

Conclusion

We recognise the various forms of connection to water, waterways, and the life within them, and the role water has in supporting the cultural, social, environmental and economic lives of all those in river communities and beyond. This is acknowledged in the growing body of literature on Blue Space (Gledhill & James, 2008) and its connection with human health (Gascon, Zijlema, Vert, White, & Nieuwenhuijsen, 2017; Völker & Kistemann, 2011) as well as social and psychological benefits (de Bell, Graham, Jarvis, & White, 2017). The connection between water and people runs deep, with Whalley (1988) stating that connections to waterscapes provide a sense of place and "a symbolism difficult to achieve with any other natural element."

We imagine a future where we are all open to doing things differently; including recognising, considering and integrating other perspectives, and transdisciplinary approaches to bring together different place-based 'ways of knowing'. To achieve this we need to apply active listening, build genuine partnerships and enable innovative solutions. Such solutions likely integrate ideas which wholly respect Indigenous knowledge and connections. Clear examples of this are available in New Zealand, through ecosystem frameworks developed through Māori knowledge; whereby aspirations and wellbeing are interdependent on ecosystem health and services (Harmsworth & Awatere, 2013). These include the Cultural Health Index, Māori Wetlands Indicators and Mauri Assessment Model. In Australia, there is also important progress towards greater knowledge sharing between water managers and Indigenous Elders around cultural flows and allocation of such flows (MLDRIN et al., 2017).

Through improved waterway and water supply health, renewed respect can be gained for the authorities responsible for maintaining our water supply and managing the competing interests around this most valuable of assets. Active engagement and involvement of the community can provide agency in this process, empowering a sense of common ownership and responsibility. As well as biodiversity and social health, healthy water also leads to economic benefit through increased recreation both on and around the waterway, ecotourism enterprises and increased property value (Mahan, Polasky, & Adams, 2000). Each of the case studies here provides examples of significant threat to community fabric through impact to waterway health—but also the opportunities for renewal for both the environment and society. Given the challenges to our waterways and water supply that is anticipated with climate collapse, the only way forward is for us all to be connected within our regions, communities and catchments to establish respect, trust, celebration and renewal.

Acknowledgements The authors wish to acknowledge Central Highlands Water, who funded the research in Maryborough. The Marysville work was part of the 'Rebuilding Together' program, funded by the Victorian and Commonwealth governments' Statewide Bushfire Recovery Plan after the 2009 wildfires. We are also grateful for access to the Jimmy Watson's Naval Club for use as makeshift office space. We would like to thank Andrew Geschke at the Arthur Rylah Institute (ARI), Department of Environment, Land, Water and Planning, Victorian State Government, who drafted the map for this chapter.

References

Australian Bureau of Statistics. (2019). 2016 Census QuickStats. Retrieved May 30, 2019, from http://quickstats.censusdata.abs.gov.au

de Bell, S., Graham, H., Jarvis, S., & White, P. (2017). The importance of nature in mediating social and psychological benefits associated with visits to freshwater blue spaces. *Landscape and Urban Planning, 167*, 118–127.

Bureau of Meteorology. (2019). Retrieved May 31, 2019, from http://www.bom.gov.au/climate/current/season/vic/summary.shtml

Central Goldfields Shire. (2019). Retrieved May 1, 2019, from http://www.visitmaryborough.com.au/arts-culture-heritage/aboriginal-wells

Central Highlands Water. (2015). *2014/15 Water quality report*. Ballarat: Central Highlands Water.

Department of Economic Development, Jobs, Training and Resources (DEDJTR). (2018). *Latrobe Valley regional rehabilitation strategy program summary*.

Department of Health and Human Services. (2015). 'Central Goldfields Statistical Profile' State of Victoria, Melbourne.

Department of Sustainability and Environment (DSE). (2013). *Advisory List of Threatened Vertebrate Fauna in Victoria*. Melbourne: Victorian Government Department of Sustainability and Environment. Retrieved from https://www.environment.vic.gov.au/__data/assets/pdf_file/0014/50450/Advisory-List-of-Threatened-Vertebrate-Fauna_FINAL-2013.pdf

Engie. (2019). Retrieved May 23, 2019, from https://engie.com.au/news/hazelwood-pondage/

Environmental Protection Authority. (2015). *Gippsland Lakes and catchment literature review*. Report.

Gascon, M., Zijlema, W., Vert, C., White, M. P., & Nieuwenhuijsen, M. J. (2017). Outdoor blue spaces, human health and well-being: A systematic review of quantitative studies. *International Journal of Hygiene and Environmental Health, 220*(8), 1207–1221.

Gem Industrial Solutions. (2019). Retrieved May 23, 2019, from www.geminstrial.com.au/?project=yallourn-river-diversion

Gledhill, D., & James, P. (2008). 'Rethinking urban blue spaces from a landscape perspective. Species, scale and the human element'. In: J. Breuste, (Eds.), Ecological perspectives of urban green and open spaces. *Salzburger GeographischeArbeiten, 42*, 151–164.

Gordon, R. (2009). *Community and psychological responses to disaster*. VPS briefing.

Graymore, M., & Schwarz. (2016). *Water Connections: exploring town water perceptions and use Maryborough water supply district*. Federation University Australia, Mt Helen.

Hames, F. (2012). *Engaging the Community in Native Fish Recovery Following Bushfire: Black Saturday Victoria 2009 Natural Values Fire Recovery Program*. Heidelberg, Victoria: Department of Sustainability and Environment.

Harmsworth, G. R., & Awatere, S. (2013). Indigenous māori knowledge and perspectives of ecosystems. In J. R. Dymond (Ed.), *Ecosystem services in New*

Zealand – Conditions and trends. Lincoln, New Zealand: Manaaki Whenua Press.

Hawe, P. (2009). *'Community recovery after the February 2009 Victorian bushfires: An Evidence Check' rapid review brokered by Sax Institute for the Public Health Branch*. Victorian Government Department of Health.

International Union for Conservation of Nature (IUCN). (2019). The IUCN Red Kist of Threatened Species. Version 2019-1. Retrieved May 30, 2019, from https://www.iucnredlist.org/

Isaak, D. J., Lucec, C. H., Rieman, B. E., Nagel, D. E., Peterson, E. E., Horan, D. L., et al. (2010). Effects of climate change and wildfire on stream temperatures and salmonid thermal habitat in a mountain river network. *Ecological Applications, 20*, 1350–1371.

Lyon, J. P., & O'Connor, J. P. (2008). Smoke on the water: Can riverine fish populations recover following a catastrophic fire-related sediment slug? *Austral Ecology, 33*, 794–806.

Mahan, B., Polasky, S., & Adams, R. (2000). Valuing urban wetlands: A property price. *Applied Landscape Economics, 76*, 100–113.

McNiven, I. J., Crouch, J., Richards, T., Gunditj Mirring Traditional Owners Aboriginal Corporation, Dolby, N., & Jacobsen, G. (2012). Dating aboriginal stone-walled fish traps at Lake Condah, southeast Australia. *Journal of Archaeological Science, 39*(2), 268–286.

Murray Lower Darling Rivers Indigenous Nations (MLDRIN), Northern Basin Aboriginal Nations (NBAN) & North Australian Indigenous Land and Sea Management Alliance (NAILSMA). (2017). Cultural Flows Field Studies Final Report. Retrieved June 6, 2019, from http://www.culturalflows.com.au/

Park Tracks. (2015). Aboriginal Cultural Heritage Trail (Waterhole Creek). Retrieved June 5, 2019, from https://walkingmaps.com.au/walk/3076

Parliament of Victoria. (2009). *Victorian bushfires royal commission report – summary*. ISBN 978-0-9807408-1-3.

Raadik, T. A. (1993). *A research recovery plan for the barred galaxias, Galaxias fuscus Mack 1936, in south-eastern Australia*. Report to the Australian National Parks and Wildlife Service, Department of Conservation and Natural Resources, Victoria.

Raadik, T. A., Fairbrother, P. S., & Nicol, M. (2009). *Barred Galaxias, Galaxias fuscus, recovery actions: Fire and drought impacts – summary report 2007–2008*. Report to Goulburn-Broken CMA. Arthur Rylah Institute for Environmental Research, Client Report.

Raadik, T. A., Saddlier, S. R., & Koehn, J. D. (1996). Threatened fishes of the world: *Galaxias focus* Mack, 1936 Galaxiidae. *Environmental Biology of Fishes, 47*, 108.
Vinson, T., Rawsthorne, M., Beavis, A., & Ericson, M. (2015). *Dropping off the Edge 2015: Persistent communal disadvantage in Australia*. Jesuit Social Services & Catholic Social Services Australia, Richmond.
Völker, S., & Kistemann, T. (2011). The impact of blue space on human health and well-being - Salutogenetic health effects of inland surface waters: A review. *International Journal of Hygiene and Environmental Health, 214*(6), 449–460.
Walker, B. (2019). *South Australia, Murray-Darling Basin Royal Commission*. Report.
Whalley, J. M. (1988). Water in the landscape. *Landscape and Urban Planning, 16*, 145–162.

16

Tidal Country and Cultures in Northern Australia

David Kelly and Michele Lobo

Introduction

This chapter explores past, present and emerging tidal cultures in Djugun/Yawuru Country (Broome, pop. 15,000) and Larrakia Country (Darwin, pop. 120,000) situated in resource-rich Northern Australia. We draw on ethnographic research, interviews, focus groups and walk-alongs with Indigenous peoples, ethnic minority asylum seekers and non-Indigenous activists. Their everyday encounters and stories illuminate the fragility and affordances of these liminal spaces and call for ethical relations with tidal country. We argue that such research that dwells and works with multiple realities in the in-between space of tidal country is essential given the advent of a new climatic regime known as the Anthropocene, which is yet to engage deeply with the situated knowledges of displaced, dispossessed and racialised bodies. These knowledges of place in white settler societies like Australia have the potential to produce a "place-based metamorphosis" (Larsen & Johnson, 2012: 1) that

D. Kelly • M. Lobo (✉)
Deakin University, Geelong, VIC, Australia
e-mail: david.kelly@deakin.edu.au; michele.lobo@deakin.edu.au

A. Campbell et al. (eds.), *Located Research*,
https://doi.org/10.1007/978-981-32-9694-7_16

transcends the land/sea binary and offers new understandings of regions in transition. As an Australian man of Irish heritage and an Australian woman of Indian heritage we come together to think about the vitality of situated knowledges in Broome and Darwin, two urban centres described as "remote" or "frontier regions" within national policy agendas that draw on western frameworks of development (Commonwealth of Australia, 2015).

Northern Australia: A Region in Transition

Broome is a small remote coastal urban centre in far North West Australia, often referred to in visitor pamphlets and promotional material as the "gateway to the Kimberley". The town services—through government administration, healthcare provision and economic activity—an expansive tropical region in North Australia that is home to Djugun and Yawuru people. Located 1800 km north east of Broome is Darwin the capital city of the Northern Territory that is home to Larrakia people (Larrakia Nation, 2006). The city is a strategic defence base and aims to be "smart, liveable, productive and sustainable" through infrastructural and place-making projects (City of Darwin, 2019, online). Both cities have trading histories with south east Asia that predate white settlement at the end of the nineteenth century (Martínez, 2006; Martínez & Vickers, 2015), but it was the European demand for pearl shell and beef that contributed to the rapid growth of Broome, as well as Darwin, as regional towns and colonial outposts in northern Australia (de La Rue, 2004). The expansion of the pearling and pastoral industry was followed by mining and more recently unconventional oil and gas extraction inland and offshore. It is in Broome, however, that activism has been widespread, epitomised by resistance to Woodside petroleum's plans to build a gas plant at Walmadan (James Price Point) on the coast (Kelly, 2019; Muecke, 2018). In 2008, following the abandonment of the plan, the Northern Territory Government welcomed INPEX, a global energy company to commence construction of an onshore plant at Bladin Point, Darwin to process gas from the Browse basin transported via an 890 km gas pipeline.

The recent expansion of offshore LNG projects in Darwin Harbour, the lifting of the moratorium on fracking for natural gas and mining in the resource-rich hinterlands of Darwin and Broome, demonstrates that Northern Australia continues to be perceived as a "frontier" that offers many economic affordances. With the first shipment of liquid natural gas from Darwin to Niigata, Japan in November 2018, the energy frontier that runs from the Browse gas Basin (off the coast of Broome) via subterranean pipelines to the processing plant at Bladin Point in Darwin Harbour now reaches Asia—the north is developing.

Developing the North is a federal policy agenda that aims to stimulate economic growth in Northern Australia and service a growing Asian region (Commonwealth of Australia, 2015). Within this agenda, the region north of the Tropic of Capricorn is explicitly described as the "last frontier", underutilised and isolated, but with "coordinated planning" has the potential to be transformed into the "next frontier". The goals outlined in the policy documents—such as the White Paper on Developing Northern Australia, Green Paper on Developing Northern Australia, Pivot North: Inquiry into the Development of Northern Australia and the 2030 Vision for Developing Northern Australia—are to be met by harnessing knowledges and incorporating strategies that would double the agricultural export industry, grow the tourist economy and leverage natural resources to build an energy export industry. The aim is to address key obstacles to investment and growth by addressing excessive bureaucracy in relation to environmental approvals of already protected sites, reducing the risk to resource companies that pay mining taxes; investigating the potential for establishing special economic zones, and building infrastructure for industry and providing incentives to attract skilled workers from southern Australia (Commonwealth of Australia, 2014). In respect to Indigenous priorities, the focus is on promoting Indigenous arts/culture and supporting enterprise employment models. The prediction is that in the next 15 years, north Australia will have a population of six million aided by new dams, new mines, new gas fields as well as becoming a food bowl for Asia supplied from new mega-farms (Commonwealth of Australia, 2014). There is little doubt that the "pristine natural environment" of tropical Northern Australia, while contributing to the expansion of the tourism industry, also offers affordances for economic development. In the next section we explore the concept of affordances to think about located research in tidal zones of Broome and Darwin.

Conceiving Tidal Affordances

A focus on what a place affords should require more than a simple myopic focus on its ability to service other distant places. In this section, we develop the concept of affordance to encapsulate the situated knowledges of local places that persist despite metanarratives of development and economic gain. The first comprehensive explanation of the term "affordance" was introduced within an ecological context and coined by ethologist James Gibson (1966, 1979). Affordance focuses on what the material environment offers human and nonhuman animals "for either good or ill" and is discerned through a range of bodily senses and available actions (Gibson, 1979: 127). The senses here figure centrally as critical systems that discern not just sensate stimuli such as touch, taste, sound and so on, but also as dialogic systems that imbue meaning when considered alongside cultural ways of being with and inhabiting place. In this chapter, we think of the objective environment through the materiality of tropical tidal places that situate metabolic exchanges between human and nonhuman bodies and circulate feelings, emotions and affects. What this leads to is an understanding of affordances of tidal environments through place-based properties that condition possibilities for action that are imbricated with affective sensations and cultural meanings.

For Tay (2009: 509) "the 'affordance' of an object is a sum total of its practical (social arrangement) and expressive (affective) order". The capacity to perform one action or another is therefore not an inherent capacity of objects and subjects, but the potential that pools when they are brought into relation. This resonates with Gibson's (1979: 129) thoughts:

> an affordance is neither an objective property nor a subjective property.... It is equally a fact of the environment and a fact of behavior. It is both

physical and psychical, yet neither. An affordance points both ways, to the environment and to the observer.

It is the in-between quality of affordance that we see as a productive conjoiner to the term "tidal", which we describe in this chapter as an agentic location, metaphor, concept and phenomena that pervades ordinary life in the Broome and Darwin. The tide does the work of concealing/revealing, aggregating/dispersing, constantly reforming relations and boundaries between things—most obviously of earth and water.

Our understanding of affordance in an ever-shifting tidal land-/seascape offers compelling trajectories for people to make relations and connections through momentary actions that enable them to dwell in the present. The measure of what is possible through such dwelling is simultaneously determined by the interplay of cultural sensibilities, stories of place and ontological groundings that foreground embodied ways of knowing place. As Edensor (2017) demonstrates through the illumination of the landscape, sensory data that the body receives and makes meaning from affords a particular emplaced cultural understanding of the scene, which, in turn, affords limitations and possibilities on what is enactable and open to making meaning.

Negotiating the Tide in Research

> They're up at Crocodile River on a moonless night,
> Trying to shine a light on his red beady eyes,
> In a dingy got no paddle on a turning tide,
> At the mercy of a salty, such is life…
> *Stephen Pigram, Crocodile River* (2013)

Another thing is those crocodiles on Cable Beach … somehow, quietly, get rid of those crocodiles

Colin Barnett. (Parke, 2015)

In 2015, I stood on the iconic Cable Beach among a crowd of gathered people, when the then Premier of Western Australia, Colin Barnett, stated that two things pose an existential threat to Broome's appeal as a tourist destination: (1) environmental protest against big oil and gas, and (2) crocodiles. The controversy over crocodiles is emblematic of a contest that has been in play since the colonial invasion of this tidal country in the Nineteenth Century; that is, the slippage between a rigid official understanding of a distant place, and local ways of being that move and live with the affordances of a dynamic environment. Stephen Pigram, a revered Yawuru artist, embellishes this localism, singing popular songs that resonate with the everyday life of local people in Broome and the West Kimberley. For instance, Will, a non-Indigenous man, reflects during an interview "Why am I here?", saying:

> If you get to see Steve Pigram play, his music, it … for me it refills my bucket with *why*. He used to sing from that time, that time where the people were in sync with the tides and things … when you get to see Steve play a gig it reminds you of why you're here.

To get rid of the crocodiles, as the then-Premier recommended, this place would have to bely the environmental conditions that produce such risk, but also an understanding of place that is anchored in the dynamism of the here-and-now. The tidal affordances of north Australia have always been integral to the development and maintenance of life here, including artistic and economic outputs across the North. Tides in Broome reach 9 metres and up to 11 in some surrounding areas. As such, the tidal zones they move through can be expansive, creating large ecosystems that depend upon its dynamism for it to reproduce its relations that are economic, cultural and lively.

In 1861, the extraction of *Pinctada Maxima*, the largest pearl in the world at that time, gave rise to a global industry at Roebuck Bay (Edmonds, 1996) and these tidal affordances began to be exploited by colonial industrialialists. By the late nineteenth century, Australia had become the largest pearling industry in the world and Broome had become the pearling capital of the industry (Martínez & Vickers, 2015). Yu, Pigram, and Shioji (2015: 251) describe how the pearling industry left an "enduring influence on the cultural and economic character of the north … changing qualities and cultural meanings of pearl shell and pearls". At the same time, the industry was shaped by the emplaced realities of the tide, where large fleets of pearling luggers required intimate knowledge of the tides to extract the shell offshore. Knowledge of tidal patterns had to be deployed in order to protect the fleet during "layup" season, where luggers would sail deep into the bay during high tide to escape the brunt of cyclones that regularly battered the beach.

But these tidal patterns that produce so much value and life, have been ignored in more recent economic projects. Since 2009, Indigenous-led environmental protest against a large gas processing hub on Aboriginal land, widely regarded as a pristine example of an intact yet vulnerable ecology, precipitated the collapse of the project in 2014 (Muecke, 2016). While Indigenous politics has been a sustained global (Canessa, 2014; Fisher, 1994; Nadasdy, 2005; Shah, 2010; Whyte, 2017) and Australian domestic (Jackson, Palmer, McDonald, & Bumpus, 2017; Toussaint, 2008; van Holstein & Head, 2018) discourse in environmental activism, the emergence of these controversies are sporadically rhythmic. The rhythm of controversy over natural resources, environmental risks and development is often rapturous here, and these events can obscure a more quotidian way of understanding place. For non-Indigenous people that come to Broome, the rhythmic conditions of everyday life must be understood if they are to meaningfully engage with and support the political/cultural projects of Aboriginal people.

My research encounters centre around a group of participants who have been involved in these environmental protests and activism around the future of Aboriginal livelihoods on country. These encounters can be narrativized, perhaps distorted into neat fieldwork vignettes, but their content brims with affective data that influence the ways in which the research was conducted. For me, a white man that parachutes into town from the South to "do" research, the rhythmic tidal affordances of place became integral to forming connections with place and relationships with participants. On many occasions, the tide's agency snapped into focus during interviews and ethnographic moments of waiting and moving.

While interviewing Elster—a well-known Yawuru Elder, nurse and activist in the Kimberley—at her home in 2015, we discuss how she imagines the good life, where she experiences it and when that was. Elster recalls growing up around the town, when pearling was the predominant industry and where state-imposed curfews and segregation for Aboriginal people was an oppressive reality, however, she says:

> when the tide was big we used to go to the Sun Pictures, we used to row our boats, Dad's dingy, to the picture show and pick mum and dad up and row them home because our house was like on a little island, we had our own little beach, that was funny. Then we made our own canoes and we used to go across to pick up all the kids and go to the airport because where the oval [near the airport] is, that was 10-foot drop, a big lagoon, that was full of water, saltwater. So, we used to go there with our canoes, pick people up and bring them home, go fishing, swimming. We had good childhood we did that till we were 19, 20 we were still playing around like little kids.

Despite apartheid conditions, Elster recollects moments in her historical narrative of place that exude a playfulness facilitated by the changing of the tide. What is possible on one moment, changes in the next, enabling her to move about country in different ways, allowing her to renew attachments to place. I interview Elster for over an hour and after turning the audio recorder off, we have another long discussion over tea and cake about places around Broome where she still goes regularly to tap into good feelings. Crab Creek is one of her favourite places to go. Elster remarks, "If you want to know what the good life is, I'll take you to Crab Creek". She gets up from the table where we sit and disappears into another room inside the house. She returns with a small book of tide tables and tide charts to check when the best time to go would be. We schedule in an appointment for a few days later to travel to Crab Creek for some fishing and swimming, only when the tide is right.

For other research participants, the tide played a role in the conduct, form and taking-place of interviews. Mitch, who is a Djugun woman and respected filmmaker and environmentalist, became a friend during the research as we shared numerous afternoons at her home chatting outside under a pergola. One night we are discussing the ways in which she "knows" a particular place and we record the conversation. I ask her about the postcard images of Broome, in particular an image of the "staircase to the moon", a natural phenomenon that occurs from March to October. As the moon rises in the east, it reflects sunlight off shallow water pooled in the depressions of the tidal flats in Roebuck Bay that resembles a staircase reaching to the moon. She responds:

> that's our fishing spot there where that photo was taken ... if you waited another half an hour the tide comes in and on it, if it's this time of the year you'll hear the salmon hitting the water and you know you're going to catch salmon. So that's the other story... Don't have to look at the calendar and we can sit here at the block and know, see this light breeze coming, tide is turning. Can you feel it?

The light breeze provides some respite to the exhausting tropical heat, and when Mitch asks if I can feel the tide turning, I think of its cooling effect. As it cools, we scratch at our skin, and Mitch smiles while saying "that's why we're scratching because the sand flies start":

> what happens is the bay is filling up. So as the mangrove trees get covered with water and the midges and the mosquitos come up in the air, they come here, they fly to all the dry places.

When I arrange to meet up with another participant Greg, a non-Aboriginal man who has lived in Broome for over 20 years, he suggests Town Beach. We sit on the grass overlooking the beach and casually chat before I am given permission to turn begin recording. Abruptly, at the 20-minute mark of what is usually at least a 45-minute interview, Greg terminates the discussion. I am confused, thinking that I have done something wrong, unethical, or simply just bored him. But as he stands up and brushes the grass from his shorts, he says "the tide has turned, wind's picked up ... I have to go home and get my kite, go for a surf". Greg was recommended to me as someone essential to speak to by about a dozen Broome locals, and just as the tide had turned, the interview was over.

Learning to Love a New Country

Nomadic wanderings that started at 8 am interspersed by bus rides that ended at 10.30 pm immersed me in serendipitous events that unfolded in the tidal country of Darwin, the capital city of the Northern Territory. These events that sometimes unfolded on beaches and beach reserves provided affordances for learning with tidal country as an Australian woman of Indian heritage. The overwhelming support from Larrakia Nation Aboriginal Corporation and the warmth of Aboriginal peoples including Larrakia or saltwater people taught me how to begin to feel the love of land and sea country as an ethnic minority migrant woman of colour. These intimate connections with country are nourishing when the subtleties of white privilege or the brutality of "in your face" racism takes my body by surprise even after 19 years in Australia.

The tidal country of Darwin with its torrential rain, tropical cyclones, stormy seas, oppressive humidity and extreme heat seems like familiar ground as I grew up in monsoonal Kolkata along the banks of the Hooghly River near the UNSECO heritage listed Sundarban mangrove forests. The affordances and dangers of tidal country in deltaic West Bengal, home to the Royal Bengal Tiger, Gangetic dolphin, saltwater crocodile as well as fishers, farmers, honey collectors, woodcutters and prawn seed collectors are celebrated in myths and contemporary stories such as Amitav Ghosh's (2004) *Hungry Tide. Adivasis* (Indigenous peoples) who were brought to clear the mangrove forests for cultivation during the colonial period continue to be sustained by tidal cultures along with Dalits (former untouchables within the Hindu caste system) and Bangladeshi Muslim migrants and refugees. These affordances of tidal country sensitised my body to the challenges that hypervisible Indigenous peoples (including those who "live rough") as well as ethnic minority humanitarian migrants and asylum seekers from countries in Africa, Asia and Middle East encounter in Darwin. Their ongoing experiences of dispossession, displacement and institutional racism, as well as their visibility in public spaces, interrupt federal, territory and local visions of Darwin as a global city and the "next frontier" of Australia.

Tidal country has historically entangled Indigenous and Asian cultures prior to white settlement. Harvesting and trading trepang (sea cucumbers) as well as pearl shell brought fishermen from Macassar who arrived with the monsoon winds and camped for a few months in coastal country (Martínez, 2006). Among Larrakia people, traditional custodians of the land, practices of hunting and fishing are ways they learn from country, feel the love from land and sea country and communicate response-ability to a more-than-human world (Larrakia Nation, 2006). As saltwater people, Larrakia have traditionally hunted for mammals such as dugong (Damidangga-la), reptiles such as saltwater crocodiles (danggala-ba) or salties and sea turtles (Dawud'li-rra). James Gaykamangu and Taylor (2014: 2), a Senior Elder and Lawman I met in Darwin underlines that "thousands of years of wisdom" about skilful and responsible fishing and hunting in tidal country are secrets learnt from parents. For example, at high tide, particularly during the wet season (November–February), it is easy to catch barramundi (Deme'bi-la) in the mangroves, but it is at low tide that stingray (Muli'babi-la) and other fish are more visible. Gayamangu underlines that enhanced skills are required during the dry season (March–October) when food is more difficult to find. It involves digging waterholes to get fish, mud crabs, mud mussels (Gunart'barrwa) periwinkles (Damu'gu-la), longbum (Danija-rra) and withchetty grubs. "Kitchen middens" along Darwin's coast (known as Garrmalang by Aboriginals) show the remains of these shellfish, crabs, mangrove worms (Da-law) and mangrove mollusc (terebralia and Geloina) collected from the intertidal zone of mangrove forest that were extensive before white settlement in 1869.

Yvonne and Sabrina are Larrakia rangers who developed connections with tidal country through family picnics at the beach. They recall how they were frugal in collecting longbum, mud mussels and mud crabs (madla):

Yvonne: As young children we used to come down here (Mindil beach) with our parents to do a lot of swimming, fishing and after that have something to eat like longbum, periwinkles and fish. Mum would make damper you know and just enjoy the day and take only what we needed you know.

Sabrina: You get fresh and salt water mussels, used to be good, but sewerage area now, sewerage everywhere, killing everything. Used to go fishing—now houses around there, we don't go anymore.

Longbum is popularly known as "poor man's tucker" or food (Bourke, 2015). Ethnographic fieldwork of contemporary Larrakia resource use around Darwin describe how longbums and "periwinkles" (i.e. Neritidae) were historically collected and consumed by small family groups (Larrakia Nation, 2006). Mangrove wood was used to light a fire to boil the shellfish. The cooked seafood was then placed on a layer of paperbark, twigs and leaves to cool—the flesh was then pulled out with pieces of shell and eaten. Shellfish collected in the 1960s and 1970s in urban areas such as Casuarina and Rapid Creek in Darwin were no longer collected in the 1990s due to pollution. Instead, Indigenous peoples collected this food from the Middle Arm of Darwin harbour and outer suburban areas of the neighbouring Palmerston township away from rapidly urbanising Darwin.

Yvonne and Sabrina regret that the quantity and quality of seafood had deteriorated. Although they enjoyed this fresh food which they used to boil in a billy can on the beach, unlike "long grassers" or "countrymen" who live rough and asylum seekers with little income, they did not depend on this food for survival. As I spent time with "long grassers" and asylum seekers who drift through Darwin rather than fly-in-fly-out like skilled workers and professionals, I listened to their stories of fishing, hunting and collecting food from mangroves. Hakeem, an asylum seeker from Myanmar, spoke of how he used Burmese style hand-made nets to catch barramundi and stingray at Lee Point along Darwin's northern coast. Mary and Aboriginal woman who "lived rough" regretted, however, that she often had to eat "dreary bush food" such as small bandicoots because seafood was scarce and too expensive to purchase from large grocery stores such as Woolworths. When Hakeem sees "long grassers" like Mary camping at Lee Point, he shares his catch. He said, "We give to them [fish], they like jellyfish".

The sandy beach at Lee Point is also a nesting ground for sea turtles. Leena, a young Larrakia ranger, works with government and private agencies to rescue injured turtles that get washed up on the beach. I spoke to her after she had spent a long day in the hot sun. She said, "I was out on the boat. But, yep we had a turtle washed up at Casuarina Beach, and yeah it seemed to be injured; a boat hit it". Turtles and dugongs (sea cows) feed on dwindling seagrass beds in the Darwin region offshore from Casuarina coastal reserve. This stretch of tidal country backed by sand dunes and cliffs is Saleem's favourite place. As a young man who arrived by boat from Myanmar and lives with his mother in community detention, he enjoys spending time at Casuarina beach even during the Muslim fasting period of Ramadan. Walking along the beach Saleem sees and learns about Old Man Rock, an offshore reef visible in a low tide that embodies the spirit of Larrakia ancestor, Darriba Nungalinya who protects the harbour. But perhaps the favourite place for many people who live rough is the tidal country with rocky cliffs and reef platforms at East Point Reserve, Fannie Bay (Bauman, 2006). While many long-term ethnic minority migrants and white settlers enjoy swimming at the nearby Alexander Lake, many homeless people enjoyed the affordances of tidal country that offered a cool sea breeze on a hot day. This gentle wind is celebrated at the annual Seabreeze Festival held at the Nightcliff foreshore. The festival draws a large crowd of long-term residents as well as newcomers, but the stone fish trap nearby which is about 100 years old and visible at low tide often escapes attention (Bauman, 2006). In the past, these funnel-shaped structures made of sticks and stones brought fish inland during high tide. When the tide ran out the fish were trapped by the rock barriers (Bauman, 2006).

Loving and Loved by Country

In understanding the history and connections to place, this chapter has unveiled how tidal country unveils temporally amorphous stories of love, life and creation. In Broome and Darwin, this liminal zone produces affordances that are well known by Indigenous peoples but must be learnt by fly-in fly-out workers and newcomers, including Anglo-Australians and ethnic/ethno-religious minority migrants, refugees and asylum seekers who make temporary or permanent homes in these tropical places. Although how one might be able to act, and the consequences of that action, depends on environmental conditions and social contexts, affordances also signal capacities to discern and act upon possibilities in the here-and-now that go beyond purely economic extractive logics. Situated knowledges highlighted in this chapter draw attention to the hidden affordances that emerge when we love and feel loved by tidal country.

Acknowledgements The research in Broome was funded by an Australian Postgraduate Award scholarship and fieldwork grants from the Faculty of Arts and Education, Deakin University. The research in Darwin was funded by an Australian Research Council Discovery Early Career Researcher Award (DE 130100250, 2013–2016). Special thanks to the friends/participants in Broome, Nulungu Research Institute, Larrakia Nation Aboriginal Corporation, Multicultural Council of the Northern Territory, Darwin Asylum Seeker Support and Advocacy Network, Darwin Community, Arts and the Northern Institute, Charles Darwin University, Australia.

References

Bauman, T. (2006). *Aboriginal Darwin*. Canberra: Aboriginal Studies Press.

Bourke, P. (2015). 'Poor man's tucker' – Historic and contemporary consumption of longbum shellfish around Darwin. *Northern Territory Naturalist, 26*, 4–12.

Canessa, A. (2014). Conflict, claim and contradiction in the new 'indigenous' state of Bolivia. *Critique of Anthropology, 34*(2), 153–173.

City of Darwin. (2019). Media release: #SmartDarwin Strategy open for comment. 15 January. Online. Retrieved May 13, 2019, from www.darwin.nt.gov.au/council/news-media/news/smartdarwin-strategy-open-for-comment

Commonwealth of Australia. (2014). *Green paper on developing Northern Australia*. Canberra: Commonwealth of Australia.

Commonwealth of Australia. (2015). *Our north, our future: White paper on developing northern Australia*. Canberra: Commonwealth of Australia.

De La Rue, K. (2004). *The evolution of Darwin, 1869–1911: A history of the Northern Territory's capital city during the years of South Australian administration*. Darwin: Charles Darwin University Press.

Edensor, T. (2017). Seeing with light and landscape: A walk around Stanton Moor. *Landscape Research, 42*(6), 616–633.

Edmonds, C. (1996). Pearl diving: The Australian story. *South Pacific Underwater Medicine Society Journal: Pearl Diving Supplement, 26*(1), 4–15.
Fisher, W. H. (1994). Megadevelopment, environmentalism, and resistance: The institutional context of Kayapó Indigenous politics in Central Brazil. *Human Organization, 53*(3), 220–232.
Gaykamangu, J., & Taylor, P. (2014). *Striving to bridge the Chasm*. Darwin: Larrakia Nation Aboriginal Corporation, Public Purposes Trust, Arnhem Land Progress Aboriginal Corporation.
Ghosh, A. (2004). *The hungry tide*. Delhi: Harper Collins.
Gibson, J. (1966). *The senses considered as perceptual systems*. Boston: Houghton Mifflin Company.
Gibson, J. (1979). *The ecological approach to visual perception*. Boston: Houghton Mifflin Company.
Jackson, S., Palmer, L., McDonald, F., & Bumpus, A. (2017). Cultures of carbon and the logic of care: The possibilities for carbon enrichment and its cultural signature. *Annals of the American Association of Geographers, 107*(4), 867–882.
Kelly, D. (2019). Apprehending felt futures in Broome, Australia. *Geoforum*. https://doi.org/10.1016/j.geoforum.2019.03.002
Larrakia Nation Aboriginal Corporation. (2006). *Saltwater people: Larrakia stories from around Darwin*. Darwin: Larrakia Nation Aboriginal Corporation.
Larsen, S. C., & Johnson, J. T. (2012). In between worlds: Place, experience, and research in Indigenous geography. *Journal of Cultural Geography, 29*(1), 1–13.
Martínez, J. (2006). Ethnic policy and practice in Darwin. In R. Ganter (Ed.), *Mixed relations Asian-Aboriginal contact in North Australia* (pp. 122–139). Crawley: University of Western Australia Press.
Martínez, J., & Vickers, A. (2015). *The Pearl Frontier: Indonesian labor and Indigenous encounters in Australia's northern trading network*. Honolulu: University of Hawa'ai Press.
Muecke, S. (2016). *The Mother's Day protest and other fictocritical essays*. London: Rowman & Littlefield International.
Muecke, S. (2018). Goolarabooloo futures: Mining and aborigines. In J. K. Watson & G. Wilder (Eds.), *The postcolonial contemporary: Political imaginaries for the global present* (pp. 208–223). New York: Fordham University Press.
Nadasdy, P. (2005). Transcending the debate over the ecologically Noble Indian: Indigenous peoples and environmentalism. *Ethnohistory, 52*(2), 291–331.

Parke, E. (2015). WA Premier tells Broome to 'quietly get rid of crocodiles' on Cable Beach to protect tourism. *ABC News*. Retrieved March 26, 2016, from https://www.abc.net.au/news/2015-10-09/wa-premier-tells-broome-to-quietly-get-rid-of-crocs/6841980

Shah, A. (2010). *In the shadows of the state: Indigenous politics, environmentalism, and insurgency in Jharkhand, India*. Durham: Duke University Press.

Tay, A. (2009). The material. In R. Kitchin & N. Thrift (Eds.), *International encyclopedia of human geography* (pp. 505–510). Amsterdam: Elsevier.

Toussaint, S. (2008). Kimberley friction: Complex attachments to water-places in Northern Australia. *Oceania, 78*(1), 46–61.

van Holstein, E., & Head, L. (2018). Shifting settler-colonial discourses of environmentalism: Representations of indigeneity and migration in Australian conservation. *Geoforum, 94*, 41–52.

Whyte, K. (2017). The Dakota access pipeline, environmental injustice, and U.S. colonialism. *Red Ink: An International Journal of Indigenous Literature, Arts, & Humanities, 19*(1), 154–169.

Yu, S., Pigram, B., & Shioji, M. (2015). Lustre: Reflections on pearling. *Griffith Review, 47*, 251–261.

17

Vulnerable Researchers: Opportunities, Challenges and Collaborative Co-design in Regional Research

Beth Edmondson

Introduction

This chapter provides a personal reflection upon some of the challenges and privileges of undertaking collaborative co-designed research focused on improving health and wellbeing in the Latrobe City and Inner Gippsland regions of Victoria. These reflections are framed by Lalor, Begley & Devane's (2006) observations regarding the importance of recognising the relational and intersectional impacts of research upon researchers. They also draw upon Fine's (1993) call for researchers to more honestly engage with 'the underside of qualitative' research that makes it difficult for researchers to discuss negative experiences, poor interactions and problematic boundaries. The projects reflected upon here involved embedded research practices that aimed to identify new approaches that express the preferences and interests of community members and to support changes in policies and practices to improve

B. Edmondson (✉)
Humanities and Social Sciences, Federation University Australia, Churchill, VIC, Australia
e-mail: beth.edmondson@federation.edu.au

A. Campbell et al. (eds.), *Located Research*,
https://doi.org/10.1007/978-981-32-9694-7_17

their wellbeing. However, these projects each derived from governmental/departmental initiatives that also contributed complex institutional, structural and individual relationships.

Co-designed research is rightly promoted as offering opportunities for vulnerable or marginalised groups and communities to become co-designers in research, rather than subjects of it. There are well-documented benefits in bringing diverse perspectives and experiences into research dialogues. Ensuring that researchers are not detached observers is integral to these processes. However, in adopting these research approaches, researchers can be exposed to the circumstantial vulnerabilities of others and their emotions in navigating them. These intersecting relationships alter the perspectives of researchers. Through them, researchers are no longer detached from the experiences of their research participants, nor from the systemic and structural conditions that underpin them.

Writing this chapter has provided opportunities to think anew about what it means to be a researcher engaged in collaborative co-designed research in the 'small connected communities' of these intersecting regions, where many people experience considerable social, economic and health related disadvantages (Damianakis & Woodford, 2012). It has also highlighted some of the unexpected vulnerabilities that might be experienced by researchers engaged in embedded regional research projects. In this region, encompassing the broader setting of the Victorian Government designated Inner Gippsland Region (comprised of four local government areas, Bass Coast Shire Council, Baw Baw Shire Council, Latrobe City Council and South Gippsland Shire Council) and the more contained settings of the Latrobe Health Innovation Zone and Latrobe City, structural and systemic power inequities underpin current patterns of individual and community disadvantage, opportunities and wellbeing. Here, numerous research and practice based initiatives have been pursued to address the complex inequities associated with social-economic opportunities, public and individual health and wellbeing, and life opportunity challenges.

Since 2016, new emphasis has been placed on collaborative co-design as enabling practice and policy innovations to improve health outcomes within this region. The Hazelwood Mine Fire Inquiry Report 2015/16 is rightly credited as marking new Victorian Government efforts to improve health and wellbeing in Latrobe City (Parliament of Victoria, 2015). These changed approaches were initiated partly in response to widespread criticisms following the Hazelwood Mine Fire, which began on 9 February 2014 and burned for 45 days. As noted in the first Hazelwood Mine Fire Inquiry Report 2014 this fire 'sent smoke and ash over the town of Morwell and surrounding areas for much of that time' (Parliament of Victoria, 2014). During this period, 'members of communities felt that they were not listened to and were not given appropriate and timely information and advice that reflected the crisis at hand and addressed their needs' (Parliament of Victoria, 2014). This fire came to be identified as 'a chronic technological disaster' that led to 'a significant and lengthy environmental and health crisis' (Parliament of Victoria, 2014).

However, it is also important to recognise that efforts to re-orient strategies for improving health outcomes across Inner Gippsland/Latrobe City also sit alongside other initiatives that had commenced earlier, such as the Inner Gippsland Children and Youth Partnership which aims to improve services that currently 'struggle to meet the needs of children, young people and families who are vulnerable' (State Government of Victoria, 2019). Many of these have projects involved 'Team Australia' partnerships whereby 'universities, industry and government' work together to 'build capabilities' (CSIRO Futures, 2018). In Inner Gippsland and Latrobe City/Latrobe Health Innovation Zone, these projects have engaged in 'deep consultation' to identify solutions that can be considered 'tried, true and trusted' by community members, regional stakeholders and service providers (CSIRO Futures, 2018).

Across these projects, some common dialogic goals were incorporated into research design and communication plans. These include ensuring that research practices, aims and methodologies provide community members with:

1. Access to being heard;
2. Intersectional links between themselves, the systems and structures that underpin their lives;
3. Opportunities to engage in new dialogues;
4. Opportunities to see their experiences with new eyes to understand their place within their community in new ways;

5. Opportunities to envision new experiences and new relationships; and
6. Evidence of their interests and preferences in recommendations and other research outcomes.

Researching health and wellbeing needs, service practices, emerging policy and practice based innovations and contributing to relevant local knowledge banks blurs boundaries between living, working and researching in regional communities. These intersectional dynamics are exacerbated when researchers are engaged in co-designed projects that aim to change practices and policies. In these projects, researchers become 'boundary spanners' as they 'negotiate system interchanges with other organizations … connect two or more systems whose goals and expectations are likely to be at least partially conflicting' and also need to 'manage the tensions at the interface between flexible, collaborative partnerships and the bureaucratic organizational structures of their partners' (de Leeuw & Wise, 2015: 44; see also Damianakis & Woodford, 2012). Regionally embedded co-designed research undertakings can present deeply personal experiences for researchers, especially in working with vulnerable people or within vulnerable regional communities (Fine, 1993; Lalor et al., 2006; Sherry, 2013).

Researchers who also live and work in the regions where their research takes place are vulnerable to experiencing emotional impacts arising from these intersections—and from their relationships with vulnerable others. In co-designed regional health and wellbeing research in Inner Gippsland/Latrobe City, such relational, role and emotional intersections are sometimes inevitable. As Emma Sherry (2013: 283) observes: 'We cannot truly separate our personal lives from our work'. These complexities and relational challenges present 'predictable surprises' that can increase the difficulties of meeting project aims and/or timelines (Bazerman & Watkins, 2004).

About the Inner Gippsland and Latrobe City Regions

The Hazelwood Mine Fire Inquiry Report 2015 identified the Latrobe Valley as a 'community that has been marginalized and neglected for decades' (de Leeuw & Wise, 2015: 10). While there was little news in this finding for researchers and service providers living and working in this region, it elevated expectations that the Victorian Government should be held responsible for addressing the cycles of disadvantage that had been identified. As a result, in 2016, the 'Victorian Government designated the Latrobe Valley as a Health Innovation Zone … to improve health outcomes in this region' (State Government of Victoria, 2017b). It also established new public entities, namely the Latrobe Health Assembly and Latrobe Health Advocate 'to promote and support innovative co-designed health and wellbeing models and programs' (State Government of Victoria, 2017b).

Across these intersecting regions, several different studies had previously found that the 'social and economic disadvantages experienced' were 'positively associated with poor health and premature death' (de Leeuw & Wise, 2015: 47). Across the Inner Gippsland Region, correlations 'between social and economic disadvantage and child abuse and neglect' had been identified as contributing to cycles of vulnerability, marginalisation and inter-generational disadvantage (State Government of Victoria, 2017a, 2017b). Identified urgent priorities included reducing:

- large numbers of children living in out of home care;
- numbers of children entering school with developmental vulnerabilities (more than 1/3 of children being developmentally vulnerable in more than one area and more than 50% of Morwell children entering school with a developmental vulnerability);
- high rates of family violence;
- high rates of people reporting high psychological distress;
- high rates of youth offending; and
- high rates of suicide.

Additional indicators of social and economic disadvantage had also been identified, such as 45% of Latrobe City residents having incomes of less than $400 per week, 58% of residents not meeting daily activity guidelines and 50% not meeting dietary guidelines for fruit and vegetables (Latrobe City Council, 2017; State Government of Victoria, 2017a; Commonwealth of Australia, 2015).

Overall, the Dropping Off the Edge Report 2015 indicated that the township of Morwell (in the heart of the Latrobe City/Latrobe Health Innovation Zone) was 'one of the six most disadvantaged postcodes in Victoria, scoring relatively high in terms of juvenile convictions, child maltreatment and domestic violence and demonstrating low levels of school readiness for children 3–5 years old' (Vinson, Rawsthorne, Beavis, & Ericson, 2015; State Government of Victoria, 2017a: 15).

About Researching in the Inner Gippsland and Latrobe City Regions

Co-designed research in this region demands direct collaborative engagement—coming to the table as equals—with diverse community members, agencies and government service providers. It is time-consuming and labour intensive. In these project related roles, researchers are constrained by the contractual obligations of each specific research project, and by complex duties of care to others, both directly and indirectly. The work involved in these projects demands careful navigation in politically and diplomatically complex relational settings as diverse stakeholders and socially interested parties negotiate, bargain, form collective goals, progress ideas and evaluate their relative merits (de Leeuw & Wise, 2015). These co-designed research practices throw the researcher into the midst of various contests for resources, being heard, and advocating for change.

To some extent, the regional research opportunities that arise for me (and my colleagues) derive quite directly from governmental interests and their associated imperatives of electoral and budget cycles. The knowledge aims at the centre of these projects are largely determined by governmental or other organisational entities—together with their funding commitments, project scoping requirements and timelines. Despite these numerous personal and professional challenges, embedded researchers can identify and solve people-place problems, extend opportunities and reduce disadvantage.

When It Goes Well

Engaging in collaborative research with marginalised and vulnerable communities can also be enormously rewarding. This is especially the case when changes follow research findings and recommendations. Co-designed research often carries aspirations of changes in practices and policies that impact vulnerable communities (and individuals within them). In these instances, regional communities and the researchers both benefit from seeking 'positive change' through co-designed evidence based 'scholarly activism' (de Leeuw & Wise, 2015: 43). As shown in Fig. 17.1, such research can enable diverse groups and individuals to work together to identify shared interests, goals, and strategies for their

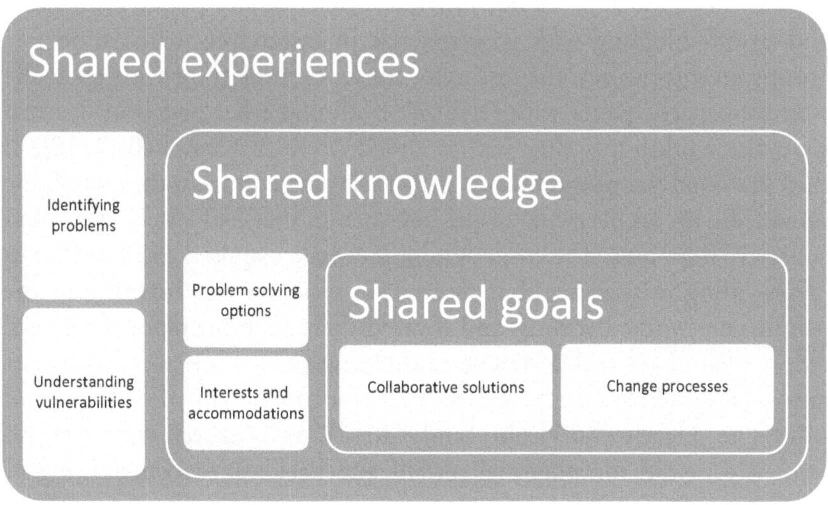

Fig. 17.1 Co-designed research: common goals and strategies for change

attainment. To do so, they re-configure relationships and structures to enable ongoing dialogues and facilitate change.

An especially joyful series of experiences were provided by the research, development and adoption of the Latrobe Health Innovation Zone Charter in 2017. As a multidisciplinary team, we were comprehensively engaged in 'boundary work, brokerage … entrepreneurship' (de Leeuw & Wise, 2015: 43). This work investigated numerous health zone models, re-visiting the detailed findings of the Hazelwood Mine Fire Inquiries, examining recent studies of community health (and wellbeing) improvement programs in diverse settings and seeking expert advice. The most important work, however, negotiated new spaces for the voices and preferences of community members and endorsements of the Latrobe Health and Wellbeing Charter by governmental stakeholders and service providers brokered new recognition of social determinants of health among key stakeholders. These rewarding experiences match de Leeuw & Wise's (2015: 44) descriptions of researchers whose work spans 'evidence, action and policy' to improve outcomes for others.

Alongside scholarly and policy-focused research, a series of community workshops were held in different towns at different times to maximise participation by community members. Targeted workshops were also held for key groups of service providers, practice and policy influencers and so on. Aligning with wise practice in researching with vulnerable groups, in this project, the research team undertook to provide all of the workshop participants with feedback outlining what had been learned from the workshop/s they had attended (Pyer & Campbell, 2012). If they provided the research team with their email address, they would be provided with an overview of the key themes that had emerged in their collective responses. The co-designing process recognised the importance of providing reassurance that researchers were actively listening to the people who worked with us and showed them that our reporting would reflect their views and preferences. Building trust was especially important as community trust levels had been eroded following the Hazelwood Mine Fire (Hazelwood Health Study, 2017). As a research team, we also wanted to avoid the potential pitfalls and backlash of contributing to research and consultation fatigue across the community. As workers-carers-residents, we were already keenly aware that a great many consultations, research projects, investigations of one kind or another had occurred following the Hazelwood Mine Fire.

Additionally, this project aimed to establish a new advisory group to support the development, implementation and endorsement of the Latrobe Health and Wellbeing Charter once it was created. Finding and creating opportunities to establish new relationships and dialogues were key project objectives. Continuous communication loops were established to meet these goals. This involved repeat cycles of listening, synthesising, listening and showing what we'd heard. These project co-design decisions were underpinned by recognition that community members, agency stakeholders, service providers, policy and program developers were all invested in improving health and wellbeing in this region. As a research team, we endeavoured to provide 'an ear' for community members 'to share their feelings and experiences', working together towards 'more reciprocal relationship[s]' (Sherry, 2013: 285). As a result, this project enjoyed an easy and uncomplicated path from project contracting, to research design and scoping, to implementation of the research strategy. Once established, continuous communication loops were easily maintained between the research team and contract related stakeholders and also between the research team and community stakeholders.

The Victorian government-endorsed Latrobe Health and Wellbeing Charter which now underpins the activities of health and health related agencies and services providers across the Latrobe Health Innovation Zone (see excerpt in Fig. 17.2) was created through this project. Every word in this charter, every element in its design emerged from the collaborative co-designed research that enabled its development. Every sentence reflects the views and expresses the voiced of many people. This was made possible only through many hours with many people, where everyone had many opportunities to think, talk, consider, reflect, revise, reiterate, confirm, listen and clarify in sustained co-design cycles. Through these processes, Latrobe City/Latrobe Health Innovation Zone communities identified core values of collaboration, integrity, innovation, access, inclusion and equity as key elements of their Health and Wellbeing Charter. They anticipate that health and wellbeing outcomes can be improved in this region when these values held in common by stakeholders, service providers, government agencies and community members.

Fig. 17.2 The Latrobe Health and Wellbeing Charter

Guiding Principles of the LHIZ Charter

Latrobe Valley communities, organisations and government who commit to the Charter will:

1. Use a community-led approach to ensure all people, including those who are marginalised and under-represented, have choice, control and a voice in developing the supports and services they need in their lives
2. Work with First Nations communities and health service providers to improve health and wellbeing
3. Do something different, try new approaches and learn from mistakes to create new and innovative health and wellbeing models and programs
4. Be fair and transparent by doing as we say to follow through on commitments to Latrobe Valley communities

5. Communicate in plain language to build trust and respect among individuals, families, communities and service providers
6. Work to improve the lives of everyone, especially the most vulnerable and disadvantaged
7. Develop equitable access to the most appropriate supports and services for all people in Latrobe Valley communities
8. Grow and nurture collaboration and coordination between agencies to develop new ways of delivering services that cut across boundaries to best meet the needs of each person
9. Use a co-design model to shift how we think about and organise health services in the Latrobe Valley, and create innovative approaches to their design and delivery. (State Government of Victoria, 2017b)

However, the underlining purpose and objectives of this project help to explain why only few major challenges were experienced in it. It involved establishing something new. It didn't need to recommend undoing anything. It didn't invade or intrude upon any established structures or systems. It created a vision/values/guiding principles/mission statement for agencies, service deliverers, program creators and policy-makers. It wasn't replacing anything that held importance to anyone.

When It Goes Badly

Not every experience in regional co-designed research runs smoothly. As Fine (1993) observed more than a quarter of a century ago, research reports and publications tend to favour positive narratives of the relationships that play out in research settings. Others, including Sherry (2013), have cogently pointed out that the relationships entered into by researchers can lead them to experience new emotional and psychological vulnerabilities. She reminds us that researchers can experience fatigue or vicarious trauma as they become 'a repository of the participants' stories' (Sherry, 2013: 281). By hearing of others' experiences, researchers are, in turn, 'left to hold or bear their stories' (Sherry, 2013: 281). These perspectives provide salutary reminders that talking, listening, analysing, sorting data, and remaining open to alternative readings of information over extended periods can be exhausting. Their respective observations encourage new and more holistic accounts of researchers' experiences to include honest recognition of the psychosocial challenges that arise.

At times, the best efforts by co-designing researchers cannot prevent disruptions or project de-railments. Sometimes, external factors, such as changed leadership or reporting structures in a key stakeholder entity can lead to changed priorities, or lead to new elements in project design, or introduce new project elements. Sometimes persistently problematic relationships can arise. Low levels of confidence in the design of a project, or in its leadership can present particular challenges for researchers—whether these arise from within the research team or from other stakeholders or community members. These problems exceed differences in personal style, or individual preferences. At times, they arise from structural or systemic factors, including contests for authoritative status or funding control between stakeholders.

Contracted co-designed regional research inevitably creates new intersections. It is, therefore, equally inevitable that these intersections will sometimes present uncomfortable experiences for researchers—as well as participants and stakeholders. These are difficult matters to include in project reports or research publications. At one level or another, co-designing researchers are compromised or precluded from airing the 'dirty laundry' of their experiences. To do so, researchers risk (or feel vulnerable to risking) exclusion by new or disenfranchised contracting, other stakeholders, community or team members.

Writing about what happens when research teams do not work well together is almost antithetical to the firmly established but nonetheless informal rules of collaborative and/or multidisciplinary and/or co-designed research. Writing about these instances at all also requires one to step away from drawing upon the routine forms of evidence or proof to support claims. I cannot, for instance, responsibly name or identify in other ways, the projects where relationships with co-researchers have fractured, or where disagreements have arisen regarding research approaches. I cannot be explicit in discussing moments when ethical considerations might have led me to withdraw from a project, or to seek interventions or other support to enable a project to 'get back on track'.

Working as a researcher in the regional community where I also live-work-care for others brings elevated responsibilities to protect the reputations of all co-researchers. In some respects, the dynamics of intersecting roles and relationships contribute levels of co-dependencies between researchers, especially when colleagues are co-researchers across multiple projects. Focusing on project goals of co-design and capacity building to empower communities and improve outcomes through changed practices and policies provides sound motivations for multidisciplinary co-designing researchers. However, these can also become detrimental to their personal and professional wellbeing.

The choice to withdraw from a research project in a regional co-design setting holds a significant public knowledge dimension as contracting parties, other stakeholders, community participant and others beyond the immediate co-researching team become aware of an individual researcher's ongoing absence. Such a decision inevitably exposes both the individual researchers and their colleagues to a range of expected and unexpected questions or observations about this change throughout the remainder of the project—or perhaps even beyond it.

Co-designed research in regions raises a number of considerations regarding the roles, relationships and wellbeing of researchers, especially when such research involves vulnerable and marginalised communities, groups or individuals. As observed throughout, co-designing regional researchers experience multiple challenges in navigating the circumstantial vulnerabilities of others, their emotions in navigating them—and also in navigating their own intersecting personal and professional roles. These embedded circumstances preclude researching as 'detached observer[s]' (Sherry, 2013: 281). As a regionally located researcher, I cannot avoid interactions with research participants (as well as co-researchers, project stakeholders and contracting parties) in other settings, such as supermarket queues, recreation grounds and cafes. These blurred role boundaries and intersecting relationships are part of being a co-designing researcher who lives-works-cares for others within a regional community. These important dynamics present psychosocial factors that demand ongoing management and accommodation strategies for individual researchers and also for understanding how located research practices can be improved and extended.

Conclusion

Research practices that aim to improve health outcomes and reduce disadvantage requires researchers to share in both individual and community experiences of personal and public vulnerabilities. It involves intersections between professional and personal roles and identities and creates new relational fluidities between the public and private spheres that we inhabit. A researcher-worker-resident-carer in a regional setting is presented with unique opportunities to gain insight, experience and understanding of policy, service delivery, agency capability, health, housing and public transport systems. Working to increase understanding of the complex needs and preferences of service users in regional settings repeatedly places researchers at the intersections of policy, public sector agencies and bureaucratic structures. It also repeatedly places us in proximity to the frustrations of others for whom current systems are not working and/or where their life opportunities seem circumstantially constrained.

Integrated and collaborative regional research approaches can reduce levels of disadvantage and lessen its impacts upon people living in regional communities. Such collaborations begin with opportunities for 'ordinary' people and 'experts' to work together. These can be supplemented by collaboration between agencies and organisations, but the success of their collaborative endeavours relies upon the shared experiences, perspectives, values and knowledge created by 'ordinary' people and 'experts' working together to develop new solutions and approaches.

References

Bazerman, M. H., & Watkins, M. (2004). *Predictable surprises: The disasters you should have seen coming, and how to prevent them*. Boston, MA: Harvard Business School.

Commonwealth of Australia. (2015). *Australian early development census*. Online. Retrieved April 3, 2019, from https://www.aedc.gov.au

CSIRO Futures. (2018). *Future of health: Shifting Australia's focus from illness treatment to health and wellbeing management*. Online. Retrieved April 3,

2019, from https://www.csiro.au/en/Do-business/Futures/Reports/Future-of-Health

Damianakis, T., & Woodford, M. R. (2012). Qualitative research with small connected communities: Generating new knowledge while upholding research ethics. *Qualitative Health Research, 22*(5), 708–718.

Fine, G. A. (1993). Ten lies of ethnography: Moral dilemmas of field research. *Journal of Contemporary Ethnography, 22*(3), 267–294.

Hazelwood Health Study. (2017). *Hazelwood health study 3rd annual report.* Online. Retrieved April 5, 2019, from http://hazelwoodhealthstudy.org.au/wp-content/uploads/2017/01/Hazelwood-Health-Study-3rd-Annual-Report_v1.2.pdf

Lalor, J. G., Begley, C., & Devane, D. (2006). Exploring painful experiences: Impact of emotional narratives on members of a qualitative research team. *Journal of Advanced Nursing, 56*(7), 607–616.

Latrobe City Council. (2017). Living well Latrobe. *Municipal public health and wellbeing plan 2017–2021.* Online. Retrieved May 3, 2019, from http://www.latrobe.vic.gov.au/Our_Community/Living_Well_Latrobe

de Leeuw, E. J., & Wise M. (2015). Population health development in the Latrobe Valley. A literature review of world best practice in building healthy communities and health systems, Hazelwood Mine Fire Inquiry. Online. Retrieved May 3, 2019, from http://hazelwoodinquiry.vic.gov.au/wp-content/uploads/2015/09/TOR-7-Health-Improvment-Report-of-E-de-Leeuw-M-Wise.pdf

Parliament of Victoria. (2014). *Hazelwood mine fire inquiry report.* Online. Retrieved April 3, 2019, from http://report.hazelwoodinquiry.vic.gov.au/

Parliament of Victoria. (2015). *Hazelwood mine fire inquiry report 2015/2016, Volume 4.* Online. Retrieved April 6, 2019, from https://www.parliament.vic.gov.au/file_uploads/11172_HAZ_MFIReport-2015_16-Volume4_FA_LR_15B0_pQfGZRfC.pdf

Pyer, M., & Campbell, J. (2012). Qualitative researching with vulnerable groups. *International Journal of Therapy and Rehabilitation, 19*(6), 311–316.

Sherry, E. (2013). The vulnerable researcher: Facing the challenges of sensitive research. *Qualitative Research Journal, 13*(3), 278–288.

State Government of Victoria. (2017a). *Children's services coordination board report.*

State Government of Victoria. (2017b). *Latrobe health and wellbeing charter.* Online. Retrieved May 3, 2019, from https://www2.health.vic.gov.au/about/health-strategies/latrobe-health-innovation-zone/latrobe-health-and-wellbeing-charter

State Government of Victoria. (2019). Children and youth area partnerships: Inner Gippsland. Online. Retrieved April 23, 2019, from https://ingipps.areapartnerships.vic.gov.au/

Vinson, T., Rawsthorne, M. L., Beavis, A., & Ericson, M. (2015). *Dropping off the edge: Persistent communal disadvantage in Australia.* Catholic Social Services Australia. Online. Retrieved May 3, 2019, from https://www.researchgate.net/publication/299389688_Dropping_off_the_edge_Persistent_communal_disadvantage_in_Australia

18

Energy, Households, and Regions: Feminist Methodologies for Transdisciplinary Research

Gordon Waitt

Introduction

This chapter advocates for feminist methods and methodologies to help study regions and regional phenomena. At a time when regions are often characterised by socio-economic disadvantages, it is helpful to consider approaches that may mitigate the imbalances of power in relationships between researchers and those researched by carefully considering questions of epistemology and methodology. The chapter discusses why a feminist inflected epistemology and methodology offer a key optic in relation to addressing energy poverty in the Global North and, more broadly, research addressing regional issues. The chapter considers feminist conceptual and methodological tools applied in a project with older low-income households living in the Illawarra, New South Wales, Australia, which face energy poverty because of rising fuel costs. For many older low-income households, everyday activities like cooking,

G. Waitt (✉)
School of Geography and Sustainable Communities, University of Wollongong, Wollongong, NSW, Australia
e-mail: gwaitt@uow.edu.au

A. Campbell et al. (eds.), *Located Research*,
https://doi.org/10.1007/978-981-32-9694-7_18

showering, lighting, heating, or cooling become a source of anxiety because of concerns over future energy bills. There is growing agreement amongst policymakers and academics that those with insufficient access to affordable energy will be hurt the most by increased energy hosts, with implications for health, wellbeing and quality of life (Day & Hitchings, 2011).

In this context, the transdisciplinary project's aim, from which the empirical example is drawn, was to better understand the everyday experiences of energy in the making and unmaking of home for older, low-income households. This type of transdisciplinary research, one that brings together social scientist and scientist, will continue to gain significance in the future given the ageing of societies in the Global North, the pressure from climate change to transition away from fossil-fuel power to more sustainable energies, uncertainty over where energy supplies would come from to supply peak demand if demand was left unchecked, and the number of houses requiring energy upgrades, given they were built before the introduction of energy efficiency building standards. Consequently, business as usual is untenable.

Through the example of energy poverty, this chapter focuses on feminist epistemology and methodologies as a lens for addressing regional inequalities in the following ways. First, the feminist concept of 'situated knowledge' draws attention to how all knowledge is partial and contextual. To strengthen methodological rigour, researchers must therefore reflect on their position in an uneven social matrix, and the implications of their positionality through the intersections of gender, class, ethnicity, and age. A method that is reflexive obviates some of the concerns that market-driven decisions around regional policy making—including conceptualisation of energy efficiency—is defined by a perspective of 'scientific objectivity'. Second, a method that 'opens-up' rather than 'closes' discussions with households about what is both socially desirable and feasible is then discussed. Specifically, the method of 'collective video storytelling' is proposed to bring together the knowledge of engineering science with the everyday experiences of older low-income residents. The melding together of lay with technical narratives is argued to respond to the call for a new modus operandi that brings together divergent epistemic foundations of energy knowledge, so creating learning from clashes in ontology, reflexivity, and dialogue (Castree & Waitt, 2017; Gordon, Waitt & Cooper, 2017).

The structure of the chapter is as follows. The next section provides an outline of the research background and its grounding in feminist methods. Section three offers a discussion of the notions of situated knowledge and positionality as applied to the specific case study. The fourth section illustrates the application of 'collective story telling'. Finally, I offer conclusions.

Background on the Method: Broadening Perspective on Regional Issues

Academics and practitioners are discussing how the social sciences need to work alongside the natural sciences to tackle issues such as climate change and energy efficiency (see Castree, 2016; Cooper, 2017). Consequently, regional studies can benefit from the incorporation and socialisation of the knowledge of engineers, scientists, and social scientists. In the context of domestic energy, scientific and technical knowledge is important because it can help answer questions that householders may have about energy use, specifically when considering different appliances (Dieu-Hang, Quentin Grafton, Martínez-Espiñeira, & Garcia-Valiñas, 2017). Whereas, social science scholars have opened-up new perspectives on everyday behaviour by drawing attention to diverse social norms, aspirations, embodied competencies, technologies, and desires (see Gibson, Farbotko, Gill, Head, & Waitt, 2013; Head, Farbotko, Gibson, Gill, & Waitt, 2013; Shove, 2003; Strengers & Maller, 2015). This work points to how everyday experiences of householders are important to understand why certain household behaviours may be more, or less, resistant to change. The starting point of a more 'transdisciplinary' enquiry towards regional studies insists on thinking about how science may speak to the different material, technical, social, and embodied relationships that constitute the problem. Hence, this chapter argues how a feminist epistemology and methodology offers a useful lens to help recruit and activate engineering science in a practical way that is alive to both the social and material elements that comprise energy use.

The project was titled, *Energy Efficiency in the 3rd Age* (EE3A), and branded Illawarra+Energy for the social marketing campaign. The team was comprised of engineers, geographers, social marketers, educators, aged-care housing providers, and community forums. The project was one of 20 trials across Australia funded under the Australian Government's Low-income Energy Efficiency Program (LIEEP). The objectives of LIEEP were twofold: to trial and evaluate different approaches to assist low-income households to lower energy consumption, and to capture and analyse data to inform future energy efficiency policy and programme approaches. To undertake this task, the EE3A project trialled three interventions: (1) a tailored social marketing programme to 830 households; (2) customised technological interventions to a subset of 185 of those 830 households that involved one or more of the following: a replacement refrigerator, replacement water-tank, reverse cycle air conditioner, floor/ceiling insulation, pipe lagging, and window/door seals; and (3) energy workshops and leadership capacity building courses.

The project design involved 11 qualitative focus group interviews ($n = 55$, 35 females and 20 males), and semi-structured interviews that combined sketches of house floor-plans and home insights with 37 households ($n = 51$, 30 females, 21 males) aged 60+ years that are home-owners or private tenants (i.e. not living in public housing) in regional NSW, Australia. All participants had a personal disposable income below $26,104 per annum, the Australian Bureau of Statistics' threshold for low income at the time of data collection (Australian Bureau of Statistics, 2011).

Focus group participants were recruited using a purposive sampling approach. This involved approaching several known networks including community social groups, service providers, and providers of independent living units and residential aged-care homes. Semi-structured interview participants were recruited as a sub-sample of a cohort of household participants recruited using random digit dialling telephone sampling of those who were taking part in a community energy efficiency programme. Focus group interviews were conducted in homes, community venues, or at the University of Wollongong campus according to participants' preference and convenience. Semi-structured interviews were conducted in participants' homes.

A semi-structured interview schedule elicited narratives regarding energy use, energy efficiency, and everyday practices. Questions to start conversations included, 'How do you use energy in the home and what do you do?' 'What do you think are some of the major contributors to electricity use in your home?' 'How did you learn about how to save energy?' Participants were encouraged to tell their own stories about important topics and practices regarding domestic energy use. Focus group interviews averaged around one hour. Ethnographic interviews averaged around 90 minutes. All focus groups and ethnographic interviews were audio recorded and professionally transcribed verbatim. The University Human Research Ethics Committee approved the study, and all participants received an information sheet and provided written informed consent. A $50 gift voucher was given to each participant to recompense for their time.

Methodological Insight: Application of Situated Knowledge and Positionality

The concepts of situated knowledge and positionality were a principle feature of the project design. Conventionally, household energy consumption is defined by a view of 'scientifically objective' knowledge. The work of the engineer is to classify, measure and ideally model household energy consumption as if it were predictable and knowable, rather than chaotic, unpredictable, and dynamic. The scientist is conventionally portrayed as the expert and speaks of charged electrons, protons, the laws of physics and physical units to quantify energy use. Following this logic, an understanding of energy efficiency may be achieved by studying the building and its constituent technologies. Priority is given to measuring the physical attributes of the building and electrical appliances. Knowledge of energy efficiency for engineers is often derived from complex computer modelling that simulates various scenarios for testing hypotheses based on mathematical abstractions composed of variables and equations and digital representations of the spaces and technologies. In doing so, much engineering laboratory work involves disinterest in everyday experiences of those living in houses. Instead, scientific objectivity can render people as static, essentialised and passive: merely cogs in a machine who respond logically to stimuli.

Consequently, the solution to the 'problem' of domestic energy overuse is through more efficient structures or appliances. For example, the introduction of the smart metre is one example of an energy policy which advocates for the use of technological advances to bring about lower energy (Strengers, 2013). Such thinking illustrates the wider discourse of ecological modernisation in which technological advances offer a solution, in this case energy efficiency, thus lowering energy bills and carbon footprints. Yet, as noted by Strengers (2013), cost alone does not explain decisions around energy consumption, and smart metres did not challenge the social practices that householders consider unchangeable. As noted by Cooper (2017), in a market-driven, governance decision-making context, an objective conceptualisation of energy efficiency may paralyse public agency by positioning science as the authoritative voice.

The concept of situated knowledge can offer a richer interplay between the physical and social sciences in energy research and regional studies more broadly. The concept of situated knowledge draws on Harding's (1986) standpoint theory, that all knowledge claims are socially situated, and hence knowledge claims can only be partial. Methodological rigour is strengthened by researchers therefore reflecting on their social situatedness in a project along the lines of disciplinary knowledge, gender, ethnicity, class, and age, among others (see Moss & Falconer Al-Hindi, 2008).

Interpreted through the concept of situated knowledge, the feminist researcher thus accepts a plurality of situated knowledges. Several important implications arise from this framework that troubles the energy science-to-policy paradigm. First, older low-income households can no longer be preconfigured as vulnerable, but instead must be repositioned as specialists in their everyday domestic energy use. Second, households can no longer be conceived as a predictable and knowable machine, but rather they are ambiguous, chaotic, and contradictory. Third, following the lead of Barry and Born (2013), energy policy is driven by a 'logic of innovation' fashioned through the conflicts that arise from the plurality of knowledge. For example, in the Illawarra+Energy project, this plurality of situated knowledges generated moments of contest that triggered negotiation, self-reflexivity, mutual learning, and new knowledge being co-generated.

In such moments, the goal was not to work towards a singular truth; engineers need not believe that appliances like refrigerators may be full of emotional meaning and sustain reciprocal human–home relationships. Nor, in our case, must residents accept energy upgrade recommendations based on faith in engineer's computer-generated models. Instead, the goal is to accept that knowledge is always partial and embedded in uneven power relationships. Through a process of submitting these uneven power relationships to debate a mutual learning process can be generated. As Harding (1992) argues, 'strong objectivity' is encouraged by researchers reflecting on their social situatedness, and the implications this has on their paradigm, power, position, and policy perspective. In practice, this means listening to participants to help identify those energy upgrades that were unrealistic with reference to practical and ethical concerns.

Methodological Insight: Collective Story Telling

This section discusses collective story telling as an experimental method to address the gulf between social sciences and science in order to better understand energy poverty. Collective story telling is positioned as an experimental method that may assist to unsettle dominant narratives around a topic, in this case, domestic energy use, but has wider application to many regional issues. In this example, video storytelling offers a technique to meld lay and scientific knowledge through combing through the narratives, voices, and visuals of our older low-income participants' homes, alongside video graphics informed by engineers (see http://www.energyplusillawarra.com.au/?page_id=84). In doing so, collective video storytelling builds on the feminist literature that argues that narratives are a technique to both document and give voice to marginalised groups; including older low-income households (see Waitt & Gorman-Murray,

2011; Waitt, Roggeveen, Gordon, Butler, & Cooper, 2016). Moreover, collective video storytelling embraces feminist arguments that narrative methods offer insights to the messiness and contradictions of everyday life. In the context of this project, everyday domestic energy narratives offered understandings as to how energy is enrolled in the making and unmaking of a house-as-home (see Galvin, 2013; Reeve, Scott, Hine, & Bhullar, 2013). Equally, energy narratives provide clues to the gendered division of household labour (Cupples, Gutatt, & Pearce, 2003), doing age (Day & Hitchings, 2011), ideas of seasonality (Hitchings, Waitt, Roggeveen, & Chisholm, 2015), and thrift (Waitt et al., 2016).

Alongside these applications of narrative, our interest was in how van Laer, de Ruyter, Visconti, and Wetzels (2014) made the case for narratives as acts of knowing, and how these may by employed to mobilise change through encouraging self-reflection. Van Laer et al. (2014) argue that the basic principle of narrative-transformation is that people should not be told what do to but instead called on to become self-reflexive about personal experiences. In this case, the employment of collective video storytelling encouraged old lower-incomes households to reflect upon personal energy practices and their wider implications.

Three questions about designing collective video stories to help guide researchers with the practical task of employing narratives to socialise science arise: How to weave together lay and scientific narratives? Who tells the narrative? Where a narrative is told?

How to Weave Together Lay and Scientific Narratives?

The starting point to making the collective video stories was 'drafting' a series of lay narratives by weaving together extracts of the stories told to us in the focus groups and semi-structured interviews. Specific attention was given to narratives that illustrate participants' description of everyday concerns about different domestic energy practices (including lighting, heating, cooking, hot-water, and refrigerators). Next, the engineers provided explanations of the way appliances operate in lay scientific terms from an energy and thermal comfort perspective. This included how the appliances within participants' houses are used and how these issues influence the overall energy consumption and comfort of each household. By bringing scientific and lay energy narratives into conversation, possibilities were offered to challenge some energy-related misunderstandings held by many participants and the community generally. The script for each of our narrative videos involved a process of editing together lay energy narratives from the qualitative research, supplemented with technical explanations of appliance energy consumption and basic building physics where relevant.

Three key findings from our analysis of participants' energy narratives guided our selection of stories for the narrative videos: resourcefulness, thriftiness, and common concerns. Resourcefulness illustrated how many older low-income people were often already involved in many energy efficiency practices out of economic necessity, notions of waste and/or to sustain generational difference. For example, one participant told us 'we just switch lights on as much as we need them' commenting that doing otherwise would be wasteful and negligent. Therefore, we endeavoured to ensure that our collective video stories acknowledged and helped reinforce the strategies and capabilities of energy efficiency already practiced by older low-income people. Importantly, the video encouraged participants to reflect on the outcomes of particular domestic energy practices rather than talking about what they might be told to do. For example, the heating narrative video features a narrative on how the participant performs certain practices of keeping warm: 'I have a blanket that I wrap myself up in. I try to warm up that way first before I put the heater on'. This helps reinforce the existing energy efficient practices that older low-income people use.

Thrift was an important emergent theme. Importantly, being thrifty in domestic energy practices had important implications for participants' comfort, health, and wellbeing. For example, one participant recalled that: 'I know a friend who used to use the street lights at night through the window instead of turning lights on. He had a fall in the bathroom and ended up in hospital'. This creates the 'tyrannies of thrift'; that is contradictions in the discourses of energy efficiency. Older low-income people who understand energy efficiency as 'thrifty' may risk their health and comfort. In response, the collective video stories introduced counter-narratives to help redefine energy efficiency as not concerned solely with using less, but to encompass maintenance of comfort and wellbeing.

Finally, participants spoke of concerns about being energy efficient. For example, as one participant said:

> I often wonder about silly things like that, what opening and closing the fridge door does to my energy usage. I don't know the answer to how to most efficiently open the fridge to get the milk out to put in my coffee and then I put the milk back in. Should I close the door in between or should I leave the door open?

Another concern related to a perception of many that reverse cycle air conditioning is an expensive, inefficient heating option: 'I've got reversible air conditioning, but I don't use that either because it's too expensive'. Technical knowledge of the engineers in our team addressed this concern by providing a narrative that illustrates how a reverse cycle air conditioner is one of the most energy efficient ways to heat a house (Fuller & Luther, 2003). It is a basic thermodynamic fact that a reverse cycle air conditioner supplies heating thermal energy to a home at a rate far greater than the rate of electricity consumed by the appliance, since additional 'free' thermal energy is captured from the outside environment. The 'coefficient of performance' of such reverse cycle air conditioning appliances means that the householder receives more than 2.5 times the amount of heating for the same amount of electricity consumption as would be provided by 'direct' electrical heating, that is, by a 'fan heater', or an oil-filled portable radiator. The collective video narrative illustrated this fact, and related technical issues in ways that were not only clear and simple but, importantly, scientifically accurate. Face-to-face interactions between participants and engineers facilitated the development of believable and relevant ways of explaining the technical issues in the plot of the narrative videos. These face-to-face interactions occurred through the engineering team engaging with participants that had energy consultations, monitoring equipment installed in their houses (to measure indoor temperatures, energy use, etc.), and had energy efficiency improvements made to their houses. This meant that the engineering team had many personal insights into the participants and their views on domestic energy efficiency, which facilitated understanding of the social and economic contexts of the participants' lives.

A further concern was that when making a purchase, buying a high-efficiency appliance is the most important consideration relating to energy efficiency. A star-rating system is used for major household appliances in Australia, such as refrigerators, air conditioners, and washing machines. Overlooked is selecting appropriately sized appliances for households. To assess properly the impact of their new appliance on their energy bills (see http://www.energyrating.gov.au), consumers should examine appliance 'energy labels' for both the appliance efficiency (star rating) and the kWh-per-annum energy consumption. The video reproduced this narrative.

Another example of where quite complex scientific/engineering issues were socialised was the case of 'active cooling' of homes, that is, using air conditioners or fans, during periods of warm or hot weather. Conveying the concept that air movement provided by a pedestal or ceiling fan can reduce the perceived temperature of air in a householder's home is important. Participants can significantly reduce their air conditioning energy consumption by combining the use of their fan and their air conditioner, which can then be set at a higher 'set-point temperature'. Therefore, animations in the narrative videos were designed to meld the scientific knowledge and recommendations of the engineers in our team, with the lay narratives of participants. In effect, the videos consider how domestic appliance narratives are produced, circulated, and given credibility. More broadly, the collective video stories contest the perceived gap between scientific and lay knowledge of domestic energy use as mutually exclusive.

The Who and Where of Storytelling?

The constructs for narrative-transportation identified by van Laer et al. (2014) informed collective video storytelling. Hence, questions of who and where a narrative is told are crucial. In each narrative video, the narrators and characters are identifiable characters relatable to older low-income people. The narrative videos feature participants' voices, faces, bodies, technologies, and homes. To create identifiable characters for older low-income viewers, the video featured research participants. Character identifiability is an important consideration in social marketing narratives by facilitating narrative-transportation and relatability (Küntay, 2002; Stern, 1994). For example, Slater, Buller, Waters, Archibeque, and LeBlanc (2003) reported that narratives featuring identifiable characters promoting a healthy lifestyle helped induce more nutritious eating habits among story-receivers. Likewise, Dal Cin, Gibson, Zanna, Shumate, and Fong (2007) reported that narratives featuring identifiable characters may increase story-receivers' intentions to smoke.

To create an imaginable plot for viewers video participants acted out everyday energy use practices. While the narrative story is read, practices are shown being performed on the screen and the audio of the narrative can be heard. For example, the cooking video features Jeannette preparing a meal as she would do normally using her own food and kitchenware, and technologies such as her fridge and oven. This synergy between narrative, places, performers, and practices is designed to help the viewer imagine the plot; making entry into the narrative world more possible (Green & Brock, 2002).

The aim of verisimilitude is to create lifelikeness, and a sense of believability in the collective narrative videos. Hence, participants read from and acted out according to a script—effectively speaking and living the collective narratives while situated in their houses. As Bal, Butterman, and Bakker (2011) identify, believability is more important than consistency or continuity in narratives. The collective video narratives were thus filmed and edited in ways to appear believable and lifelike, creating a sense of verisimilitude.

As feminist scholars emphasise, the spatial dimension of storytelling matters. Hence, the video narrators are speaking in their own homes, using their own appliances and materials, or in local community spaces and places, as relevant to the topic of the video. This contributes to creating identifiable characters (van Laer et al., 2014) and brings to the fore the spatial elements of subjectivity. Acknowledgement of the spatial dimensions points out how energy use and its relationship with thermal comfort is more than measuring kilowatts and temperature, because the various practices that consume energy helps to sustain the reciprocal relationships between people and place. By explicitly situating the narrative videos in participants' homes and communities, they avoid the common narrative approach of 'erasing' consideration of the spatial, and underscore the importance of situated knowledge in social marketing narrative videos.

Conclusion

This chapter argues that feminist epistemology and methodology offer a productive optic for transdisciplinary research studying regional issues by employing the example of domestic energy consumption. In sum, feminist epistemology and methodology may facilitate transdisciplinary research through how reflexivity extends and contests conventional approaches to knowledge about domestic energy consumption. In the context of energy research, the reflexivity and positionality of feminist thinking challenges the conventions that position the researcher as expert, and older, low-income households as vulnerable. Instead, attention turns to becoming reflexive about unequal power relationships and the politics behind these relationships of power. Hence, integral to the research is the everyday experiences of participants alongside the scientific knowledge of domestic energy consumption. Consequently, feminist epistemology and methods encourage moving discussions of domestic energy away from solely kilowatts consumed to employing ideas and methods that enable documenting what energy consumption enables people to do, including sustain a sense of self place and self. A transdisciplinary project design that embraces feminist epistemology and methodology has application beyond questions of access to domestic energy, but also in a more general sense to problem solving within regional locations, including, but not restricted to, the topics of climate change, fresh water, land, food, and mineral resources.

The example of collective video storytelling offers one experimental narrative approach within transdisciplinary research by the melding of lay and scientific knowledges. Acknowledging these different perspectives on domestic energy use and energy efficiency can help open up conversation by encouraging viewers to reflect upon their everyday experiences through stories that feature identifiable characters, recognisable social practices, and socialisation of science in robust and accessible plots. Future regional research can be benefitted from feminist concepts of 'situated knowledge' and 'positionality' through the encouragement of methods that combine reflexive thinking and narratives to encourage ongoing debate and help recognise and mitigate against the imbalances of social power in research design and relationships.

References

Australian Bureau of Statistics. (2011). Household expenditure survey and survey of income and housing: User guide 2003–04, cat. no. 6503.0. Online. Retrieved May 28, 2019, from http://www.ausstats.abs.gov.au/ausstats/subscriber.nsf/0/5BCF25BEABB47CF3CA25715600159C04/$File/65030_2003-04.pdf

Bal, P. M., Butterman, O. S., & Bakker, A. B. (2011). The influence of fictional narrative experience on work outcomes: A conceptual analysis and research model. *Review of General Psychology, 15*(4), 361–370.

Barry, A., & Born, G. (Eds.). (2013). *Interdisciplinary: Reconfigurations of the social and natural sciences*. London: Routledge.

Castree, N. (2016). Geography and the new social contract for global change research. *Transactions of the Institute of British Geographers, 41*(3), 328–347.

Castree, N., & Waitt, G. (2017). What kind of socio-technical research for what sort of influence on energy policy? *Energy Research and Social Science, 26*, 80–86.

Cooper, A. (2017). Building physics into the social: Enhancing the policy impact of energy studies and energy social science research. *Energy Research and Social Science, 26*, 80–86.

Cupples, J., Gutatt, V., & Pearce, J. (2003). Put on a jacket, you wuss': Cultural identities, home heating and air pollution in Christchurch. *New Zealand. Environment and Planning A, 39*(12), 2883–2898.

Dal Cin, S., Gibson, B., Zanna, M. P., Shumate, R., & Fong, G. T. (2007). Smoking in movies, implicit associations of smoking with the self, and intentions to smoke. *Psychological Science, 18*(7), 559–563.

Day, R., & Hitchings, R. (2011). "Only old ladies would do that": Age stigma and older people's strategies for dealing with winter cold. *Health & Place, 17*(4), 885–894.

Dieu-Hang, T., Quentin Grafton, R., Martínez-Espiñeira, R., & Garcia-Valiñas, M. (2017). Household adoption of energy and water-efficient appliances: An analysis of attitudes, labelling and complementary green behaviours in selected OECD countries. *Journal of Environmental Management, 197*, 140–150.

Fuller, R. J., & Luther, M. B. (2003). Using small reverse cycle air conditioners in relocatable classrooms – A case study. *Energy and Buildings, 35*, 619–629.

Galvin, R. (2013). Impediments to energy-efficient ventilation of German dwellings: A case study in Aachen. *Energy and Buildings, 56*, 32–40.

Gibson, C., Farbotko, C., Gill, N., Head, L., & Waitt, G. (2013). *Household sustainability: Challenges and dilemmas in everyday life*. Cheltenham, UK: Edward Elgar.

Gordon, R., Waitt, G., & Cooper, P. (2017). A social marketer, a human geographer and an engineer walk into a bar: Reflections on doing interdisciplinary projects. *Journal of Social Marketing, 7*(2), 366–386.

Green, M., & Brock, T. (2002). In the mind's eye: Transportation-imagery model of narrative persuasion. In T. Brock, J. Strange, & M. Green (Eds.), Narrative impact: Social and cognitive foundations (pp. 315–342). New Jersey: Lawrence Erlbaum Associates.

Harding, S. (1986). *The science question in feminism*. Ithaca, NY: Cornell University Press.

Harding, S. (1992). After the neutrality ideal: Science, politics, and strong objectivity. *Social Research, 59*, 567–587.

Head, L., Farbotko, C., Gibson, C., Gill, N., & Waitt, G. (2013). Zones of friction, zones of traction: The connected household in climate change and sustainability policy. *Australasian Journal of Environmental Management, 20*, 351–362.

Hitchings, R., Waitt, G., Roggeveen, K., & Chisholm, C. (2015). Winter cold in a summer place: Perceived norms of seasonal adaptation and cultures of home heating in Australia. *Energy Research and Social Science, 8*, 162–172.

Küntay, A. (2002). Development of the expression of indefiniteness: Presenting new referents in Turkish picture-series stories. *Discourse Processes, 33*(1), 77–102.

van Laer, T., de Ruyter, K., Visconti, L. M., & Wetzels, M. (2014). The extended transportation-imagery model: A meta-analysis of the antecedents and

consequences of consumers' narrative transportation. *Journal of Consumer Research, 40*(5), 797–817.
Moss, P., & Falconer Al-Hindi, K. (2008). Feminisms, geography and knowledge in Feminisms. In P. Moss & K. Falconer Al-Hindi (Eds.), *Geography: Rethinking space, place and knowledges* (pp. 1–27). Lanham, MD: Rowman & Littlefield.
Reeve, I., Scott, J., Hine, D., & Bhullar, N. (2013). 'This is not a burning issue for me': How citizens justify their use of wood heaters in a city with a severe pollution problem. *Energy Policy, 57*, 204–211.
Shove, E. (2003). *Comfort, cleanliness and convenience: The social organisation of normality*. Oxford: Berg.
Slater, M. D., Buller, D. B., Waters, E., Archibeque, M., & LeBlanc, M. (2003). A test of conversational and testimonial messages versus didactic presentations of nutrition information. *Journal of Nutrition Education and Behavior, 35*(5), 255–259.
Stern, B. B. (1994). Authenticity and the textual persona: Postmodern paradoxes in advertising narrative. *International Journal of Research in Marketing, 11*(4), 387–400.
Strengers, Y. (2013). *Smart energy technologies in everyday life: Smart utopia.* Sydney: Palgrave Macmillan, Angus & Robertson.
Strengers, Y., & Maller, C. (2015). *Social practices, intervention and sustainability: Beyond behaviour change*. New York: Routledge.
Waitt, G., & Gorman-Murray, A. (2011). It's about time you came out. Sexualities, mobility and home. *Antipode, 43*, 1380–1403.
Waitt, G., Roggeveen, K., Gordon, R., Butler, K., & Cooper, P. (2016). Tyrannies of thrift: Governmentality and older, low-income people's energy efficiency narratives in the Illawarra, *Australia. Energy Policy, 90*, 37–45.

19

Out in the Regions: Queer Film Festivals, Community-Building and the Cultivation of Creative Talent in Regional Victoria

Chloe Benson

Introduction

Queer Film Festivals (QFFs) have tended to be linked with major capital cities in Australia, most notably Sydney (Queer Screen) and Melbourne (Melbourne Queer Film Festival or MQFF)—locations that more readily facilitate long-term financial viability and professionalisation. This reflects wider global trends amongst QFFs, as well as the symbolic metropolitan and non-metropolitan divide that has long-permeated the queer imaginary (Halberstam, 2005; Weston, 1995). Yet, events like the Bendigo Queer Film Festival (BQFF) and the Geelong Pride Film Festival (GPFF) highlight the contributions that queer cinema can bring to regional communities. Although film festivals are a distinct type of arts festival, they can be located within the broader context of queer festivals in rural and regional locations. Like other regional queer festivals, such as Daylesford's Chillout Festival, these events challenge the exclusion of rural and regional spaces from the queer imaginary (Gorman-Murray, Waitt, &

C. Benson (✉)
Federation University Australia, Churchill, VIC, Australia
e-mail: cbenson@federation.edu.au

A. Campbell et al. (eds.), *Located Research*,
https://doi.org/10.1007/978-981-32-9694-7_19

Gibson, 2008). As Ford demonstrates, regional QFFs can also play a 'unique role in bringing about changes in attitudes within the general community' (Ford, 2014) and diversifying representation outside metropolitan areas. For Ford (2013), this is crucial because 'entrenched homophobia' and misunderstanding of LGBTQIA+ (i.e., lesbian, gay, bisexual, trans, queer, intersex, asexuality and other sexualities, sexes and genders that aren't included in these few letters) people remains common in regional areas.

Acknowledging these issues of inclusion and accessibility for rainbow communities in regional Victoria, the LGBTI Equality Roadshow, led by Gender and Sexuality Commissioner Ro Allen, has been touring throughout regional and rural Victoria in an effort to 'reduce discrimination and promote inclusion of LGBTI people in everyday rural and regional life' (State Government of Victoria, "About the Roadshow", 2019a). A further $1 million has been committed to supporting the event over the next two years, with the Minister for Equality, Martin Foley, stating: 'the roadshow has consistently proved the need to engage with LGBTI communities that live outside Melbourne's metropolitan bubble.... We need to make sure the equality agenda doesn't stop at the end of the tram tracks' (Wade, 2018). The Rural and Regional Program seeks to pursue equality outside Melbourne via a range of long-term improvements for regional LGBTQIA+ communities, including: 'improved mental health outcomes; greater population retention and economic inclusion; capacity building for communities to empower themselves, for example, developing leaders; building lasting networks between LGBTI community, service providers and government agencies; [and] greater broader community understanding and support for inclusion and diversity' (State Government of Victoria, "LGBTIQ Equality", 2019b). Creative endeavours have a key role to play in the attainment of these aims, as the Victorian Government's Pride Events and Festivals Fund highlights. This fund is intended to 'promote LGBTI arts and culture, celebrate diversity and acceptance, and build resilience against LGBTI discrimination' (State Government of Victoria, "Pride Events", 2019c). Successful grant recipients throughout 2018 and 2019 have included a range of regional events including Ballarat Frolic Festival and Gippsland's Rainbow Formal.

Both BQFF and GPFF have also been awarded grants to support their contributions to LGBTQIA+ arts and culture.

Regional QFFs like GPFF and BQFF showcase important LGBTQIA+ cinema otherwise difficult to access. The power of such representation is well documented; as Dyer (2002: 1) has surmised, 'how we are seen determines in part how we are treated; how we treat others is based on how we see them; such seeing comes from representation'. But in addition to this representation or on-screen visibility, RQFFs can bolster community and facilitate the development of creative capital through the initiation of funding and mentoring opportunities for emerging creatives. The earliest QFFs were typically grassroots start-ups, marked by a 'corrective and self-affirming nature ... [and] positive imagery politics' (Loist, 2012: 158–159); in these contexts, film was used for consciousness raising and liberationist purposes. These political origins remain relevant, yet QFFs are also evolving as they endeavour to '[promote] social empowerment while still remaining financially viable' (Richards, 2016: 6). This chapter considers these various challenges alongside the generative potential of regional QFFs, both as contexts of exhibition and community as well as incubators for emerging filmmakers and storytellers.

A Regional Context

An urban-bias pervades research undertaken in Australian festival studies (Gibson, Waitt, Walmsley, & Connell, 2010). In their analysis of close to 3000 rural and regional festivals in Australia, Gibson and Stewart (2009) address this issue, observing the immense diversity and significance of festivals in regional Australia. This diversity is unsurprising, given the varied landscape that constitutes Australia's regions. Broadly, regional Australia encompasses 'all of the towns, small cities and areas that lie beyond the major capital cities' (Regional Australia Institute, 2014). That is, the regions encompass *a lot* of Australia. It makes sense, then, that a one-dimensional view of regional Australia is neither feasible nor useful (Regional Australia Institute, 2014). To help elucidate the diversity of regional Australia, the Regional Australia Institute identifies four types of regional communities: Regional Cities, Connected Lifestyle Areas,

Industry and Service Hubs and Heartland Regions. Both Bendigo and Geelong, the towns that are home to the RQFFs discussed in this chapter, fall into the category of Regional Cities, which tend to have higher populations, a diverse economy and employment profiles comparable to major metropolitan centres (Regional Australia Institute, 2014). Whilst regional festivals are not limited to these types of larger locations, large regional towns, lifestyle regions and popular tourist spots do tend to host the most regional festivals (Gibson & Stewart, 2009).

Rural and regional festivals span various niche interests and range vastly in terms of their size and duration (Gibson & Stewart, 2009). They stimulate communal and economic benefits for communities and successfully bring diverse groups of people together; 'they blend attitudes, enlargen social networks and encourage improvements in social cohesion ... catalys[ing] community in the name of fun' (Gibson & Stewart, 2009: 5). Challenging a binary conception of the inner city as the site of creativity and the surrounding sites beyond the city as places dominated by consumption and a lack of innovation or creativity, Gibson and Brennan-Horley (2006: 469) assert that 'it is time to break out of the mould of thinking about creativity as a bohemian, inner city phenomenon'. Ford (2014), founder of the Queer Fruits Film Festival in Lismore, echoes this sentiment as she reflects on the concentration of QFFs in metropolitan areas and their relative absence in the regions:

> Regional or rural communities have historically had a tendency (or been viewed) as living in the same areas and mingling with the same folk, not taking easily to outsiders, or new ideas and with a certain old-fashioned mindset. Remote rural and regional locations have also been seen as places where odd and unwelcome things happen to outsiders.... [But] The view of the old-fashioned rural environment in comparison to the futuristic metropolis is now becoming transformed and in many places redundant.... Regional areas now can be experienced as intellectual and cultural havens, where distance may be used as an advantage, providing space (physically and intellectually) to create new forms.

Ford (2014) identifies this long-held view of regional locations, as well as economic and organisational constraints, as contributors to the uneven geographical distribution of QFFs. As noted at the start of this chapter, QFFs tend to be associated with urban settings. In Australia, 'there are very few [QFFs] ... that are wholly regional and that exist outside the major cities and metropolitan areas' (Ford, 2014) and understanding those events that do or have existed is made difficult by the limited scholarship focussed on queer cinema and its uses within non-metropolitan settings.

There are a range of reasons for both the limited critical focus on queer cinema in the regions and the limited number of QFFs in regional locales. In broad terms, the urban/rural divide is used across queer politics and narratives to represent a significant symbolic divide. The city and its promise of community and acceptance have been identified as a crucial component of the gay imaginary and is a prominent influence for queer migration (Weston, 1995). Given the focus on urban queer culture and its perceived greater freedoms, 'there has been little attention paid to date to the specificities of rural queer lives' (Halberstam, 2005: 32). Moreover, many narratives of gay and lesbian identity emphasise urban queer lifestyles in positive terms whilst rendering rural queer life negatively:

> While the story of coming out tends to function as a temporal trajectory within which a period of disclosure follows a long period of repression, the metronormative story of migration from 'country' to 'town' is a spatial narrative within which the subject moves to a place of tolerance after enduring life in a place of suspicion, persecution, and secrecy. Since each narrative bears the same structure, it is easy to equate the physical journey from small town to big city with the psychological journey from closet case to out and proud. (Halberstam, 2005: 34–35)

To some extent, research supports such a narrative and its inclinations. LGBTQIA+ people dwelling in regional and rural areas may feel less connected to their communities and experience increased rates of discrimination, which leads to migration to metropolitan areas, identified as a notable characteristic of internal migration amongst LGBTQIA+ individuals (Frost & Meyer, 2012; Henrickson, Neville, Jordan, & Donaghey, 2007; Weston, 1995). Whilst such studies may indicate why QFFs have been more commonly associated with major cities, they also signal the need for increased community-building efforts in regional locations. More nuanced and productive narratives of LGBTQIA+ lives in regional locations are needed, and QFFs provide one means of unearthing and delivering these stories.

Contextualising the Queer Film Festival

In his book *The Queer Film Festival: Popcorn and Politics*, Stuart Richards (2016: 5–6) provides a concise and useful definition of the queer film festival as 'a series of film screenings that primarily focus on queer themes', adding that 'the primary purpose of these festivals is to provide a space for the exhibition of films that would otherwise struggle to secure a large audience'. QFFs were born out of struggle and they continue to seek increased visibility of and for LGBTQIA+ people. QFF scholar Skadi Loist (2012: 158) traces the political origins of these events back to the 1960s and the founding of 'several strands of identity-based film festivals—all with a corrective and self-affirming nature'.

Mirroring the political impetus of black and women's film festivals, the earliest QFFs aimed for the inclusion of homosexuality in mainstream culture and 'were marked by positive imagery politics' (Loist, 2012: 158–159). The Stonewall Riots were a crucial catalyst for QFFs and their visibility politics. The riots that erupted in New York during a 1969 police raid on the Stonewall Inn sparked a turning point in the gay liberation movement and by the late 1970s, amidst 'post-Stonewall vigor and enthusiasm' QFFS began to form (Zielinski, 2008: 980). In 1977, *The Gay Film Festival of Super-8 Films* was first instituted. Now known as *Frameline*, this event is the longest-running QFF globally. *Frameline* was founded by the Persistence of Vision, a 'loosely run collective' thriving in a subcultural context where 'self-expression was everything and money was beside the point' (Frameline, 2012: 19–21). The festival's beginnings were modest: a pinned-up sheet and a rented projector in an old community centre (Frameline, 2012: 18), yet its aims were ambitious: 'Our goal has been to provide a forum for our art and at the same time provide a pool of talent, energy and equipment to help each other' (Frameline, 2012: 3).

Since this time, QFFs have proliferated globally. An interactive Google Map, established by Loist, lists 326 LGBT/Q film festivals in 346 locations operating globally since 1977. Events continue to crop up that are yet to make their way onto Loist's map, such as the relatively new Geelong Pride Film Festival. This map makes use of colour-coded pins to designate those events that are still running and to demarcate different stages of the QFF's global development and corporatisation. Loist's map not only highlights the proliferation of QFFs globally, it demonstrates the increased professionalisation and financial sophistication of larger, contemporary events, which have evolved from their grass roots origins 'to become part of an elite film institution with an influential position in queer cinema' as well as their specific cities or regional locales, where they generate 'creative value' and play important social functions (Richards, 2016: 1–2). This cataloguing of QFFs is important for, as Richards (2016) points out, contemporary festivals must secure sustainable financial stability if they are to realise their social missions, and this can require negotiation of complex economic and political issues. In programming events, for instance, organisers must consider the needs and interests of their communities or audiences, as well as the expectations of their sponsors, including to what extent the content they screen should progress political aims or provide more marketable (and perhaps comfortable) entertainment (Gamson, 1996; Richards, 2016). Richards (2016: 245) demonstrates that as social enterprises, QFFs can strike a balance between these competing programming challenges, signalling the professionalisation of QFFs, which have largely moved away from their underground beginnings, to become part of a 'commercial creative industry'. Major cities now pursue QFFs because they lend locales a certain cache; the gloss of being 'global' and 'creative' (Zielinski, 2012). In this context, regional festivals may face an 'uneven playing field' when it comes to securing screening fees, or in trying to compete with the prestige that larger, international events can offer potential filmmakers (Ford, 2014). Thus, the corporatisation of events may present additional challenges to smaller festivals, particularly those emerging in regional and rural areas.

In many ways, contemporary events vary notably from the grass roots origins of *Frameline*, particularly in economic terms. Yet, the tenets of unity and visibility that underpinned this event in its earliest incarnation have remained key political aims of many festivals on the contemporary queer circuit. The type of content screened at festivals can be markedly different than mainstream offerings, in terms of content (including who is represented and how), as well as form. Experimental, low-budget and short works are less likely to be distributed outside festivals. Thus, QFFs continue to screen content that would be unlikely to find exhibition elsewhere. Moreover, as Rhyne (2006: 618) explains, these events 'do not simply acquire and screen the films they show; they actually create the economic conditions that enable their production'.

The unifying and representative power of QFFs also exceeds the content they screen. As Richards (2016: 6) encapsulates nicely, these events encompass 'So much more than films'. The impact of viewing queer films in a cinema with a largely queer audience remains an important distinguishing characteristic of the QFFs exhibitive function (Rich, 2006). Broadly, this point is true of all film festivals. Conducting focus groups with members of the Glasgow Film Festival audience, Dickson (2015) identified that festival attendees articulated their experiences of the event in primarily spatial and corporeal terms. For instance, audiences commented on the impact of their environment and fellow audience members during screenings in unconventional venues. Significantly, audiences also commented on the role of increased proximity at film festivals: 'In the festival context, reduced social distance and being visibly identified as "a festival-goer", "film lover", "a devotee" or "not a bam" emerge as key points of gratification' (Dickson, 2015: 716).The physical location of events and gathering of bodies makes spatial pleasures an important dynamic of all film festival experiences. This dynamic is amplified in the context of identity-based film festivals, where belonging and a shared sense of community is an essential facet of events and spectatorship is shaped distinctively by context (Rich, 2006; Searle, 1997; Siegel, 1997).

QFFs foster an environment in which LGBTQIA+ people and their allies can attend and enjoy curated screenings amongst their community safely. In part, the QFFs community-building function stems from its status as an exhibition context, where films are screened in public places and communities literally come together to gather in front of a screen. But in addition, QFFs often involve other events and communal spaces beyond the confines of the cinema—such as question and answer sessions, public forums, pop-up festival bars, official affiliations with other establishments (e.g., restaurants and bars), art exhibitions, workshops, networking opportunities and social activities like speed dating. These activities, alongside the programmed films, all contribute to the QFFs potential to foster a sense of community and belonging. In these ways, QFFs prioritise LGBTQIA+ recognition and visibility on *and* off screen.

Regional Queer Film Festivals in Action

As this overview of QFFs highlights, these events have the potential to overlap with a range of the prioritised outcomes for regional LGBTQIA+ people outlined by the Victorian State Government. In particular, through their capacity building, networking, and showcasing of LGBTQIA+ stories, QFFs offer great opportunities for improving LGBTQIA+ health and wellbeing and for increasing acceptance, inclusion and community empowerment. More broadly, QFFs also align with the general functions, aims and benefits of regional festivals, which build community, generate creative capital and contribute cumulatively to economic growth (Gibson & Stewart, 2009). To best demonstrate this potential in action in the regions, the remainder of this chapter will focus on two examples of QFFs: Bendigo Queer Film Festival (BQFF) and Geelong Pride Film Festival (GPFF).

The Bendigo Queer Film Festival has been a staple of Bendigo's events calendar for 16 years. In collaboration with Melbourne Queer Film Festival, BQFF was founded by Chris Walters in 2004. This connection with MQFF has continued and the larger, metro festival remains BQFF's programmer and key sponsor. Relative to BQFF, Geelong Pride Film Festival has a much shorter history. In 2018, the festival was launched as Geelong Queer Film Festival. In late 2018, the event changed its name, opting to move away from the term 'queer' in favour of the Geelong *Pride* Film Festival. This choice was made in an effort to reflect the festival's commitment to 'pride and dignity' and a 'mission for regional, cultural and generational inclusivity' (GPFF, 2018). In 2019, GPFF ran its second festival under this new branding. Like BQFF, GPFF is also connected to and supported by MQFF, who develop and collaborate on the event's programming.

This collaboration between MQFF and each regional event resonates with Ford's (2014) description of showcase programmes, where films 'programmed by an urban festival curator … [are] presented as a "touring" program to a region by an established city-based film festival'. Ford (2014) differentiates between these touring events and 'wholly regional' festivals, which are 'established and produced independently in a regional area'. Neither BQFF nor GPFF fit easily within the classifications that Ford (2014) delineates. Whilst MQFF plays a role in programming each festival, BQFF and GPFF are larger events more embedded in their regions than touring programmes. This is most salient when they are contrasted with the touring programme events MQFF do offer via their regional roadshows, which have traversed Australia nation-wide. In 2018, for instance, MQFF embarked on a NT Roadshow, taking screenings to Alice Springs. Commenting on this endeavour and the role of community to MQFF, festival director Spiro Economopoulos explained: 'actually experiencing the community element… That's important enough in Melbourne … but I think it's even more applicable in regional areas like Alice Springs' (Russell, 2018). Roadshow screenings tour highlights of MQFF to wider areas, thereby offering rural and regional LGBTQIA+ people and their allies a chance to experience both this 'community element' and a showcase of new queer cinema offerings. This is significant as public spaces for affirming identity and community are particularly important to regional LGBTQIA+ people (Ford, 2014). Both GPFF and BQFF offer comparable opportunities for communal gathering and sharing by screening a selection of films from MQFF in their local communities. At both events, festival attendees gather in local venues to enjoy films in an inclusive space, and to network and partake in community discussion. For instance, in 2019, GPFF screened the documentary *TransMilitary* alongside a Q&A with co-founder of Transgender Victoria, Sally Goldner. This session created a space for the community, LGBTQIA+ and otherwise, to gather, connect and engage in dialogue. Holding this event at a larger and more central venue than the cinema where most of the festival's screenings take place, and ensuring the provision of closed captions and Auslan, helped to remove barriers to participation. In Bendigo, comparable opportunities for community-building are fostered. For instance, in 2019, the event launched its festivities with a special performance by the All Male Choir, LOW REZ and included a Special Guest Breakfast event to accompany its screening of *TransMilitary*. At both events, launch and opening night parties, forums and foyer mingling, all facilitate spaces for connecting.

Like the touring model, both GPFF and BQFF also play pivotal roles in bringing LGBTQIA+ content to regional locations where niche films may be less accessible. With fewer independent or arthouse cinemas, and a smaller pool of ticket buyers, many regional cinemas will screen only limited runs of smaller, niche films (if they secure exhibition rights at all). Streaming platforms are changing this picture, both in terms of LGBTQIA+ content and independent cinema more broadly. Special interest streaming platforms like Tello, Revry and Dekko offer subscribers access to a wealth of LGBTQIA+ entertainment. A number of mainstream streaming platforms, such as Netflix and Stan, as well as the free SBS On Demand service, also offer users content catalogued as LGBTQIA+ interest or, in the case of SBS, Rainbow Pride. Other digital forms such as podcasts and web series are further improving accessibility to content produced by and for LGBTQIA+ people. Yet, accessibility issues persist. For instance, the sheer volume of content available online may present challenges, particularly for those not already immersed in LGBTQIA+ culture. That is, how do young people seeking such content learn of its existence and whereabouts outside of established LGBTQIA+ networks, communities or press?

Access to digitally distributed content may also be curtailed by a lack of reliable Internet infrastructure, restrictive data limits and poor digital literacy, particularly for youth in regional and rural areas. Both GPFF and BQFF help audiences to overcome some of these potential barriers by playing an important role in exhibiting films that would otherwise struggle to secure a cinema screening in their respective regions. For instance, as noted earlier, both events showcase MQFF's annual Australian shorts package, a session that always offers diversity. In 2018, the session included films focussed on the Australian Sharpie scene, a feel-good transgender school story, an alien abduction tale, and the Australian women's football league, amongst others. Both events are also international in their programming. For instance, the 2019 GPFF boasted films from Kenya, Switzerland, the UK, the US and Australia in its impressive eight session programme. Shorts, documentaries, classic films and new releases also feature in the events' festival and special screening offerings. Whilst distinctions exist between the BQFF and GPFF programmes and the special screenings that they offer through the year, there is some overlap in the programmes of each event as a result of their partnership with MQFF. Whilst this may limit some of the potential for audience overlap or sharing between events, it also reinforces the capacity of these Victorian events to showcase local offerings around the state via multiple screenings. It also dramatically eases some of the challenges associated with regional festival programming identified by Ford (2014), such as securing and paying for films.

Yet, Ford (2014) suggests that the parallels between GPFF, BQFF and the touring programme model can also pose challenges to regional curation. According to Ford (2014), the distinction between wholly regional and touring festivals is particularly significant when reflecting on programming and curating strategies, because the lower socio-economic context of regional areas 'leads to audience expectations for films that may have less demoralising content than that a metropolitan film audience may tolerate'. The programming partnership between BQFF and GPFF may constrain each event's capacity to address these issues raised by Ford (2014), as the events draw largely on their relationship to the larger metro festival when it comes to selecting and securing films for their audiences. However, although MQFF continues to play an important role in programming and securing the films each event offers, each festival also exercises a level of curation. For instance, in 2019, the BQFF and GPFF programmes both shared a number of films (*Wild Nights with Emily, Australian Shorts, Tucked, Mapplethorpe* and *TransMilitary*) and offered a range of films not screened at the other event (*Ice King, Untold Histories, A Kid Like Jake, Thomas Banks' Quest for Love, Rafiki, Mario, Dykes, Camera, Action!, Dyketactics* and *Ginger*).

In contrast to the visiting programme model of events like the NT Roadshow, BQFF and GPFF are also branded and promoted as their own unique and distinctive local festivals. Each event is also operated and lead by its own organising committees. In these ways, GPFF and BQFF foster cooperation and collaboration within their regions. Like cultural festivals more broadly, they also contribute to the development of important skills in their local communities, such as leadership and management (Gibson et al., 2010); skills particularly relevant to bolstering LGBTQIA+ wellbeing and cultural capital and enhancing the broader creative arts scene and cinephillia within each region. The planning and running of festivals from an organisational perspective alone can be seen to address a range of the improvements that the Victorian State Government seeks for regional LGBTQIA+ communities, namely: economic inclusion, capacity building and network creation. Both events are also committed to engaging and impacting their wider local communities, both in the service of improving LGBTQIA+ inclusion and in fostering creative and economic benefits. Both BQFF and GPFF reference the 'wider community' or 'whole community' in their mission statements, and promote their screenings as inclusive and open to all. By bringing LGBTQIA+ stories to regional screens and operating as a visible presence throughout the year, GPFF and BQFF promote acceptance in the wider population. The press coverage the festivals receive is notable in this regard. For instance, in Geelong, the GPFF has been endorsed widely; including being advertised and editorialised in the local street magazine, *Forte*; promoted at and by the local Independent cinema, The Pivotonian, including the venue 'teaming with the theme' in the lead up to the event by adapting the colour of their logo on social media; and showcased in a special, pop-up shopfront installation on Little Malop Street, a busy thoroughfare in Geelong's central business district. In addition to fostering a sense of inclusivity and pride across Geelong's LGBTQIA+ community, this representation highlights the ways that QFFs can provide powerful LGBTQIA+ visibility in the wider community (Ford, 2014) and help to facilitate both economic inclusion and collaborative networks.

Both BQFF and GPFF also play important roles in nurturing creative talent in the regions. Broadly, the QFF's contributions to creative incubation can be classified into two categories. The first is associated with the QFF's exhibitive function. Securing a release in mainstream cinemas can be incredibly difficult for queer filmmakers. QFFs provide content that would otherwise struggle to find an audience, or a place to be exhibited

(Rhyne, 2006; Richards, 2016). The network in Victoria—that is, between BQFF, GPFF and MQFF—generates greater opportunities to screen smaller films, such as those circulated between the events for inclusion in the Australian shorts session. Secondly, QFFs can offer prizes, funding and mentoring that further assists in supporting emerging queer filmmakers. For instance, each year MQFF awards a series of cash prizes to films screened in the Australian shorts session, including: best director, best emerging Australian filmmaker, and best cinematography. Despite only commencing in 2018, GPFF has already demonstrated its capacity to foster local talent in the region. In 2018, GPFF collaborated with the Geelong Council to launch a $500 short film commission for LGBTQIA+ early career filmmakers. Applicants were encouraged to pitch their best concept for a short film that 'explores a positive LGBTQIA+ theme in the Geelong region' and 'includes an iconic Geelong location or symbol' (GPFF, 2018). In 2019, the commission is underway again, but the framework has evolved to support short film scripts rather than concepts. Bendigo has also established means of fostering local creative talent. In concomitance with the Bendigo Film Festival each year, the Queer Country art exhibition is also held to showcase and celebrate queer artists from Bendigo and regional Victoria. The exhibition provides regional artists with a platform to 'express themselves, showcase their creativity, offer another perspective and educate the broader community about GLBTI arts and culture' (BQFF, 2019). The exhibition showcases art in any media and whilst works addressing LGBTQIA+ themes and issues are prioritised, all subject matters are considered for inclusion. In addition to providing a stimulating backdrop for the launch of BQFF, this exhibition helps to nurture and showcase the work of rural and regional artists.

Conclusion

Both BQFF and GPFF highlight that RQFFs are viable in regional areas and that their various initiatives and activities can foster important benefits that align with the priorities for regional LGBTQIA+ wellbeing identified by the Victorian State Government. As this chapter has demonstrated, the tendency to overlook regional areas in the study of queer cinema and film festivals risks reifying limiting narratives that dichotomise regional and urban queer experiences. In order to better understand the potential impact of events like BQFF and GPFF, we need to begin by acknowledging their existence and challenging the dichotomising urban/regional binary that often prevails in queer narratives and studies. This requires recognition of the difference within and between regional locales, including both the unique strengths and needs of regional LGBTQIA+ people. Yet, the ways that we conceptualise and study QFFs also has a role to play here. QFFs are typically underpinned by similar political aims, but all of these events will inevitably be a product of their geopolitical location; the experience and motivations of their founders; their various financial opportunities and constraints; and their histories and community embeddedness. Thus, in studying regional festivals, we must be mindful of this specificity and avoid conflating all regional events in opposition to all metropolitan events. GPFF and BQFF demonstrate that regional events with metropolitan affiliations can and do succeed in regional Australia, but that their success and impact also requires further analysis. This includes more thorough examination of the ways regional, rural and urban geographies are conceptualised (or not) in QFF studies and the ways that the economic and cultural terrain of regional locations varies and thereby impacts the feasibility and function of various events.

Neither BQFF nor the GPFF fit Ford's (2014) description of a wholly regional event, yet they also don't adhere strictly to the touring city-based festival model. Instead, these events are regionally focused and based. Each event is organised and run by local community members in Geelong and Bendigo, but they simultaneously tap into a wider network across Victoria via their affiliation with MQFF. In these ways, BQFF and GPFF signal the need for new ways of conceptualising and theorising QFFs. Although Richards (2016) has demonstrated that QFFs typically don't work in a networked, circuitous manner, the links between MQFF, BQFF and GPFF offer a unique case study for revisiting this idea. Further analysis of these festivals and their links offers opportunities to examine not only their connections to one another but also their connections to wider LGBTQIA+ community networks. Both of these types of connections require further consideration. For instance, the ways that these festivals overlap, differ, support and disrupt one another warrant additional attention. The ways each event functions as a hub that connects community members, local businesses and LGBTQIA+ organisations within and across regional towns are also something in need of further consideration.

References

BQFF. (2019). *Queer Country*. Retrieved from www.bendigoqueerfilmfestival.com.au; http://www.bendigoqueerfilmfestival.com.au/queer-country.html

Dickson, L.-A. (2015). 'Ah! Other bodies!': Embodied spaces, pleasures and practices at Glasgow Film Festival. *Participations: Journal of Audience and Reception Studies, 12*(1), 703–724.

Dyer, R. (2002). *The matter of images: Essays on representation*. London: Routledge.

Ford, A. (2013). Regional and Queer: Refusing to be Invisible, Creating queer space in a non-queer world. Presented at the 2nd European Geographies of Sexualities Conference, Lisbon.

Ford, A. (2014). Curating a regional, queer film festival. *Fusion,* 4. Retrieved from http://www.fusion-journal.com/issue/004-fusion-the-town-and-the-city/curating-a-regional-queer-film-festival/

Frameline. (2012, August 24). *25th Anniversary San Francisco International Lesbian & Gay Film Festival Program*. ISUU. Retrieved from https://issuu.com/frameline/docs/25thsanfrancisco-international-lgbt-film-festival/20

Frost, D. M., & Meyer, I. H. (2012). Measuring community connectedness among diverse sexual minority populations. *Journal of Sex Research, 49*(1), 36–49.

Gamson, J. (1996). The organizational shaping of collective identity: The case of lesbian and gay film. *Sociological Forum, 11*(2), 231–261.

Gibson, C., & Brennan-Horley, C. (2006). Goodbye Pram City: Beyond inner/outer zone binaries in creative city research. *Urban Policy and Research, 24*(4), 455–471.

Gibson, C., Waitt, G., Walmsley, J., & Connell, J. (2010). Cultural festivals and economic development in nonmetropolitan Australia. *Journal of Planning Education and Research, 29*(3), 28–293.

Gibson, C. R., & Stewart, A. (2009). *Reinventing rural places: The extent and impact of festivals in rural and regional Australia*. Wollongong: University of Wollongong.

Gorman-Murray, A., Waitt, G., & Gibson, C. (2008). A queer country? : A case study of the politics of gay/lesbian belonging in an Australian country town. *Australian Geographer, 39*(2), 171–191.

GPFF. (2018). *Geelong Pride Film Festival Facebook Page*. Retrieved from www.facebook.com: https://www.facebook.com/GeelongPrideFilm/

Halberstam, J. (2005). *In a queer time and place: Transgender bodies, subcultural lives*. New York: New York University Press.

Henrickson, M., Neville, S., Jordan, C., & Donaghey, S. (2007). Lavender Islands. *Journal of Homosexuality, 53*(4), 223–248. https://doi.org/10.1080/00918360802103514

Loist, S. (2012). A complicated queerness: LGBT film festivals and queer programming strategies. In J. Ruoff & J. Ruoff (Eds.), *Coming soon to a festival near you: Programming film festivals* (pp. 157–172). St Andrews: St Andrews Film Books.

Regional Australia Institute. (2014). *Talking point: The foundations of regional Australia*. Regional Australia Institute.

Rhyne, R. (2006). The global economy of gay and lesbian film festivals. *GLQ: A Journal of Lesbian and Gay Studies, 12*(4), 617–619.

Rich, B. R. (2006). The new homosexual film festivals. *GLQ: A Journal of Lesbian and Gay Studies*, 620–625.

Richards, S. (2016). *The Queer Film Festival: Popcorn and politics*. New York: Palgrave Macmillan.

Russell, S. A. (2018). From Alice Springs to Geelong, Queer Film goes regional. Retrieved from https://www.sbs.com.au/topics/sexuality/fast-lane/article/2018/04/04/alice-springs-geelong-queer-film-goes-regional

Searle, S. (1997). *Queer-ing the screen: Sexuality and Australian film and television*. St Kilda: Australian Teachers of Media in association with Australian Film Institute Research and Information Centre and Deakin University School of Visual, Performing and Media Arts.

Siegel, M. (1997). Spilling out onto Castro Street. *Jump Cut, 41*, 131–136.

State Government of Victoria. (2019a). *About the roadshow*. Retrieved from www.engage.vic.giv.au; https://engage.vic.gov.au/lgbti/about-us

State Government of Victoria. (2019b). *LGBTIQ equality rural and regional program*. Retrieved from https://engage.vic.gov.au/lgbtiq

State Government of Victoria. (2019c). *Pride events and festivals fund*. Retrieved from https://www.vic.gov.au/pride-events-and-festivals-fund

Wade, M. (2018, September 12). Victoria invests $1 million in regional LGBTI roadshow over next two years. *Star Observer*. Retrieved from http://www.starobserver.com.au/news/nationalnews/victoria-news/victoria-invests-1-million-lgbti-roadshow-next-two-years/171719

Weston, K. (1995). Get thee to a big city: Sexual imaginary and the great gay migration. *GLQ: A Journal of Gay and Lesbian Studies, 2*, 253–277.

Zielinski, G. (2008). Queer Film Festivals. In J. C. Hawley (Ed.), *LGBTQ America today: An encyclopedia* (pp. 980–984). Westport: Greenwood Press.

Zielinski, G. (2012). On the production of heterotopia, and other spaces, in and around lesbian and gay film festivals. *Jump Cut, 54*, 31.

20

The Dialectics of Community and Government

Sally Weller

Introduction

Contemporary forms of government celebrate the idea of partnership between government and communities. The idea of community is "constitutive of modern politics, a keyword whose meaning turns on questions of membership, shared meanings, identity and imagination" (Watts, 2004: 196). Community is an unquestionably positive word, one that "never seems to be used unfavorably, and never to be given any positive opposing or distinguishing term" (Williams, 1973: 76). This chapter, however, views community as a contradictory idea with complex dialectics: a word that is "profoundly ambiguous and open-ended in class terms" (Gough, 2002: 417). The chapter sees the word and the idea of community as a site of struggle over meaning.

In discourse analysis, the word community is a "privileged" signifier: a word that binds together different objects, subjects, and practices into particular "systems of meaning" that constitute and organise social

S. Weller (✉)
University of South Australia Business School, Adelaide, South Australia
e-mail: sally.weller@unisa.edu.au

A. Campbell et al. (eds.), *Located Research*,
https://doi.org/10.1007/978-981-32-9694-7_20

relations (Howarth & Stavrakakis, 2000). Yet the signifier "community" is especially ambiguous. In its positive form, community is a feel-good word implying that the residents of a place are a part of some larger whole. The invocation of the idea of community carries the positive connotation of shared associations and care for others at the local level. This positive form supports and is supported by a chain linking it with notions like empowerment, engagement, local capacity building, self-help, adaptation, and resilience. The word *community* also offers a common currency that bridges different understandings of social justice, accommodating both the redistributive demands of place-based community and the recognition-based demands of communities of identity (Fraser & Honneth, 2003). This meaning of community resonates with (capitalist) economic development objectives by standing as short-hand for the locally embedded relationships, networks, and bonds—based on trust, norms, and cooperation—that are essential underpinnings of the incentives to economic effort in capitalist economies (Storper, 2005: 16).

Activists working at the grassroots of community rework this meaning to promote democratic forms of organisation at the local level in an effort to tackle the ill-effects of poverty in practical and empowering ways. In this oppositional and polemical mode, community is a synonym for "the people" and a key site of resistance to negative developments associated with the rule of markets (Escobar, 1995). In the hands of radical thinkers like Gibson-Graham (1996), community-based engagement and learning is a vehicle for imagining a world based on an ethic of care; a world beyond capitalist relationships.

Alternatively, community can be understood as a component of technologies of control in contemporary administrative systems based on the idea of governance. Here the word community is enrolled in "top down" government policies that mobilise volunteers to undertake socially useful work at the neighbourhood scale. These interventions steer the energies of community activists towards the government's projects and priorities, often with the intention of replacing redistributive funding. Volunteering in community projects also supports "active" welfare policies by orienting unemployed and under-employed people to productive activity. Gough (2002: 418) views the government's promotion of community as a socialising strategy with a "class-disciplining" effect that fosters among the residents of target locations an "acclimatisation to the permanence of their (class) social position". Schofield (2002: 197) observes that the notion of a self-governing community is one of the "defining articulations" of forms of governance associated with "neo-liberal" rule; that is, a mode of rule that leverages the positive associations of community engagement to, in practice, replace redistributive funding with localised "make-do" voluntary provisioning. In the United Kingdom, government-sponsored community empowerment projects have been tokenistic, foster conservative social relations, and enable governments to ration funds by setting up competition between community groups (Atkinson, 1999). The intermittent and precarious nature of community grants programmes socialises local leaders in poor communities into precarious lives characterised by periods of low-paid work interspersed with periods of volunteering between grants. This system is legitimised by its social valorisation as community service. Perversely, policies that seek to "empower" poor communities to tackle the local effects of (market) policy settings rarely question the underlying policy frameworks that created the poor communities in the first place. Since places are sites of complex articulations of state power with economic power, it is important to view the actions of government in "the community" as allowing, disallowing, or appropriating different forms of community economy (Hirsch, 1989).

The deployment of community-based forms of governing is a spatially selective state strategy (Jones, 1997). As Gough (2002) observes, in advanced economy contexts, the word *community* is invoked most forcefully with reference to poor or "disadvantaged" places. These are places where local-level mobilisation and empowerment is seen as a means to improve individual and collective wellbeing. Yet the adverse place identifications that justify government intervention in the community are often externally imposed by government and the media, and supported by the evidence of multiple comparative metrics. The external gaze reinforces a negative image, creates the false impression of internal homogeneity, and stifles local voices. Government policies then demand that this negative view is replaced by positive place-branding and forward-looking local vision (see Weller, 2019). As these false naming processes deny a genuine local voice, they "move(s) the discourses out of subjects", and produce a "deficit of democracy" (Adams, 2009: 154, see also Mouffe, 2000).

This chapter considers the incorporation of local community groups in government projects and forms of governance. It aims to tease out a view of how community action might interact with government in ways that retain a positive sense of community and avoid becoming extensions of government. To that end, the next section discusses the positioning of community in different modes of government. The third section uses the example of Australia's Latrobe Valley to demonstrate the issues at stake in the incorporation of community organisations in the frameworks of government. The penultimate section discusses the role of regional experts. The conclusion then discusses the implications of these developments for society and democracy.

Four Modes of Social Regulation

There are four basic forms of social regulation: markets, hierarchies, stakeholder networks and solidaristic networks (Torfing, Peters, Pierre, & Sørensen, 2012). All social regulation in any place is a complex mix of these four basic forms. Each offers a different means of allocating scarce resources, and in each the word community has different meanings and implications. These different understandings highlight not only the ambiguity of the idea of community, but also the politically charged nature of its meaning. The foundational assumptions of this chapter are that "community" needs to be understood from these different perspectives simultaneously and that its meaning is continuously contested. Whilst at some times and places, the meaning might become fixed in common sense (i.e., hegemonic) understanding, that stability is always temporary.

Market-base regulation is a "hands-off" form of regulation in which the allocation of resources is based on the interaction of supply and demand as mediated by prices. Prices are assumed to shift fluidly until supply meets demand. This creates an individualised and competitive understanding of the social world, albeit one in which individuals consent self-interestedly to follow the implicit rules of the trading game. In market-based social regulation, "community" is nothing more than a hollow aggregation of individual preferences, implying that the whole is not greater than the sum of its parts.

Hierarchical regulation is the authoritative rule of government. This is a "top-down" form of governing enacted through a dedicated bureaucracy in the collective interest: here "the community" is a universal abstraction representing the collective interest. This justifies policy settings that are designed centrally and delivered in local areas with little regard for local interests.

Regulation through stakeholder-based groups overcomes some of the limitations of markets and hierarchies by bringing together representatives of groups with specific interests to identify negotiated, practical "what works" solutions to localised issues. Here communities are always plural, since the word is effectively a synonym for institutional interest groups like firms or chambers of commerce or unions. This mode of regulation is suited to complex and politically volatile contexts where stakeholders reach negotiated compromises that accommodate incommensurable worldviews. Optimistically, these interactions might gradually moderate opposing positions.

The fourth form of social regulation is based on solidarity. It is the form of regulation of most interest in this chapter. Here government works through and with local networks and institutions in "joined up" governance arrangements (Keast, 2011). These leverage peoples' allegiances to pre-existing social institutions like churches, women's associations, or environmental groups. Here community means a community-based network in which people are bound together by shared worldviews. Regardless of whether they are interest- or issue-based, the shared perspective focuses interest on a relatively narrow set of issues in ways that enable members to process the complexities of the wider environment. Voluntary and locally based groups encourage the development of agreed attitudes, norms of conduct, and lived practices. This shared ethos has a disciplining role in the Foucauldian sense. Etzioni (1995: ix) explains:

> Communities are social webs of people who know one another as persons and have a moral voice. Communities draw on interpersonal bonds to

> encourage members to abide by shared values …. Communities gently chastise those who violate shared moral norms and express approbation for those who abide by them.

By this definition, communities are communities of practice that produce a sense of belonging and empowerment. Membership is expected to generate converging worldviews.

This form of community produces the idea of a self-regulating society in which multiple communities of practice produce a "landscape of practice" as each adjusts to others in day-to-day encounters (Wenger-Trayner & Wenger-Trayner, 2014; see also Ostrom, 2010). These accommodations reinforce the commonalities that produce local cultures. This perspective sees each community of practice as constituting a differentiated subsystem (or ecological niche) within wider society that reproduces itself recursively. Landscapes of practice promise to promote collective over individual interests in ways that enable individuals and place-based communities to thrive. This process works laterally, in the horizontal dimension of network interaction, rather than through the vertical hierarchies of government. However, the model depends on the different sub-communities being able to communicate with one another, to understand each other's needs and to make accommodations, which, in turn, requires that they "speak the same language" (Laclau, 2014).

In contemporary modes of governance based on partnership, governments seek to govern through networks of community groups, to incorporate them into hierarchical frameworks of governance and to make them participants in the formal business of government. For their proponents, this engagement delivers to communities a genuine voice in matters of government. But critics raise questions about the scope of inclusion and the implications for political citizenship. Landscapes of practice assume that everybody is "inside" communities; otherwise this idea of community creates outcasts. A dialectical view sees both possibilities as true simultaneously.

In recent times, governments across the western world have been exploring new ways to combine these four forms of social regulation. The resulting "metagovernance" strategies soften the boundaries between vertical hierarchies and horizontal networks, and in the process "tend(s) to break down the arbitrary divide that has been constructed between government and governance" (Whitehead, 2003: 7). Metagovernance alters the relationship between governments and places; it also re-calibrates relationships among sectors, places, and communities. The challenge then is to identify the sorts of articulations—between communities of practice, stakeholder-based governance, market exchange and hierarchical forms of governance—that advance rather than stifle local interests. These are likely to be place- and time-specific and depend on how spatially selective state strategies interact with the capacities of local communities. The work of experts and intermediaries who bridge multiple forms of social organisation at the regional scale are crucial in this equation. The next sections examine these interactions in relation to the Latrobe Valley in Victoria.

Governing Through "Community"

Since the early 2000s, the Victorian Government has looked to community participation to address the problems of regional communities assessed as "disadvantaged" relative to State-wide averages. In 2003, it established a Department of Victorian Communities to manage that process. Drawing on Putnam's (2000) version of the notion of social capital, policy looked to social inclusion through community participation to offset the ill-effects of living in disadvantaged locations. At the time, respected social analysts Bryson and Mowbray (2005) accused the new policy orientation of delivering a low-cost "spray-on" solution that at best created the appearance of addressing disadvantage. They saw the evidence supporting community-based solutions as ignoring the counter-evidence that community participation is associated with comfortable material circumstances, and that conversely, lack of participation is an effect (and not a cause) of economic privation. Cass and Brennan (2002) saw the roll-out of policies advocating active community participation as a ruse to mask the punitive intent of the contemporaneous shift to individualised contractualism in "active" welfare and unemployment policy. In Everingham's (2001) assessment, the new "politics of community" attributed the problems of disadvantaged areas to a deficit of "community" rather than to the structural tendency of market capitalism to increase spatial difference by concentrating wealth in already-advantaged places (see Harvey, 2000). In the government-sanctioned understanding, community becomes an "agent of government" that "reinforces a broader trend towards social exclusion" of people who lack the resources to engage in community activities (Everingham, 2001: 107). In the years since these critiques, however, community partnership has become a standard and mainstream mode of social regulation.

The Latrobe Valley has been a site of entrenched disadvantage since the 1990s when the government privatised the dominant electricity production sector. This shift from hierarchical to market-based forms of social regulation failed spectacularly to smooth the impacts of plant rationalisation and job losses. On the contrary, a market-based form of social regulation has consolidated the Valley's entrenched reputation for inter-generational disadvantage. The persistent economic crisis since privatisation has been reinforced by a sense of political disenfranchisement. The Kennett government in the 1990s—operating in a decisively hierarchal mode—merged the local governments of the Valley's three main townships—Moe, Morwell, and Traralgon—into a single City of Latrobe. Ever since, long-standing rivalries have combined with the social and physical distance between the towns to inhibit the development of a shared place-identity (Weller, 2012). This fragmentation encourages governing through multiple separate "communities".

To compensate for the failure of markets, hierarchical government has had a heavy footprint in the Valley. Numerous state and federal government agencies have offices across its townships, and governments fund many private services, both of which provide much-needed, full-time jobs and contribute to the local economy. Yet these services follow policies set centrally, in Canberra and Melbourne. Overlapping jurisdictions, an uncertain division of responsibilities and programmes, and conflicting objectives that are not suitable for local conditions produce tensions and coordination problems. Bureaucratic inertia fuels local frustrations. Long-standing policy and service delivery failures since privatisation have produced a sense of resentment and mistrust in "top-down" government agencies and policies (Duffy & Whyte, 2017; Weller, 2017a). The recent Latrobe Valley Task Force initiative is an attempt to redress this deficit by creating a direct link from the Valley to the Premier's Office. Otherwise, the principal avenue through which local people can influence top down policy is by voting in elections (see Weller, 2017a).

To address the failures of hierarchies and markets, the Victorian government has created multiple stakeholder-based groups to provide advice on specific challenges of regulation. The 2011–2013 Latrobe Valley Transition Committee (LVTC), for example, was a stakeholder-based grouping that included the relevant government departments at state, federal, and local levels as well as local experts, representatives of business, unions, and community groups. Much of its work involved evaluating evidence—research or commissioned reports from universities, consultancy firms or government departments, econometric modelling and community consultations—and taking into account local conditions, past experiences and knowledge of "best practice" in the relevant policy area (Weller, 2019). Similarly, the Federal Government's Regional Development Australia (RDA) committees bring together regional stakeholders with different and sometimes opposing agendas to craft agreed regional objectives and priorities.

Stakeholder groups constitute semi-autonomous vehicles of governance designed to identify techno-managerial "fixes" to specific presenting policy problems (see Lovering, 1999). Bringing different levels and perspectives to the same table in a "horizontal" peer network aspires to neutralising power imbalances. Yet these groups always operate in what Whitehead (2003) calls the shadow of hierarchy. The government's control of membership, agendas, and administration ensures that these arrangements produce embellishments of the policies favoured by government. People selected to serve on committees are likely to be people who are "on the same page" with government policy priorities (Weller, 2019; Weller & Tierney, 2018). The recommended "fixes" support market-based social regulation by encouraging local solutions that "take responsibility" for local problems. Solutions advocating adaptive and resilient responses therefore in effect recommend acquiescence to the rule of market forces. Since in the Australian context, there is limited capacity for regions to take responsibility in a political or economic sense, these fixes collapse into a rhetoric justifying and legitimising deindustrialisation and service retrenchment.

Solidaristic modes of social regulation build on the local social ties and commitments of the many vibrant communities of practice in the Latrobe Valley and Gippsland areas. Many local communities in Gippsland are organised around political activism and seek to operate as regional change agents; the community-based group Gippsland Climate Change Network (GCCN) is an example (Green & McGlam, 2020). Another community-based organisation, the Voices of the Valley (VoV), has been an important contributor to debates about a "just" transition from coal dependence in the Valley. These groups promise genuine participation, build local capabilities, empower people whose voices are often not heard, and reinforce community-oriented worldviews (see Gibson-Graham, 2003). A productive example is the health-related collaborations that developed after the 2014 Hazelwood fires, which succeeded in changing both the government's policy settings and its modes of engagement, go(ing) beyond working within existing structures to create new ways of working together (Edmondson, 2020). Such groups have the potential to subvert the objectives of government programmes by redefining them in terms that are both meaningful and beneficial for communities (Larner & Craig, 2005).

However, it is possible that the reverse could emerge, and the community groups become agents of government policy. Some local groups have a semi-professional dimension through contributions to government inquiries, engagement in media debates and participation in government-auspiced stakeholder-based committees. These contributions are most effective when community groups present their cases in the language government understands. But the risk is that learning to speak in the government's language influences the range of terminologies, definitions, and understandings that constitute a worldview. The compromises community groups make to secure government funding requires an at least "on paper" alignment to existing policy objectives. Competition for funding further encourages groups to focus on government objectives and induces them to accept the rationing inherent in the competitive process. In this form, engagement with government becomes a disciplining device that steers local activists towards approaches that are compatible with government policy directions. For Laclau (2014), this outcome is inevitable.

When the government deploys the rhetoric of "partnership", and explicitly endeavours to deliver services through community-based organisations, it is simultaneously minimising the cost of paying employees, replacing workers with (mostly women) volunteers, and positioning community groups at the lowest, least authoritative tier in the vertical hierarchy of government structure. Governing through community networks further consolidates the neglect of the needs of people who are not engaged with, or are excluded from those networks.

Experience from other places suggests that as the partnership model evolves, community activism becomes a threshold condition for government intervention. After community organisations in the Valley lobbied successfully for government action after the Hazelwood fires, the Environment Protection Authority proposed that in future, it would take action on environmental hazards only as a response to a community-based movement for change (EPA, 2016). This would constitute a retreat from intervention based on transgressions of the law. Similarly, the state government's intention to "roll out" the Latrobe Valley partnership model in other places suggests that to qualify for redistributive funding in the future, places will have to first demonstrate an active community. If lack of community activism is an expression of disadvantage, this move threatens to further isolate poorer communities.

Knowledge and the Rule of Experts

Local knowledge and leadership are pivotal to the coherence and effectivity of community groups. But this dimension is also contradictory. On the one hand, members of community groups can be understood as "holders" or "keepers" of knowledge of local lived practice. This role is accepted among Indigenous communities but is just as much a feature of non-Indigenous local communities. This knowledge is associated with everyday processes of subject-formation, which are intimately bound up with the local context, since the situated production of knowledge contributes to different ways of knowing (Haraway, 1990). Local knowledge holders are considered better able—compared to outsiders—to identify the nature of regional problems and to recommend viable solutions.

Place is important because spatial frames shape problem definitions, diagnosis and prognosis, and therefore the types of knowledge that are relevant. If local knowledge is tacit, grounded in local lives, and conveyed through unspoken norms and ways of living, then its custodians are likely to be lay people who have lived in the area for a long time. Ideally, these experts would assume the role of intermediaries and be engaged to translate local understandings into the language of government. Yet such people are not necessarily the community representatives invited to participate in government partnerships. The tacit form of knowledge held by these knowledge holders is quite different to the codified knowledge of research reports and quantitative data. Since this form of knowledge is not amenable to quantification or validation, it is often excluded from, or its value discounted, in the government's evidence-based policy processes (Lahiri-Dutt, 2004).

On the other hand, in both stakeholder and partnership modes of social regulation, governments rely on a different type of local expert. These people are chosen to represent the interests of specific communities and through that role become speakers for the interests of the wider community at large. These individuals—who include public servants and private sector relationship managers, local university personnel with community engagement responsibilities and people in funded advocacy positions—develop knowledge of interactions across different community groups. They are also, through their external associations, people who introduce knowledge of national and global best practice exemplars into local debates. As suggested by Putnam's (2000) notions of "bridging" and "linking" capital, these people are empowered by their roles as connectors of knowledge resources. Governments invest in grooming these emerging leaders, so in local practice, the same active and committed people tend to find their way onto multiple committees and groups. Favoured spokespersons have the capacity to communicate with policy makers in policy language, play the policy problem-evidence-solution game, and promote visions of the future that support the official policy vision. In regional areas, the social ties these community leaders often have with policymakers suggest a class dimension. These leaders become licensed to extend from representing a particular network of interests to speaking on behalf of the wider geographically defined "community".

Of course, the degree to which community groups and community spokespersons are embedded in the structures of government varies depending on the orientation of the group and the government's priorities. Some groups are more deeply involved with networks of power than others. In the case of the Latrobe Valley, the Victorian government's active destabilisation of long-standing place definitions tends to diminish the salience of local knowledge based on lived experience (Weller, 2019), and opens a space for experts in community relationship management whose version of local knowledge is more heavily inflected with globally circulating best practice tropes. Many community-based organisations in Gippsland have close links to like-minded groups in Melbourne and beyond, with the effect that they operate as local instantiations of global networks of expertise who reinterpret circulating understandings through the lenses of local conditions (Weller, 2017b).

In the Latrobe Valley, as in other places, the answer to the question "On whose behalf does the local expert speak?" is never assured. Given persistent local political and inter-place rivalries, it is difficult to imagine any person or organisation—apart from the elected Mayor of the City of Latrobe—speaking for all of the people who live in the Valley. In fact, the crucial "other" of the structural incorporation of *particular* communities into the government's policy architecture is the simultaneous exclusion of both elected local government representatives and dissenting community groups. As the state government has reoriented to governing in partnership with self-nominating community groups, it has simultaneously wound back engagement with the elected councillors of the City of Latrobe, who have in past been vocal critics of the State government.[1] Oppositional groups like the Moe Progress Association—groups that refuse to accept the government's requirement that people adapt resiliently to disadvantage and poverty—are seldom called on to work in partnership with government. What excluded organisations seem to have in common is an insistence on the right to re-distributional funding to address material disadvantages.

[1] This exclusion has been made possible by the physical redefinition of the Valley to a larger area in which the three participating local governments are mere stakeholders (Weller, 2019).

Whither Democracy?

This raises deeper questions about what stakeholder- and partnership-based governance frameworks mean for democratic society. The idea of "landscapes of practice" (Wenger-Trayner & Wenger-Trayner, 2014) proposes that complex social systems can become self-regulating and self-reproducing as the multiple communities of society adjust constantly to one another and renegotiate their multiple boundaries. Yet this view does not account for the political rifts that fragment the polity in actual places, as has been the case in the Latrobe Valley over the years. Landscapes of practice assume that individuals are associated with one or more communities of practice and that the different communities are able to communicate effectively with one another. But neither assumption is necessarily true. There is no discussion of the people who are excluded—that is, of people who are poor, or socially isolated, or just busy making ends meet. The landscapes of practice worldview allocates social power to people who bridge and intermediate among multiple groups (e.g., university researchers), but in its current realisation, this takes power at the expense of democratically elected representatives and less compliant constituencies. When the landscapes of practice idea focuses attention on "horizontal" interactions among groups, it conceals the nature of their enrolment into structures of government and hides entirely the ways "community" is being deployed to replace elected representatives.

This has worrying implications for the future. Rancière (2010: 33) insists that whenever a network creates an "inside" group, it creates simultaneously an excluded, but constitutive, outside, the "part of those that have no part". Sennett (1998: 138) suggests that in stressful times, the insider communities tend to become "morally self-righteous" and "draw very tight boundaries around themselves", so making the lived practice of community deeply divisive. In Rancière's view, democracy exists only when those who have no part are able to break through and demand that their voices are heard. From his perspective, the technologies of metagovernance being deployed in the Latrobe Valley are fundamentally anti-democratic, since the people who have a sanctioned part are selected by government on the basis of their expertise in cooperating with government.

It is important therefore that researchers are alert to the wider implications of government engagement, to the disciplinary costs of accepting government funding, and to the exclusionary tendencies of a notion of community based on specific communities of practice. To avoid these pitfalls, local engagements should endeavour to:

1. Promote respect and recognition of those who have no part. To avoid the fetishizing the experiences of impoverished people, this means highlighting the distanciated relationships, between and across scales and geographics, that produce and sustain local disadvantages (Roy, 2011).
2. Encourage recognised communities of practice to engage with and to represent the interests of their neighbours who "have no part" in the dominant world.
3. Encourage, rather than stifle, the possibility of rupture of the dominant government-sponsored worldview (see Derickson, 2017).
4. Ensure their work is helping, not hindering, the long-term interests of the community. For example, promoting "resilience" would only advance long-term interests when it complements, rather than replaces, re-distributional funding.
5. Ensure that community groups are not inadvertently providing a disciplining function for government. This means that when acting as intermediaries linking community-based organisations to government priorities, through activities like submission and grant application writing, that participants resist attempts to steer community-of-practice towards the government's frameworks of understanding.
6. To protect discursive spaces that are respectful of other worldviews, other ways of knowing, and other ways of being. For Mouffe (2011), the only way to this is to promote continuous political struggle.

The plea of this chapter is that community groups and researchers working with community groups become more aware of their contradictory roles in enacting or influencing government strategies. The idea of "collective hope" (Edmondson, 2020) is helpful, so long as it incorporates the people who "have no part" and preserves the right to democratic representation. The rule of thumb, in Derickson's (2018: 558) view, is "a refusal to participate in the creation of knowledge that contributes to the erasure and marginalisation of subjects, subjectivities, and politics that can't be seen in the totalized unitary or generalised explanatory abstractions".

Acknowledgement The research on which this chapter is based was supported by the Australian Research Council (Grant FT110100854).

References

Adams, P. (2009). *Geographies of media and communication*. Chichester: John Wiley & Sons.

Atkinson, R. (1999). Discourses of partnership and empowerment in contemporary British urban regeneration. *Urban Studies, 36*(1), 59–72.

Bryson, L., & Mowbray, M. (2005). More spray on solution: Community, social capital and evidence based policy. *Australian Journal of Social Issues, 40*(1), 91–106.

Cass, B., & Brennan, D. (2002). Communities of support, or communities of surveillance: And enforcement in welfare regimes. *Australian Journal of Social Issues, 37*(3), 247–262.

Derickson, K. (2017). Taking account of the 'part of those that have no part'. *Urban Studies, 54*(1), 44–48.

Derickson, K. (2018). Masters of the universe. *Environment and Planning D: Society and Space, 36*(3), 556–562.

Duffy, M., & Whyte, S. (2017). The Latrobe Valley: The politics of loss and hope in a region of transition. *Australasian Journal of Regional Studies, 23*(3), 421–434.

Edmondson, B. (2020). Researching opportunities and disadvantage in regions. In A. Campbell, M. Duffy, & B. Edmondson (Eds.), *Located research: Regional places, transitions and challenges* (pp. 319–334). Palgrave Macmillan.

EPA. (2016). *Discussion paper on environmental regulation*. Melbourne: Environment Protection Authority.

Escobar, A. (1995). *Encountering development: The making and unmaking of the third world*. Princeton, NJ: Princeton University Press.

Etzioni, A. (1995). *The spirit of community*. London: Fontana.

Everingham, C. (2001). Reconstituting community: Social justice, social order and the politics of community. *Australian Journal of Social Issues, 34*(3), 209–230.

Fraser, N., & Honneth, A. (2003). *Redistribution or recognition?: A political-philosophical exchange*. London: Verso.
Gibson-Graham, J. K. (1996). *The end of capitalism (as we knew it): A feminist critique of political economy*. Oxford: Blackwell.
Gibson-Graham, K. (2003). An ethics of the local. *Rethinking Marxism, 15*(1), 49–74.
Green, M., & McGlam, G. (2020). The significance of collaborative approaches and 'collective hope' in Gippsland during a time of regional transition. In A. Campbell, M. Duffy, & B. Edmondson (Eds.), *Located research: Regional places, transitions, and challenges* (pp. 223–251). Palgrave Macmillan.
Gough, J. (2002). Neoliberalism and socialisation in the contemporary city: Opposites, complements and instabilities. *Antipode, 34*(3), 405–426.
Haraway, D. (1990). *Simians, cyborgs, and women: The reinvention of nature*. New York: Routledge.
Harvey, D. (2000). *Spaces of hope*. Edinburgh: Edinburgh University Press.
Hirsch, P. (1989). The state in the village: Interpreting rural development in Thailand. *Development and Change, 20*(1), 35–56.
Howarth, D., & Stavrakakis, Y. (2000). Introducing discourse theory and political analysis. In D. Howarth, A. Norval, & Y. Stavrakakis (Eds.), *Discourse theory and political analysis: Identities, hegemonies and social change*. Manchester: Manchester University Press.
Jones, M. (1997). Spatial selectivity of the state? The regulationist enigma and local struggles over economic governance. *Environment and Planning A, 29*, 831–864.
Keast, R. (2011). Joined-up governance in Australia: How the past can inform the future. *International Journal of Public Administration, 34*(4), 221–231.
Laclau, E. (2014). *The rhetorical foundations of society*. London: Verso.
Lahiri-Dutt, K. (2004). 'I plan, you participate': A southern view of community participation in urban Australia. *Community Development Journal, 39*(1), 13–27.
Larner, W., & Craig, D. (2005). After neoliberalism? Community activism and local partnerships in Aotearoa New Zealand. *Antipode, 37*(3), 402–424.
Lovering, J. (1999). Theory led by policy: The inadequacies of the 'new regionalism' (illustrated from the case of Wales). *International Journal of Urban and Regional Research, 23*(2), 379–395.
Mouffe, C. (2000). *The democratic paradox*. London: Verso.
Mouffe, C. (2011). *On the political*. London: Routledge.
Ostrom, E. (2010). Beyond markets and states: Polycentric governance of complex economic systems. *The American Economic Review, 100*(3), 641–672.

Putnam, R. (2000). Bowling alone: America's declining social capital. In L. Crothers & C. Lockhart (Eds.), *Culture and politics: A reader* (pp. 223–234). New York: Palgrave Macmillan.
Rancière, J. (2010). *Dissensus: On politics and aesthetics*. New York: Bloomsbury Publishing.
Roy, A. (2011). Slumdog cities: Rethinking subaltern urbanism. *International Journal of Urban and Regional Research, 35*(2), 223–238.
Schofield, B. (2002). Partners in power: Governing the self-sustaining community. *Sociology: the Journal of the British Sociological Association, 36*(3), 663–683.
Sennett, R. (1998). *The corrosion of character: The personal consequences of work in the new capitalism*. New York: WW Morton & Company.
Storper, M. (2005) Society, community, and economic development. *Studies in Comparative International Development, 39*(4) 30–57.
Torfing, J., Peters, B., Pierre, J., & Sørensen, E. (2012). *Interactive governance: Advancing the paradigm*. Oxford: Oxford University Press.
Watts, M. (2004). Antinomies of community: Some thoughts in geography, resources and empire. *Transactions of the Institute of British Geographers, 29*, 195–216.
Weller, S. (2012). Regional dimensions of the transition to a low carbon economy: Australia's Latrobe Valley. *Regional Studies, 46*, 1261–1272.
Weller, S. (2017a). The political economy of regional transformation in the Latrobe Valley. *Australasian Journal of Regional Studies, 23*(3), 110–118.
Weller, S. (2017b). Fast parallels? Contesting mobile policy technologies. *International Journal of Urban and Regional Research, 41*(5), 821–837.
Weller, S. (2019). Just transition? Strategic framing and the challenges facing coal dependent communities. *Environment and Planning C: Politics and Space, 37*(2), 298–316.
Weller, S., & Tierney, J. (2018). Evidence in the networked governance of regional decarbonisation: A critical appraisal. *Australian Journal of Public Administration, 77*(2), 280–293.
Wenger-Trayner, E., & Wenger-Trayner, B. (2014). Learning in a landscape of practice: A framework. In E. Wenger-Trayner, M. Fenton-O'Creevy, C. Kubiak, S. Hutchinson, & B. Wenger-Trayner (Eds.), *Learning in landscapes of practice: Boundaries, identity, and knowledgeability in practice-based learning*. London: Routledge, 2015.
Whitehead, M. (2003). 'In the shadow of hierarchy': Meta-governance, policy reform and urban regeneration in the West Midlands. *Area, 35*(1), 6–14.
Williams, R. (1973). *Keywords: A vocabulary of culture and society*. Oxford: Oxford University Press.

Index[1]

[1] Note: Page numbers followed by 'n' refer to notes.

A. Campbell et al. (eds.), *Located Research*,
https://doi.org/10.1007/978-981-32-9694-7

N

O

P

The manufacturer's authorised representative in the EU is Springer Nature Customer Service Centre GmbH, Europaplatz 3, 69115 Heidelberg, Germany. If you have any concerns regarding our products, please contact ProductSafety@springernature.com

Printed and bound by CPI Group (UK) Ltd, Croydon, CR0 4YY
25/09/2026
02222881-0006